Tuned Liquid Column Dampers for Structural Control

Structural vibrations from excitations such as wind, earthquakes, waves and pedestrian movements can be controlled using tuned liquid dampers such as tuned liquid column dampers (TLCDs). They benefit from low installation cost, ease of maintenance and amenability to tuning. TLCDs offer high volumetric efficiency and a specific and quantifiable damping mechanism.

This book outlines passive, active, semi-active and hybrid forms of TLCD. It systematically covers basic operational principles, mathematical modelling, analysis methodologies of damper-structure systems, in frequency and in time domain, design issues and experimental studies, and structural implementations. There is also detailed coverage of innovative design variations that enhance applicability and performance.

This book is aimed at researchers in structural control, advanced students and vibration specialists in industry.

Tanmoy Konar is an engineer in the West Bengal Police Housing and Infrastructure Development Corporation, India, with extensive practical experience, especially in the design of buildings. He has regularly worked on different configurations of liquid dampers and has developed several novel damping mechanisms for the same.

Aparna Dey Ghosh is a professor at the Department of Civil Engineering, Indian Institute of Engineering Science and Technology, Shibpur, India. Prof. Dey Ghosh has a comprehensive knowledge of passive structural vibration control systems and her work on the tuned liquid column damper has served as the cornerstone for further research on the subject. She also has experience as a consulting design engineer in the power plant and steel plant sectors.

Biswajit Basu is a professor at Trinity College Dublin, Ireland. Prof. Basu is an internationally acclaimed researcher with expertise in a wide range of subjects, such as, nonlinear hydrodynamics, time-frequency analysis and signal processing, stochastic dynamics and control, oceanic flow modelling and simulation, offshore renewable energy and algorithms for quantum computing.

Tuned Liquid Column Dampers for Structural Control

Tanmoy Konar, Aparna Dey Ghosh, and
Biswajit Basu

CRC Press
Taylor & Francis Group
Boca Raton London New York

CRC Press is an imprint of the
Taylor & Francis Group, an **informa** business

Contents

Foreword ix
Preface xi

Chapter 1   Introduction 1

    1.1   Structural Vibration Control 1
    1.2   Dynamic Vibration Absorber 11
    1.3   Tuned Liquid Damper 16
    1.4   Tuned Liquid Column Damper 22
    References 24
    MATLAB® Code 1.1 28
    MATLAB® Code 1.2 32

Chapter 2   Tuned Liquid Column Dampers: Working Mechanism and
Design Considerations 37

    2.1   Background 37
    2.2   Working Mechanism and Mathematical Modelling 38
        2.2.1   Working Mechanism 38
        2.2.2   Equations of Motion of the Structure-TLCD System 39
        2.2.3   Frequency Domain Representation of the Equations
of Motion of the Structure-TLCD System 44
    2.3   Design Considerations 46
        2.3.1   Mass Ratio 46
        2.3.2   Length Ratio 46
        2.3.3   Coefficient of Head-Loss of the Orifice(s) 47
        2.3.4   Tuning Ratio 49
        2.3.5   Other Design Considerations 50
    References 51
    MATLAB® Code 2.1 54

Chapter 3   Different Configurations of the Tuned Liquid Column Damper 59

    3.1   Introduction 59
    3.2   TLCD Configurations for Multi-Directional
Vibration Control 60
        3.2.1   Bi-Directional TLCD (BTLCD) 60
        3.2.2   Toroidal TLCD (TTLCD) 63
        3.2.3   Omnidirectional TLCD (OTLCD) 68
    3.3   TLCD Configurations to Enhance Control Performance 71
        3.3.1   Liquid Column Vibration Absorber 71
        3.3.2   Tuned Liquid Column Ball Damper 74
        3.3.3   Tuned Liquid Multi-Column Damper 77

        3.3.4   Circular TLCD 80
        3.3.5   S-Shaped TLCD 84
        3.3.6   Tuned Liquid Column and Sloshing Damper 86
        3.3.7   Tank-Pipe Damper 89
    3.4   TLCD Configurations to Improve Tunability 93
        3.4.1   Sealed TLCD 93
        3.4.2   Compliant Liquid Column Damper 97
        3.4.3   Pendulum-Type Liquid Column Damper 101
        3.4.4   Tuned Liquid Column Damper-Inerter 103
    3.5   TLCD Configurations for Vertical Vibration Control 105
        3.5.1   Adaptive Tuned Liquid Column Damper 105
        3.5.2   Vertical Sealed TLCD 106
        3.5.3   Vertical TLCD (VTLCD) 107
    References 108
    MATLAB® Code 3.1 112
    MATLAB® Code 3.2 116
    MATLAB® Code 3.3 120

Chapter 4   Control of Wind-Excited Structures by TLCD 125

    4.1   Introduction 125
    4.2   Frequency Domain Studies 129
    4.3   Time Domain Studies..136
    4.4   Experimental Studies 140
    4.5   Optimality Issues...143
    References 147
    MATLAB® Code 4.1 150

Chapter 5   Seismic Vibration Control of Structures by TLCD 155

    5.1   Introduction 155
    5.2   Frequency Domain Studies 156
    5.3   Time Domain Studies..160
    5.4   Experimental Studies 165
    5.5   Optimality Issues...169
    5.6   Studies Considering Soil–Structure Interaction Effects 172
    References 176

Chapter 6   TLCD for Control of Structures Subjected to Other Types of
             Excitations 181

    6.1   Introduction 181
    6.2   Control of Wave-Induced Excitation by TLCD....................181
    6.3   Control of Pedestrian-Induced Excitation by TLCD.............192
    6.4   Miscellaneous...194
    References 196

Chapter 7   Active and Semi-Active Tuned Liquid Column Damper 200

    7.1   Introduction 200
    7.2   ATLCD with a Compliant Mechanism 201
        7.2.1   Equations of Motion of ATLCD-C 201
        7.2.2   Controller Design Algorithms................................203
    7.3   ATLCD with Propellers 207
        7.3.1   Equations of Motion of ATLCD-P..........................208
        7.3.2   Controller Design and Implementation 209
    7.4   Gas-sealed TLCD with Active Pressure Control 209
        7.4.1   Equations of Motion of STLCD-APC......................209
        7.4.2   Controller Design and Implementation 211
    7.5   TLCD with Semi-Active Orifice Control 211
        7.5.1   Equations of Motion of TLCD with Semi-Active
                Orifice Control 212
        7.5.2   Controller Design and Implementation 213
        7.5.3   Irregular Structures and Controllers Based on
                Neural Networks 216
    7.6   Semi-Active MR-TLCD 217
        7.6.1   Equations of Motion with Semi-Active MR-TLCD......217
        7.6.2   Controller Design and Implementation 219
    7.7   Semi-Active Spring-Connected TLCD 219
        7.7.1   Equations of Motion with Semi-Active
                Spring-Connected TLCD..220
        7.7.2   Controller Design and Implementation 222
    7.8   Hybrid Dampers with Semi-Active TLCD............................226
    References 227

Chapter 8   Practical Implementation of TLCD..230

    8.1   Introduction 230
    8.2   Higashi-Kobe Bridge 234
        8.2.1   Overview 234
        8.2.2   Description of Damper and Performance 235
    8.3   Hotel Cosima 237
        8.3.1   Overview 237
        8.3.2   Description of Damper and Performance 238
    8.4   Prospect Communication Tower 240
        8.4.1   Overview 240
        8.4.2   Description of Damper and Performance 241
    8.5   One Wall Centre 242
        8.5.1   Overview 242
        8.5.2   Description of Damper and Performance 243
    8.6   Random House Tower 244
        8.6.1   Overview 244
        8.6.2   Description of Damper and Performance 247

8.7   Comcast Center 247
        8.7.1   Overview 247
        8.7.2   Description of Damper and Performance 249
8.8   New Songdo City First World Towers 250
        8.8.1   Overview 250
        8.8.2   Description of Damper and Performance 251
8.9   Sky Gate 253
        8.9.1   Overview 253
        8.9.2   Description of Damper and Performance 254
8.10   B2 Tower...255
        8.10.1 Overview 255
        8.10.2 Description of Damper and Performance 257
8.11   Australia 108..258
        8.11.1 Overview 258
        8.11.2 Description of Damper and Performance 260
References 260

Index.. 265

Foreword

Structural vibration control is a very important field as there has been a growing concern in recent years to better protect structures – particularly tall buildings, long span bridges, and stay cables – from not only extreme events but also from loads that cause serviceability problems in an economical and sustainable manner. The field has gained significance mainly because, with advancements in material science, we are able to build more slender structures that possess low intrinsic damping. Apart from seismic base isolation, which has had its own trajectory of development, a wide variety of energy dissipation systems for structures under dynamic loading have evolved. Many of these, such as metallic dampers, friction dampers, fluid viscous dampers and tuned mass dampers, have been researched extensively and successfully implemented in several structures in different parts of the world. With the passage of time, the more economical counterpart of the tuned mass damper, namely the tuned liquid damper, has also witnessed significant development after its original innovations in mainly Japan (by Professor Fujino and his group for tall buildings, long span bridges, and stay cables) and many other countries. A type of tuned liquid damper, the tuned liquid column damper (TLCD), has become the fowcus of researchers due to its many advantages over the other systems. The existing books on structural vibration control have elucidated these different dampers, but mostly concisely. Books that deal exclusively with a single damping device from the point of view of structural vibration control are lacking. In this context, the current book that concentrates only on the TLCD provides a very valuable resource material to researchers and practicing engineers on the subject.

The book is very well written and well-organized. It initially provides a broad overview of the different categories of structural vibration control systems, along with an elaboration on the tuned liquid damper, which should prove helpful to those relatively new to the field. The next two chapters clearly explain the working mechanism of the TLCD, both in its original and modified configurational forms. What is commendable here is that the authors have presented the matter related to a large number of different TLCD configurations in a logical and systematic format so that the reader should have no difficulty in grasping the reasons why these developments have taken place, and can deduce future directions of research as well. In each of the subsequent chapters, studies relevant to a particular type of dynamic excitation for which the TLCD has been utilized to control the resulting structural vibrations have been addressed. While passive TLCDs dominate, a single chapter written with rare insight has been devoted to the works that have so far been carried out on active, semi-active, and hybrid TLCDs. The book concludes with an interesting assimilation of real installations of TLCDs worldwide that acquaint the readers with the potential of TLCDs in modern structural engineering, particularly tall buildings, and long-span bridges. I compliment the authors for writing such a fine book that will benefit the structural vibration control field and the broader engineering community significantly.

Satish Nagarajaiah, Ph.D.
NAI (USA), Dist. M.ASCE, M.SXI, F.SEI
Professor of CE and Professor of ME (& MSNE Courtesy)
Rice University, Texas, USA
Senior Editor – MSSP | Editor – SCHM
Nov. 11, 2024

Preface

The concept that vibrations of structures, whether onshore or offshore, fixed or floating, can be reduced through means external to the structure germinated in ancient times. Several historic structures on which the concept of base isolation was applied in some form have weathered the vagaries of nature and stand testament to the efficacy of this technique. In the more recent past, the use of anti-rolling tanks to stabilize ships was another innovative concept of structural vibration control that served to pioneer the development of tuned liquid dampers and of tuned liquid column dampers, which is the topic of this book.

With the advancement of the subject of structural dynamics, a clearer understanding of energy dissipation mechanisms emerged and that led to the development of a wide variety of vibration control methodologies and devices. The success of the early implementations of these systems in mitigating wind, earthquake and wave-induced vibrations of structures, especially in the era of building taller and more flexible structures with lighter materials, spurred further research on the subject.

The advent of performance-based design necessitated new ways to address vibration-related, both safety and serviceability, issues and promoted the development of a number of innovative structural vibration control systems. The progress in the field of control and sensor technology further helped in the realization of a wide variety of active, semi-active and hybrid control systems. The passive control systems, however, with their intrinsic reliability, relative simplicity and cost-effectiveness, continue to remain the most popular and suitable for applicability to different types of structures.

The tuned liquid column damper (TLCD) was proposed as a damping device for tall structures a little more than 30 years ago, and though initial research on the device was not taken up by many, in the past decade, there has been a remarkable surge in literature on different aspects of this damper. The authors have themselves worked on the topic for more than 20 years and now feel that there is a need for a book that assimilates all the different aspects of investigations that have so far taken place on the device. It is interesting to note that despite substantial research efforts having been directed towards this niche topic, the TLCD has been covered rather briefly in most existing books on structural vibration control. But we have now come to the stage when a few pages will not suffice to guide a student or researcher on the topic.

This book is generally intended towards students and researchers in structural dynamics and vibration control. It should also prove very useful to structural engineers trying to implement innovative concepts of vibration control, which given the mandate of building economic, resilient and sustainable structures might soon become a necessity rather than a choice. This book starts with a background on the dynamic vibration absorber and tuned liquid dampers in general. It then elucidates the working principle of the TLCD and its design parameters. The large number of configurational variations of the original TLCD, developed to overcome its limitations, are discussed next. The following chapters systematically deal with specific works available in the literature on the use of the TLCD for suppression of vibrations due to wind, seismic, wave and other excitations. As far as possible, frequency domain, time domain and experimental studies have been organized in separate sections. A separate chapter is dedicated to active, semi-active and hybrid TLCDs. The treatment of each type of TLCD, whether passive, active or semi-active, is complete, from modelling, formulation, results and discussion. The final chapter presents some landmark real-world installations of the TLCD, supplying details of the structure, damper and its performance. MATLAB® programs have been provided where necessary. The authors hope that this book will enthuse many more researchers to explore the TLCD, which holds a lot of potential in protecting vibrating structures in an efficient and cost-effective manner.

1 Introduction

1.1 STRUCTURAL VIBRATION CONTROL

Vibration, in general terms, is the back-and-forth motion of an elastic system from its equilibrium position. Vibration is induced in a structure when it is "excited," that is, when it receives mechanical energy from external sources. During the process of vibration, this mechanical energy is alternately transferred from potential to kinetic form and vice versa. A vibratory system thus essentially includes a means for storing potential energy (spring or stiffness element) and a means for storing kinetic energy (mass or inertia element). Further, every real-life vibratory system has the capacity to dissipate a certain amount of energy, by the conversion of the mechanical energy causing the vibration, to heat and/or acoustic energy. This is termed as damping. Since the inherent damping ability of a structure is complex and stems from various mechanisms, such as friction in micro-cracks, straining and yielding of materials, it is generally modelled in an equivalent sense. The mathematical representations of damping are therefore in terms of equivalent viscous damping, friction damping, hysteretic damping and so on. Among them, the equivalent viscous damping model is the most commonly used for a structure. In this, the damping element is represented by a dashpot, and the damping force, acting opposite to the motion, is proportional to the velocity of the vibrating body.

Figure 1.1 shows a simple vibratory system with single-degree-of-freedom (SDOF). Here, the mass, the stiffness and the damping are denoted by M, K and C, respectively. $F(t)$ represents the external excitation. The displacement of the system with respect to its equilibrium position is denoted by $x(t)$. The equation of dynamic equilibrium of the system is given in Eq. (1.1) as

$$M\ddot{x}(t) + C\dot{x}(t) + Kx(t) = F(t). \tag{1.1}$$

In Eq. (1.1), an overdot represents a derivative of the respective variable with respect to time, t.

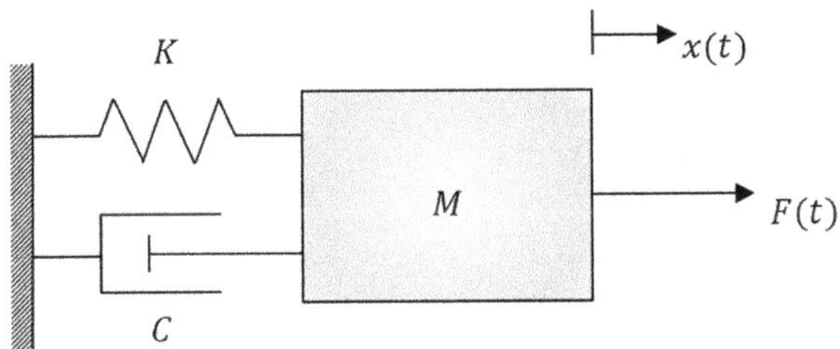

FIGURE 1.1 A single-degree-of-freedom vibratory system.

DOI: 10.1201/9781003377894-1

For a civil engineering structure, vibration is induced due to the mechanical energy received by it during natural events, such as strong wind, earthquakes and ocean waves; or, as a result of anthropogenic activities, such as bomb blasts, vehicle movement and pedestrian movement (see Figure 1.2). In general, vibration in civil engineering structures is undesirable. This is for two reasons. First, vibrations may result in serviceability issues, such as causing occupant discomfort by exceeding floor acceleration limits (Lamb and Kwok, 2019; Yamada and Goto, 1975), damage to architectural finishes and hindrance in the functioning of equipment. Second, the dynamic stresses developed during vibration may lead to fatigue failure of structures. Thus, vibration can impair the function as well as the life of a structure or its components.

The economic and humanitarian cost of vibration-induced structural dysfunction or failure is enormous. The humanitarian toll of the recent major earthquakes is given in Table 1.1, which mostly occurred due to the collapse of building structures under earthquake-induced vibration. The total number of human lives lost due to earthquakes between 2000 and 2019 is estimated to be 721,318 (CRED, 2019). The cost of repairing the buildings in Los Angeles city in the aftermath of the 1994 Northridge earthquake was estimated to be US\$ 2.3 billion (Eguchi et al., 1998). Strong wind or storm is the other major cause of losses resulting from vibration-induced structural failure. From 1980 to 2018, tropical cyclones were responsible for a loss of US\$ 2,111 billion (MunichRE, 2018). The collapse of the Tacoma Narrows Bridge, USA, in 1940, is considered to be one of the major incidents of structural failure due to wind-induced vibration (see Figure 1.3). In recent times, wind-induced vibrations beyond the level of human comfort have been reported in the Storebaelt suspension bridge (Denmark) in 1998, the Østerøy suspension bridge (Norway) in 1999 and the Volgograd bridge (Russia) in 2010 (Yu et al., 2022). The London Millennium footbridge was closed within a few days from its inauguration in June 2000, as the bridge experienced excessive pedestrian-induced lateral vibrations (Dallard et al., 2001). The recent chemical detonation that occurred on August 4, 2020, in the Port of Beirut, Lebanon, caused an earthquake of a 3.3 magnitude on a Richter scale and a massive blast wave (Al-Hajj et al., 2021). This led to massive damage to the surrounding structures. All these facts and figures highlight the importance of the study of structural vibration and its control.

A structure should thus be engineered to withstand, along with the static loads, the expected effect of vibrations during its design life, without compromising its safety and serviceability. The traditional approach of structural design utilizes the

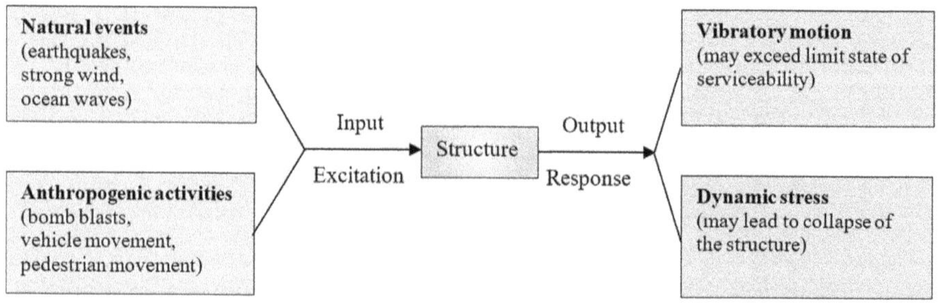

FIGURE 1.2 Causes and effects of structural vibration.

TABLE 1.1
Recent Major Earthquakes and their Humanitarian Toll

Name	Year	Number of Deaths (CRED, 2022, 2019; Hussain et al., 2023)
Kahramanmaraş (Türkiye) earthquake	2023	57,000 (approx.)
Haiti earthquake	2021	2,575
Nepal earthquake	2015	8,969
Haiti earthquake	2010	222,570
Sichuan earthquake	2008	87,476
Kashmir earthquake	2005	73,338
Andaman (Indian Ocean) earthquake	2004	226,408
Bam (Iran) earthquake	2003	26,716
Bhuj (Gujarat) earthquake	2001	20,005

FIGURE 1.3   The failure of Tacoma Narrows Bridge, USA, in 1940 ("https://commons. wikimedia.org/wiki/File:Tacoma-narrows-bridge-collapse.jpg," 2010).

strength, stiffness, ductility and energy dissipation capacity or damping of structures to resist the impact of vibration. Since the beginning of the 20th century, intensive efforts have been devoted to the development of design codes and guidelines for the construction of structures that can successfully withstand vibration, especially earthquake and wind-induced vibrations (Baker, 2007; Guevara-Perez, 2008). Despite these efforts, making structures self-sufficient to sustain expected vibrations without hampering the structural integrity and serviceability is not always feasible from an economical and practical perspective. Moreover, due to rapid urbanization and scarcity of suitable land, people are compelled to extend their buildings and other structures in the vertical direction by employing advanced construction techniques including the use of lightweight and high-strength materials. Consequently, these tall structures possess higher flexibility and lower damping, making them more susceptible to vibration. Further, the trend in the design of modern-day structures is to set specific performance objectives for the structures (Shu et al., 2019). The code-guided procedures may not always meet these objectives. In all such cases, vibration control technologies offer a solution to reduce the vibrational response and meet the performance expectations of the structures.

The concept of vibration-resilient construction can be traced back to ancient human settlements. Several building construction techniques and configurations that provided improved performance under dynamic loading were developed, even though the means of analysing them came much later. The use of robust architectural forms, for instance, buildings with symmetric plans and elevation; and resilient structural configurations, such as buildings with bands and braces, are some examples of indigenous techniques that were used to improve the seismic performance of historic structures (Sinha et al., 2004). The application of the sandbox technique for seismic isolation of buildings and structures was also practised (Daka et al., 2018). During the Iron Age, around the Atlantic sea border of Europe, a particular form of building, known as the roundhouse, with a circular plan and a conical roof was developed to enhance safety against strong wind (Pryor, 2003). There are many ancient structures that are still standing, where vibration-resilient construction techniques were used. Some prominent examples include the Ramappa (Kakatiya Rudreshwara) temple, Telangana, India (Daka et al., 2018) (see Figure 1.4); the Apollo temple, Bassae, Greece (Stiros, 2020); the Mausoleum of Cyrus, Pasargadae, Iran (Bayraktar et al., 2012); and the Athena temple, Paestum, Italy (Carpani, 2017).

Modern-era research on structural vibration control was initiated more than a century and half ago with the development of contemporary base isolation techniques. The main concept here is to provide flexibility at the junction of the foundation and

FIGURE 1.4 The Ramappa (Kakatiya Rudreshwara) temple, Telangana, India (Welch, 2021).

the superstructure of a structural system to prevent the transmission of vibration from the former to the latter. An implementable base isolation system not only requires sufficient horizontal flexibility, but other properties such as adequate damping as well, to prevent excessive bearing displacements, recentring ability and resistance to sustain design vertical loads. Stevenson (1868) is considered the inventor of the modern base isolation technique. He was a consultant to the company entrusted by the British Government with the work of building several lighthouses in Japan. Given the seismically active nature of the region, one of the main challenges was to keep the lamp apparatus stable to ensure an uninterrupted light signal. To meet the challenge, Stevenson developed a rolling-bearing device named "aseismatic joint," which was, in fact, a seismic isolation system. The rolling-bearing device, which consisted of spherical rollers in niches, was provided below the large lamp tables of the lighthouses (see Figure 1.5a and b).

A couple of years later, on February 15th, 1870, Touaillon patented a technique for base isolation of building structures to control earthquake-induced vibrations. He used spherical rollers in niches between the superstructure and the foundation (see Figure 1.6a). The striking resemblance between Touaillon's base isolation system and Stevenson's rolling-bearing device is self-evident. Only 2 weeks after Touaillon,

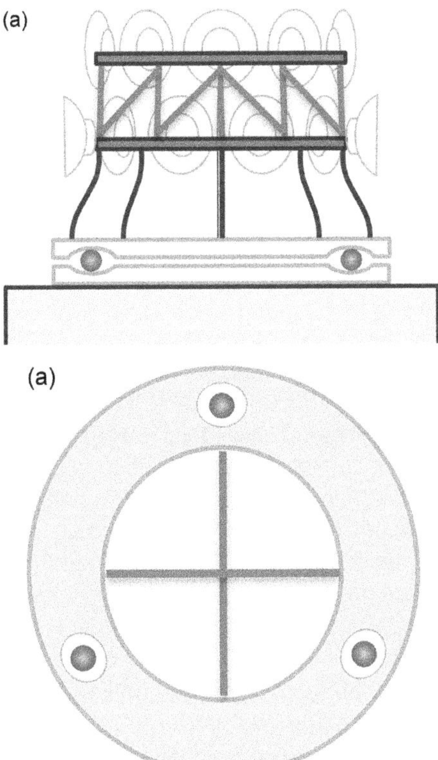

FIGURE 1.5 Schematic of lamp table provided with the rolling-bearing device developed by Stevenson (1868). (a) Sectional elevation and (b) plan.

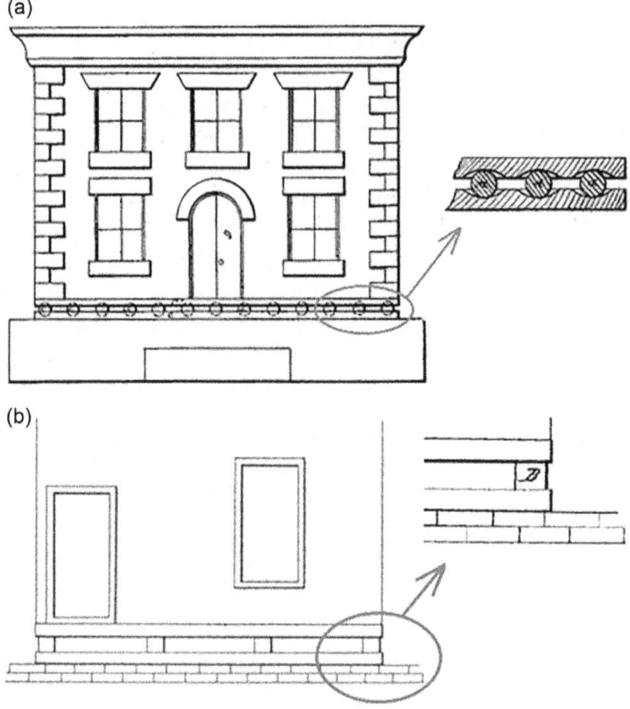

FIGURE 1.6   Base isolated buildings (a) proposed by Touaillon (1870) and (b) proposed by Cooper (1870).

on March 1st, 1870, Cooper patented a system that involved the use of natural rubber bearings between the superstructure and the foundation, with the aim of dampening the shock from earthquakes (see Figure 1.6b). Despite these early developments, even a hundred years later, in the 1970s, there were only a few seismic isolated buildings, partly because traditional structural engineers were still hesitant to accept the concept of deliberately introducing flexibility at the base of buildings (Makris, 2019). However, from the early 1980s, the scenario changed and base isolation technology began to gain popularity. Since then, numerous instances of installation of base isolation systems in new structures as well as in existing structures as a retrofitting measure have been reported. The William Clayton Building, Wellington, New Zealand (1981); the Foothill Communities Law & Justice Centre, San Bernardino, USA (1984); the Salt Lake City and County Building, USA (built in 1894, retrofitted in the mid-1980s); the San Francisco City Hall, USA (built in 1912, retrofitted in 1995); the San Francisco Airport terminal building, USA (2000); the Los Angeles City Hall, USA (built in 1913, retrofitted in 2001); the Shimizu Corporation Tokyo Headquarters, Japan (2012); the Benicia-Martinez Bridge, California, USA (2007); the Nunoa Capital Building, Santiago, Chile (2016); and the Adana Integrated Health Campus, Turkey, (2017) are some of the significant structures with base isolation systems (Lago et al., 2019; Lewis, 2017; Luca and Guidi, 2019; Makris, 2019). A typical present-day base isolator, in use in the Utah State Capitol Building, USA, is shown in Figure 1.7.

FIGURE 1.7 Typical base isolator used in the Utah State Capitol Building, USA (Renlund, 2008).

Although effective, base isolation is mainly applicable for earthquake-induced vibration control of short to mid-period structures (Lago et al., 2019). Base isolation is normally considered unsuitable for long-period structures, such as structures having a fundamental period of more than 4 s. Further, there are concerns regarding implementation, maintenance and cost.

In addition to base isolation, there are several other structural vibration suppression techniques. Those techniques mainly work on three principles. The first is the alteration of dynamic characteristics of the structure, such as stiffness modification, redistribution of modal mass and aerodynamic modification. The second one involves the application of a response-dependent control force. The third one requires the incorporation of supplemental damping devices in the structure. During vibration, a supplemental damping device absorbs a portion of the input mechanical energy and dissipates the absorbed energy by converting it into heat and/or acoustic energy, which is ultimately released into the environment. In the mathematical model of a structure, the supplemental damping is represented to act in parallel with the intrinsic damping of the structure.

Supplemental damping devices are divided into two major groups, namely material-based dampers and inertia-based dampers. The metallic damper, friction damper and magnetometric damper are examples of the first category. Here, the dampers are provided with materials having energy dissipative properties, and the dampers are attached to the structural frame in a distributed pattern. On the other hand, an inertia-based damper restrains the structural vibratory motion through an inertial force developed by a secondary mass connected to the structure. Such a system functions effectively when the natural frequency of the secondary mass is tuned, that is, kept nearly equal to the natural frequency of the structural mode to be controlled. Thus, inertia-based dampers are often referred to as tuned dampers. Inertia-based dampers also have a damping element to dissipate the energy they receive through dynamic interaction with the host (primary) structure. The idea of the inertia-based damper has stemmed from Frahm's (1911) absorber, or, the dynamic vibration absorber (DVA), which is elaborated upon in Section 1.2. The inertia-based dampers are broadly of two types. First one, the tuned mass dampers (TMDs), in which the auxiliary mass is a solid, and the second type, tuned liquid dampers (TLDs), wherein the auxiliary mass comprises of liquid, normally water, in a partially filled container. TLDs are further divided into two major groups, namely (a) the tank sloshing dampers and (b) the tuned liquid column dampers (TLCDs). Within the larger family of structural vibration control techniques, the position of TLCD, which is the focus of this book, is indicated in Figure 1.8.

Based on the power requirement of the dampers, they are broadly divided into four groups, namely passive, active, semi-active and hybrid dampers.

A passive damper is characterized by the non-requirement of any electrical power to operate and is activated by the input vibrational energy from the structure to which the damping device is attached. The control force developed by a passive damper is a function of the response of the structure at the location of the damper. A structural system, without any energy dissipation device, and with a passive damper, is schematically represented in Figure 1.9a and b, respectively. Passive control devices are inherently stable and reliable. As compared to other supplemental damping devices, passive dampers have a simple working principle. They are economical and easy to design, construct, install and maintain as well. Further, the possibility of power failure during a disaster makes passive dampers a popular choice for application in civil engineering structures. However, passive systems are often referred to as systems with limited intelligence because they are unable to adapt to the excitation and response of the host structure and are thus

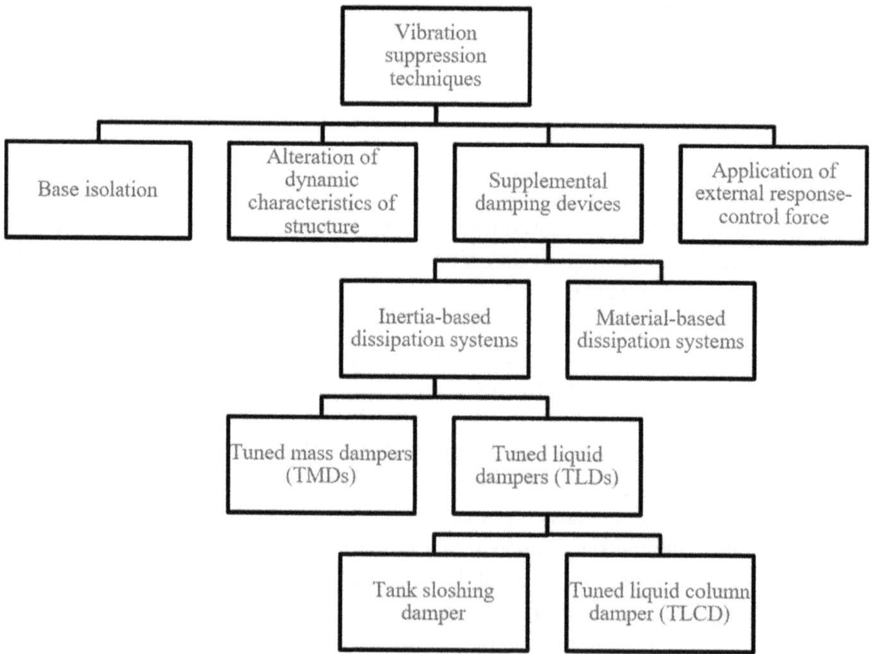

FIGURE 1.8 Position of the TLCD within the larger family of structural vibration control techniques.

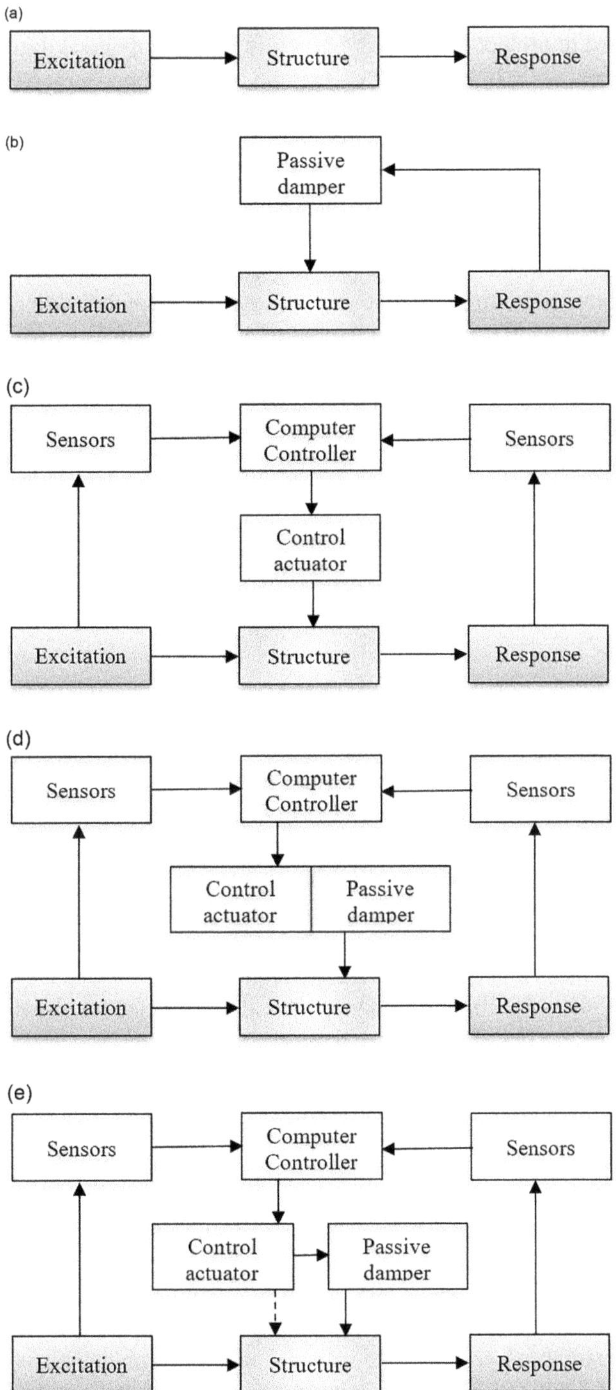

FIGURE 1.9 Working mechanism of different types of supplemental damping systems. (a) Uncontrolled structure, (b) structure with passive damper, (c) structure with active damper, (d) structure with semi-active damper and (e) structure with hybrid damper.

characterized by a limited control capacity. A passive damper system may be optimally designed to protect the primary structure from a stipulated dynamic loading, but its efficiency will obviously either degrade or remain uncharacterized for other types of dynamic loading.

An active damper is equipped with a control actuator that runs on an external power supply and, depending on the feedback of the structural response and the external excitation received through sensor(s), generates a control force that attenuates the motion of the structure. The working mechanism of an active damper system is depicted in Figure 1.9c. The primary advantages of active dampers are (a) enhanced vibration control effectiveness, (b) instant adaptability to excitations, (c) applicability to a wide spectrum of excitation characteristics and (d) effectiveness for multi-hazard vibration mitigation. However, active control systems involve high capital and maintenance costs. The scale of civil engineering structures being large, the magnitude of the dynamic loading experienced by them is also significant. Thus, for application in civil engineering structures, active dampers require a substantially large power source and actuating equipment to generate the required control force. Moreover, an active damper system would cease to function in case of a power failure, which is not uncommon in the event of an earthquake or cyclone. This is one of the main disadvantages of this type of damper. Active control systems are also comparatively more complicated and require greater effort to design. Further, active dampers induce a shift in the dynamic behaviour of the structure by adding or removing energy from it, this may result in an unwanted or even unstable condition (Christenson, 2001). Due to all these reasons, active control systems have to be used with caution and with sufficient resources in civil engineering structures.

A semi-active damper is essentially a controllable passive dissipation device. The properties of the semi-active damper are regulated by an external power source, based on the information acquired regarding the excitation and/or structural response. The components of the system thus include sensors for measuring the input and/or output, a computer that processes the measurement to generate a control signal and a mechanism to regulate the behaviour of the passive energy dissipation device. A schematic of the working mechanism of a semi-active damper is presented in Figure 1.9d. Since in this system instead of directly applying a force to the structure, the actuator is used to control the properties of the passive damper, the power requirement is small and is met through batteries. Hence, a semi-active damper is expected to provide control of the structural response to some degree even in the event of a power failure. This is an important merit of semi-active dampers. Despite being more complex and costly as compared to passive dampers, semi-active systems are still relatively easy to manufacture, install and maintain and offer considerable promise. However, the control capacity of these devices is restricted to the capacity of the corresponding passive devices with limited improvement (Saaed et al., 2015).

When different passive, active and semi-active dampers are combined in a single damper system, a hybrid control system is formed (see Figure 1.9e). Here, the individual damper units are grouped into series or parallel combinations in order to have the optimum advantage and to alleviate some of the restrictions and limitations that exist, when each system is functioning alone. Usually, hybrid control systems have passive devices that are designed to achieve a major part of the targeted structural response reduction, while the active or semi-active ones are used to adjust tuning, if required, and enhance the overall structural response reduction. Hybrid devices have a larger capacity and greater efficiency than a passive system and cost less than an active system. Hybrid control systems are also found to be very efficient in protecting structures from different types of excitation with dissimilar intensity and frequency content (Wu, 2011). They are more reliable and require less energy than active devices since there is no need for large control forces, but they still require significant energy (Christenson, 2001). In case of a power disruption, the passive component of the hybrid system continues to function, unlike an active system, and hence, hybrid dampers are "fail-safe." However, the implementation of hybrid control strategies is relatively complex because it incorporates multiple control methods in a single system.

Examples of vibration control systems under each of the aforementioned four categories are listed in Table 1.2. Each category has a respective variety of TLCD. The TLCD working mechanism, its configurational variations, design and performance under different loads, and implementational issues are discussed in the other chapters of the book. This chapter provides an overview of the DVA, including its operational principles and a general introduction to the two basic and conventional types of TLDs, namely the tank sloshing damper and the TLCD.

1.2 DYNAMIC VIBRATION ABSORBER

The dynamic vibration absorber is considered to be the basis of the modern-day inertia-based dampers. As TLCD belongs to the class of inertia-based dampers, a brief deliberation on the DVA is presented in this section. In the early 20th century, Frahm developed the DVA as a means of mitigation of the resonant vibrations that arise in bodies subjected to periodic loading and obtained a patent on this technology (Frahm, 1911a). Later, Ormondroyd and Den Hartog (1928) and Den Hartog (1947) provided significant theoretical insight into the DVA, including the procedure for proper selection of the absorber parameters.

The DVA consists of a vibratory system, normally referred to as an auxiliary system, having a mass and a spring or stiffness element. The absorber mass, which is much smaller than that of the primary system, is attached to the primary system through the spring element. Figure 1.10 shows a schematic diagram of a primary

---

TABLE 1.2

Major Groups of Structural Vibration Control Systems with Examples

Passive	Active	Semi-active	Hybrid
• Metallic damper	• Active tendon systems	• Piezoelectric dampers	• Hybrid bracing
• Friction damper	• Active bracing system	• Semi-active stiffness dampers	• Hysteretic-viscous hybrid damper
• Tuned mass dampers	• Active tuned mass damper	• Semi-active tuned mass damper	• Hybrid mass dampers
• Tank sloshing damper	• Active tank sloshing damper	• Semi-active tank sloshing damper	• Hybrid tank sloshing damper
• Tuned liquid column damper	• Active tuned liquid column damper	• Semi-active tuned liquid column damper	• Hybrid tuned liquid column damper

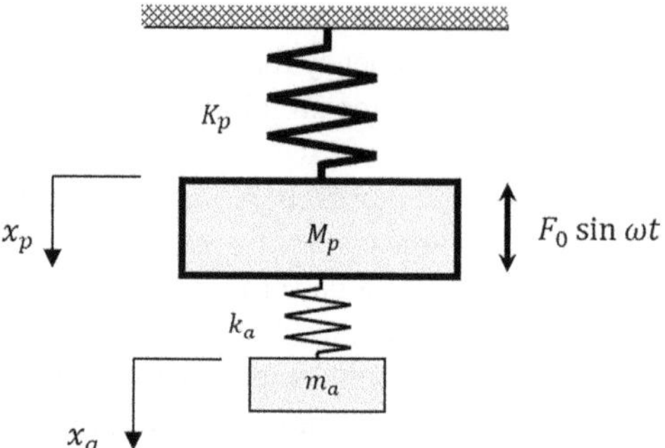

FIGURE 1.10   Primary system with attached dynamic vibration absorber (Frahm's absorber).

system with an attached DVA. The notations K_p and M_p, respectively, denote the stiffness and mass of the primary system, which is subjected to a steady harmonic force of constant frequency. The stiffness and mass of the absorber are k_a and m_a respectively. The coupled equations of motion of the primary-auxiliary system are given by

$$\begin{bmatrix} M_p & 0 \\ 0 & m_a \end{bmatrix} \begin{Bmatrix} \ddot{x}_p \\ \ddot{x}_a \end{Bmatrix} + \begin{bmatrix} K_p + k_a & -k_a \\ -k_a & k_a \end{bmatrix} \begin{Bmatrix} x_p \\ x_a \end{Bmatrix} = \begin{Bmatrix} F_0 \sin \omega t \\ 0 \end{Bmatrix}. \tag{1.2}$$

In Eq. (1.2), x_p and x_a, respectively, denote the displacement of the primary and auxiliary systems. $F_0 \sin \omega t$ represents the steady alternating force, with F_0 and ω, respectively, being the amplitude and the frequency, acting on the primary system. An overdot represents a derivative of the respective variable with respect to time, t.

Under steady-state conditions, the solutions of Eq. (1.2) may be written as

$$x_p = X_p \sin \omega t \text{ and } x_a = X_a \sin \omega t. \tag{1.3}$$

In Eq. (1.3), X_p and X_a are the amplitudes of the steady state displacement of the primary system and the auxiliary system, respectively.

The substitution of Eq. (1.3) in Eq. (1.2) leads to the following linear algebraic equations:

$$\left. \begin{array}{l} X_p\left(-M_p\omega^2 + K_p + k_a\right) - X_a k_a = F_0 \\ -X_p k_a + X_a\left(-m_a\omega^2 + k_a\right) = 0 \end{array} \right\}. \tag{1.4}$$

On solving Eq. (1.4), the expression for X_p is obtained as

$$X_p = \delta_{\text{st}} \frac{1 - \left(\dfrac{\omega}{\omega_a}\right)^2}{\left[1 - \left(\dfrac{\omega}{\omega_p}\right)^2 + \dfrac{k_a}{K_p}\right]\left[1 - \left(\dfrac{\omega}{\omega_a}\right)^2\right] - \dfrac{k_a}{K_p}}. \tag{1.5}$$

In Eq. (1.5), $\omega_p\left[= \sqrt{K_p/M_p}\right]$ and $\omega_a\left[= \sqrt{k_a/m_a}\right]$ denote the natural frequency of the primary system and of the DVA, respectively. The term δ_{st} is the static deflection of the primary system and is given by F_0/K_p.

The amplitude of the steady-state displacement of the primary system, X_p, is equal to zero when the numerator of Eq. (1.5) is zero, and this occurs when the frequency of the applied force is the same as the natural frequency of the absorber, that is, when, $\omega_a = \omega$. This implies that under this condition, the spring force would be equal and opposite to the harmonic force acting on the primary system.

The DVA was originally conceptualized to suppress excessive vibrations of machines or machine parts under the action of harmonic forces acting at frequencies that were close to the resonant frequency of the vibrating bodies (Den Hartog, 1947). However, unlike the primary system in Figure 1.10, civil engineering structures are multi-degree-of-freedom systems with an inherent energy dissipation capacity. Moreover, the environmental loads that they are subjected to are random and are composed of, not a single, but a range of frequencies. In order to adapt the DVA for application to these structures, three fundamental modifications have been made to create inertia-based energy dissipation devices. First, the frequency of the auxiliary system is designed to be nearly equal to the dominant frequency of the primary structural system to maximize the transfer of vibrational energy from the primary system to the auxiliary system through dynamic interaction (Konar and Ghosh, 2023a; McNamara, 1977), the reason why inertia-based dampers are also referred to as tuned dampers. Second, the damper systems are equipped with their own energy dissipation mechanism to damp out the vibrational energy that they receive from the primary system (McNamara, 1977). This eliminates the possibility of reverse flow of energy, that is, from the auxiliary system to the primary system. This rules out any undesirable beating phenomenon in the vibrational response of the structure. Third, the auxiliary system is connected to the structure near the location of the maximum structural displacement occurring in the mode of vibration to be controlled (Konar and Ghosh, 2023a; Moon, 2010). In general, civil engineering structures, such as buildings, chimneys, towers and bridge pylons, have maximum displacement near their top in the dominant vibrational mode. Hence, when inertia-based dampers are used in civil engineering structures, they are usually attached near the top of the structures.

The conventional TMD is the simplest form of inertia-based damper, wherein the auxiliary system is identical to that of a DVA, except for the presence of a dashpot element parallel to the spring element. A schematic of a structure-TMD system is shown in Figure 1.11. Here, the structure is represented by a single-degree-of-freedom (SDOF) system, having mass, stiffness and damping represented by M_s, K_s and C_s respectively. The mass, stiffness and damping of the TMD are denoted by m, k and c, respectively. The term $F(t)$ denotes the external force acting on the structure. The equations of motion of the structure-TMD system are given by

$$
\begin{bmatrix} M_s & 0 \\ 0 & m \end{bmatrix} \begin{Bmatrix} \ddot{u} \\ \ddot{v} \end{Bmatrix} + \begin{bmatrix} C_s + c & -c \\ -c & c \end{bmatrix} \begin{Bmatrix} \dot{u} \\ \dot{v} \end{Bmatrix} + \begin{bmatrix} K_s + k & -k \\ -k & k \end{bmatrix} \begin{Bmatrix} u \\ v \end{Bmatrix}
$$

$$
= \begin{Bmatrix} F(t) \\ G(t) \end{Bmatrix}. \tag{1.6}
$$

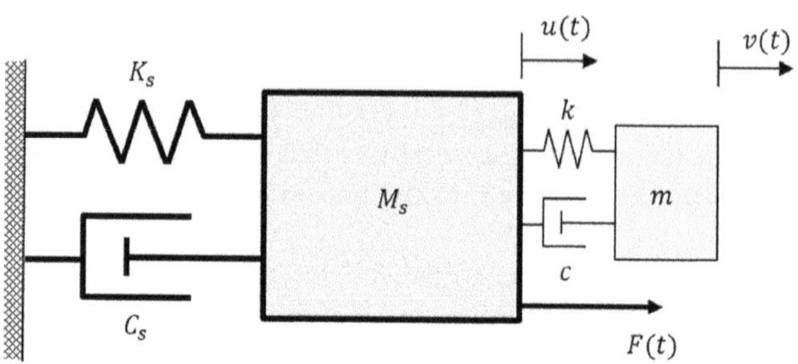

FIGURE 1.11   Schematic diagram of a structure-TMD system.

In Eq. (1.6), the displacements of the structure and the damper are represented by $u(t)$ and $v(t)$, respectively. $G(t)$ is equal to $\mu F(t)$ when the structure is base-excited and is equal to zero for other cases. The ratio of the damper mass to the structural mass, termed the mass ratio, is symbolized by $\mu = m/M_s$.

Summation of the individual differential equations in Eq. (1.6) yields

$$M_s\ddot{u} + C_s\dot{u} + K_s u = F(t) + G(t) - m\ddot{v}. \tag{1.7}$$

When $F(t)$ is a harmonic force or a stationary random input, Eq. (1.7) may be rewritten in the form of an energy or power balance equation as

$$M_s\langle\ddot{u}\dot{u}\rangle + C_s\langle\dot{u}^2\rangle + K_s\langle u\dot{u}\rangle = \langle[F(t)+G(t)]\dot{u}\rangle - m\langle\ddot{v}\dot{u}\rangle. \tag{1.8}$$

Here, $\langle\cdot\rangle$ stands for the mathematical expectation for the case of stochastic input and the time average in one cycle for the case of harmonic excitation. In case of steady-state response, the values of $\langle\ddot{u}\dot{u}\rangle$ and $\langle u\dot{u}\rangle$ are equal to zero, and Eq. (1.8) leads to

$$C_s\langle\dot{u}^2\rangle = \langle[F(t)+G(t)]\dot{u}\rangle - m\langle\ddot{v}\dot{u}\rangle. \tag{1.9}$$

In Eq. (1.9), $C_s\langle\dot{u}^2\rangle$ denotes the energy dissipated due to intrinsic damping of the structure, and $\langle[F(t)+G(t)]\dot{u}\rangle$ is the energy input from the external excitation, which is always positive. The term $m\langle\ddot{v}\dot{u}\rangle$ represents the energy flow from the structural system to the damper. This energy flow plays a fundamental role in the functioning of the TMD as a structural vibration control device. The greater the energy flow, the larger is the reduction in the root-mean-square (rms) response of the structure by the damper. The maximum flow of energy is obtained when the displacement of the damper mass lags that of the structural system by a phase angle of $\pi/2$ (Soong and Dargush, 1997). In the present case, the relative acceleration of the damper mass is in phase with the velocity response of the structural system, and the energy flow is equivalent to the effective dissipative energy, which enhances the total effective damping (C_{eq}) in the structural system. Here, C_{eq} may be expressed as (Soong and Dargush, 1997)

$$C_{eq} = C_s + m\frac{\langle\ddot{v}\dot{u}\rangle}{\langle\dot{u}^2\rangle}. \tag{1.10}$$

Den Hartog (1947) considered a case similar to that expressed by Eq. (1.6), with the only variation that the primary system was undamped, that is, $C_s = 0$. The vibration absorber in this system was referred to as a "damped vibration absorber." As civil engineering structures have a low damping ratio, normally in the range of 0.5%–5%, Den Hartog's damped vibration absorber would give a fair indication of the control effectiveness of the TMD, and the inertia-based damper, in general. Similar to the system considered by Den Hartog, let us consider that the primary system is subjected to a sinusoidal excitation with frequency ω. Thus, $F(t) = F_0 \sin \omega t$ and $G(t) = 0$. Here, the effectiveness of the damped vibration absorber is measured in terms of the dynamic amplification factor R, defined by the ratio of the maximum structural displacement under $F(t)$ to the static displacement of the structure under F_0. The expression for R, as derived by Den Hartog (1947), is

$$R = \frac{u_{max}}{u_{static}} = \sqrt{\frac{\left(\gamma^2 - \beta^2\right)^2 + \left(2\zeta\gamma\beta\right)^2}{\left[\left(\gamma^2 - \beta^2\right)\left(1 - \beta^2\right) - \gamma^2\beta^2\mu\right]^2 + \left(2\zeta\gamma\beta\right)^2\left(1 - \beta^2 - \beta^2\mu\right)^2}}. \quad (1.11)$$

In Eq. (1.11), γ denotes the frequency tuning ratio, which is the ratio of the natural frequency of the vibration absorber to that of the primary system. The forced frequency ratio β is defined by the ratio of the frequency of excitation to the natural frequency of the SDOF primary system, and ζ is the damping ratio of the vibration absorber.

The variation in R with β for $\mu = 0.05$, $\gamma = 1$ and for different ζ is presented in Figure 1.12. It can be observed that when the vibration absorber is not provided with a damping element, that is, when $\zeta = 0$, the response amplitude is infinitely large at two resonant frequencies of the combined primary-auxiliary system. On the other hand, when the damping of the auxiliary system tends to infinity, the two masses are virtually

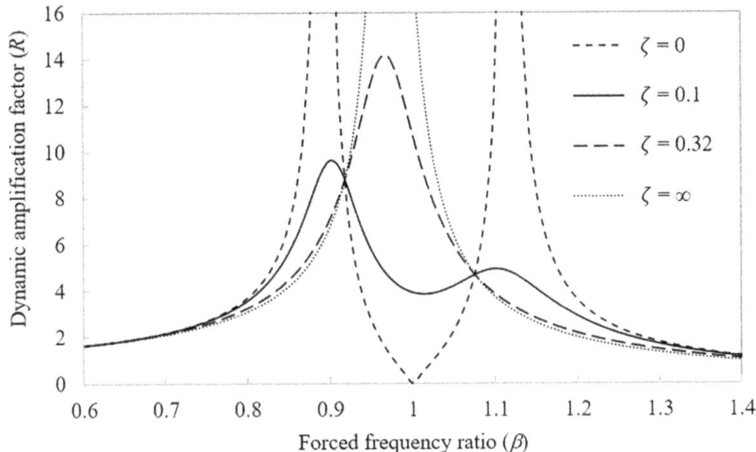

FIGURE 1.12 Variation in dynamic amplification factor (R) with forced frequency ratio (β) for mass ratio (μ) = 0.05, frequency tuning ratio (γ) = 1, and for different damping ratio of the vibration absorber (ζ).

fused to each other, and the result is an SDOF system with a mass $(1 + \mu)$ times the mass of the primary system. In this case, the amplitude at the resonant frequency is infinitely large again. Therefore, somewhere between these extremes, there must be an optimum value of ζ for which the peak of the dynamic amplification factor is minimized.

With this in view, for a structure-TMD system, the damper is designed so as to bring the resonant peak of the amplitude down to its lowest possible value, so that smaller amplifications over a wider frequency bandwidth, with β close to unity, can be attained. To achieve this objective, γ and ζ are to be optimized. The optimum values of these parameters, γ_{opt} and ζ_{opt}, depend upon the type of excitation, the mass ratio and the damping ratio of the structure. A significant amount of research has been devoted to the determination of γ_{opt} and ζ_{opt} for TMDs. In this regard, the solution for γ_{opt} and ζ_{opt} expressed as

$$\gamma_{opt} = \frac{1}{1 + \mu} \tag{1.12}$$

and

$$\zeta_{opt} = \sqrt{\frac{3\mu}{8(1+\mu)^3}} \tag{1.13}$$

derived by Den Hartog (1947) has served as a cornerstone for subsequent developments for the damped vibration absorber.

Overall, the TMD has been widely studied (Elias and Matsagar, 2017; Ghosh and Basu, 2007; Holmes, 1995; Rana and Soong, 1998; Soto and Adeli, 2013), and there are a number of real-world installations of TMDs in civil engineering structures as well. Some landmark structures equipped with a TMD system include the CN Tower, Toronto, Canada, the John Hancock Tower, Boston, USA, the Crystal Tower, Osaka, Japan, the Burj Al-Arab, Dubai, UAE and the Taipei 101, Taiwan.

Although the TMD is a successful vibration control device, it has its own limitations. The principal disadvantages of the TMD are (a) requirement of a threshold excitation for activation; (b) provision for smooth movement of the damper mass, especially at low excitation levels, leading to relatively high operational and maintenance costs; (c) requirement of space for TMD movement, which sometimes is difficult to provide; and (d) placement of a heavy solid mass near the top of the structure. These shortcomings of the TMD can be overcome by using the TLD.

1.3　TUNED LIQUID DAMPER

The TLD is an inertia-based vibration control device, in which liquid (generally water), fully or partly, constitutes the inertial mass of the damper and also contributes to the energy dissipation mechanism. Here, the restoring force (stiffness) is provided by gravity. In a TLD, the frequency of liquid motion is tuned to the frequency of the structural mode that is required to be controlled. The energy dissipation in a TLD is generally due to the sloshing of the contained liquid, with or without breaking of the sloshing waves, or by energy loss that takes place by the flow of liquid through orifices, and in some cases, due to the relative motion between the liquid and submerged objects (Konar and Ghosh, 2023a). Thus, unlike TMDs, TLDs do not require expensive external damping elements. The absence of solid moving parts and the associated maintenance also contribute to the overall cost-effectiveness of these devices. TLDs are easy to install, both in new structures and in existing structures for retrofitting. Moreover, the non-requirement of a threshold external excitation to activate the device, ease of adjustment of damper frequency and the possibility of bi-directional vibration control with a single damper are the other significant advantages of the TLD, especially in comparison to TMD. There are certain disadvantages of the device, which are discussed later in the section.

The idea of using the motion of liquid stored in a container to reduce the excessive oscillation of the container supporting system was pioneered in the 1860s (Froude, 1861). The concept was first utilized in the free-surface anti-rolling tanks developed to suppress the rolling motion of ships (Watts, 1885, 1883) (see Figure 1.13a). The nutation damper, developed in the 1960s, to control the spinning of satellites was composed of an annular container partly filled with liquid (Ayache and Lynch, 1969; Cartwright et al., 1963) (see Figure 1.13b). However, the study on the utilization of the TLD in controlling

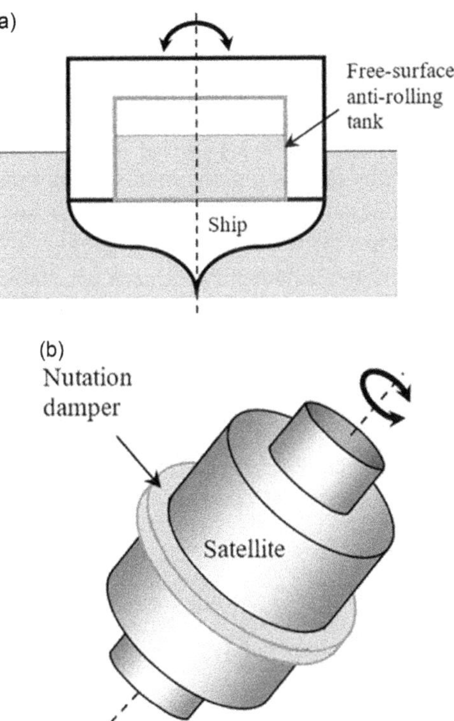

FIGURE 1.13   (a) Free surface anti-rolling tank in a ship, (b) nutation damper in a satellite (Konar and Ghosh, 2023a).

structural vibrations was initiated in the latter half of the 1970s, by Vandiver and Mitome (1978), who demonstrated that liquid sloshing in an appropriately designed container, attached to a fixed-base offshore platform could mitigate the vibration of the platform. Subsequently, Sayar and Baumgarten (1981, 1982) examined the nonlinear behaviour of the liquid sloshing damper through experimental and analytical studies. Lee and Reddy (1982) conducted a parametric study on liquid-filled containers meant for the vibration suppression of offshore platforms under wave-induced excitation and also put forward the possible methods of frequency tuning between the platform and the damper. Bauer (1984) proposed a damper consisting of a rectangular tank fully filled with two immiscible liquids to reduce wind-induced vibrations of buildings. To attenuate the oscillatory motion of structures, El-Sherif et al. (1985) suggested the use of partially filled containers with sloshing liquid. Kareem and Sun (1987) investigated the effect of liquid-containing tanks on base-excited multi-storied structures. Since the late 1980s, TLDs secured considerable attention from the engineering community for their various advantages discussed earlier (Fujino and Fujii, 1989; Modi and Welt, 1988; Noji et al., 1988; Sun et al., 1989).

As indicated in Figure 1.8, there are two major configurations of the TLD, namely the tank sloshing damper and the TLCD. However, the term TLD is conventionally used to refer to the tank sloshing damper. This is possibly because TLCD was developed much later than the tank damper, when certain difficulties of the latter became apparent to the researchers. Hence, it is important to understand the working of the tank sloshing damper to better comprehend the TLCD.

In a tank sloshing damper, liquid resides in a partially filled tank and dissipates energy through the sloshing of the damper liquid. Here, the fundamental sloshing frequency of the liquid is tuned to the frequency of the structural mode to be controlled. The liquid mass corresponding to the first sloshing mode acts as the inertial mass, while gravity provides the restoring (spring) force (see Figure 1.14) for the damper. The higher sloshing modes are generally ignored as their participation is much less in comparison to the fundamental (Das et al., 2023; Konar, 2024; Konar and Ghosh, 2022a). The liquid mass associated with the higher sloshing modes is then assumed to act with the impulsive liquid mass. Usually, containers of circular or rectangular shape are used for tank sloshing dampers. However, tanks of other shapes, such as spherical (Chen et al., 2016; Georgakis, 2011), annular (Modi and Welt, 1988; Tamura et al., 1992), rectangular with sloped bottom (Gardarsson et al., 2001; Zhang, 2020) and other complex shapes (Love and Tait, 2015), have also been studied. The objective of these works has been to improve the applicability of the

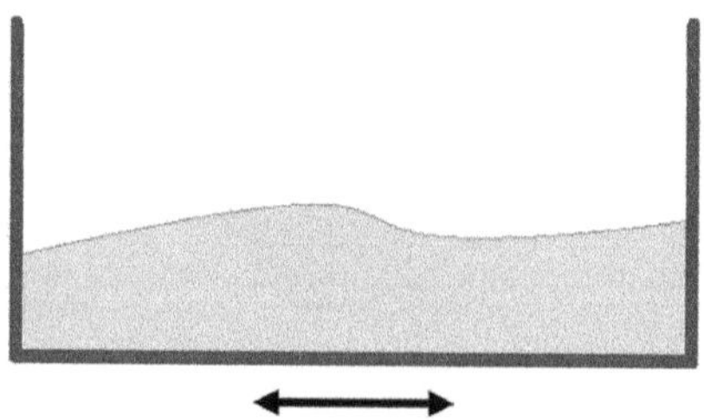

FIGURE 1.14   TLD in the form of a tank sloshing damper.

tank sloshing damper for some specific cases. For example, the annular tank damper is suitable for a cylindrical-shaped primary structure, such as a chimney. A rectangular tank damper with sloped bottom and a tank damper with a complex shape may be adopted when regular circular or rectangular-shaped tanks cannot be accommodated due to geometrical and space constraints. On the other hand, tank dampers of spherical shape have been developed for the multi-directional vibration suppression of wind turbine structures.

The fundamental frequency of liquid sloshing in a circular tank sloshing damper (Abramson et al., 1961), ω_{ac}, and in a rectangular tank sloshing damper (Lamb, 1932), ω_{ar}, may be obtained from

$$\omega_{ac} = \left[\frac{3.682g}{B} \tanh\left(\frac{3.682d}{B} \right) \right]^{1/2} \tag{1.14}$$

and

$$\omega_{ar} = \left[\frac{\pi g}{B} \tanh \frac{\pi d}{B} \right]^{1/2}. \tag{1.15}$$

In Eqs. (1.14) and (1.15), d is the depth of liquid in the damper tank at rest. B denotes the internal diameter of the damper tank for circular-shaped sloshing dampers, and the internal length of the damper tank along the direction of expected excitation for rectangular-shaped sloshing dampers. The acceleration due to gravity is denoted by g.

The liquid mass corresponding to the first mode of sloshing in a circular tank sloshing damper (Abramson et al., 1961), m_{ac}, and in a rectangular tank sloshing damper (Graham and Rodriguez, 1952), m_{ar}, may be obtained from

$$m_{ac} = M_F \frac{B}{4.4d} \tanh\left(\frac{3.682d}{B} \right) \tag{1.16}$$

and

$$m_{ar} = M_F \frac{8B}{d\pi^3} \tanh\left(\frac{\pi d}{B} \right). \tag{1.17}$$

In Eqs. (1.16) and (1.17), M_F represents the total mass of liquid in the damper tank.

Depending upon the type of sloshing waves produced under lateral excitations, tank sloshing dampers can be categorized into three groups, namely shallow tank dampers, deep tank dampers and intermediate-depth tank dampers. For practical purposes, the differentiation between these three types of tank dampers is made based on the ratio d/B. In a shallow tank damper, the lateral excitation generates travelling sloshing waves with the physical phenomenon of wave steepening or wave breaking. This happens because the water particles in the upper part of the moving wave travel faster than the water particles in the lower part. Saint-Venant equations or nonlinear shallow water wave equations are used to represent this condition. These equations are valid as long as the ratio d/B of the tank damper is within 0.1 (Antuono et al., 2014; Frandsen, 2005; Fujino et al., 1992). However, some studies have proposed slightly different limiting values of d/B, such as 0.125 (Li et al., 2012), 0.15

(Kareem, 1993) and 0.1–0.12 (Dean and Dalrymple, 1991), for shallow tank dampers. In contrast, in the fundamental mode, laterally excited deep tank dampers produce standing sloshing waves. In a deep tank damper, $d/B \geq 0.5$, and there is no motion of liquid due to sloshing waves near the bottom of the damper tank (Frandsen, 2005; Konar and Ghosh, 2022a). Tank dampers having $0.1 < d/B < 0.5$ are known as intermediate-depth tank damper, in which transitional waves are produced under lateral excitations (Frandsen, 2005; Konar, 2023).

The equivalent viscous damping ratio of a circular tank sloshing damper of shallow or intermediate depth may be obtained from the empirical formula (Ibrahim, 2005; Mikishev and Dorozhkin, 1961)

$$\zeta_{TSD} = \frac{4.86}{\pi} \sqrt{\frac{\nu}{B^{3/2} g^{1/2}}} \left[1 + \frac{0.318}{\sinh(3.68 d/B)} \left(1 + \frac{1 - (2d/B)}{\cosh(3.68 d/B)} \right) \right]. \qquad (1.18)$$

where ν is the kinematic viscosity of the damper liquid. For a deep cylindrical tank damper, Eq. (1.18) takes the form

$$\zeta_{TSD} = \frac{4.86}{\pi} \sqrt{\frac{\nu}{B^{3/2} g^{1/2}}}. \qquad (1.19)$$

In the case of rectangular tank sloshing dampers having $d/B > 1$, the equivalent viscous damping ratio may be obtained from the expression (Ibrahim, 2005)

$$\zeta_{TSD} = \sqrt{\frac{\nu}{B^{3/2} g^{1/2}}}. \qquad (1.20)$$

In the case of shallow rectangular tank sloshing dampers, the equivalent viscous damping ratio may be determined from (Fujino et al., 1992; Tait et al., 2004)

$$\zeta_{TSD} = \frac{1}{2(\eta + d)} \sqrt{\frac{\nu}{2\omega_{TSD}}} \left[1 + \left(\frac{2d}{B} \right) + S \right]. \qquad (1.21)$$

In Eq. (1.21), η and S denote the maximum amplitude of liquid sloshing and the surface contamination factor, respectively. The value of S varies between 0 and 2 (Miles, 1967). For a fully contaminated liquid surface, the value of S may be taken as unity (Konar and Ghosh, 2023b; Lepelletier and Raichlen, 1988; Miles, 1967).

Sometimes, in cases such as that of the deep tank damper, the inherent damping of a sloshing damper is not sufficient to dissipate the energy it receives from the structure during vibration. In those cases, sloshing dampers are provided with different flow damping devices, such as screens, nets, baffles and poles for enhanced damping (Konar and Ghosh, 2022b, 2021).

Real-life implementation of the tank sloshing damper started during the late 1980s in Japan, with the first being in the 42 m tall Nagasaki Airport Tower, Japan in March 1987 (Fujii et al., 1990; Tamura et al., 1995). Subsequently, sloshing dampers have been installed in structures, such as the 158 m tall Gold Tower, Utazu, Japan; the 149 m tall Shin-Yokohama Prince Hotel, Yokohama, Japan; One King

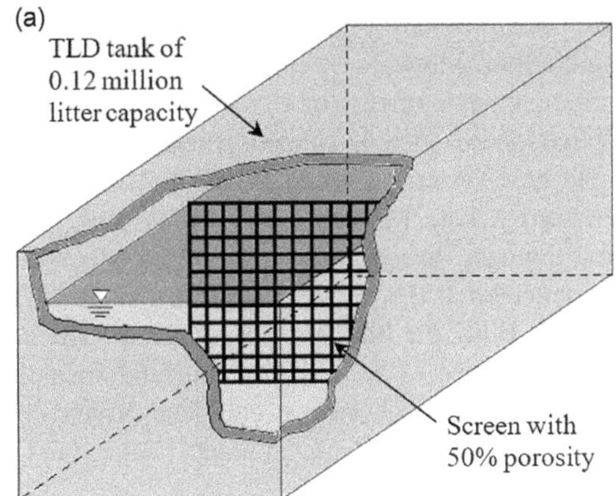

FIGURE 1.15 The 42, Kolkata, India. (a) Completed building and (b) schematic of the TLD used.

West of Toronto, a 176 m tall residential-commercial structure built upon the historic Dominion Bank building; One Rincon Hill of San Francisco, two 60 storied and 50 storied buildings; the 252 m tall Highcliff Apartments at Hong Kong; the Bai Chay Bridge, Vietnam and The 42, Kolkata, India, having a height of 250 m (see Figure 1.15a and b) (Konar and Ghosh, 2023a, 2022c).

Despite significant success as a vibration control device, tank sloshing dampers have three major disadvantages. The first is regarding the quantification of damping.

Although there are several approaches available, none of them can specifically predict the effective damping ratio of the sloshing damper over a wide range of excitation. Secondly, since only the convective liquid mass in a sloshing damper participates as the inertial mass, the remaining liquid mass that is impulsive in nature does not participate in the damping mechanism. This implies that the volumetric efficiency of the damper is low. The third disadvantage is the nonlinear behaviour of the sloshing damper. Liquid sloshing phenomena are inherently nonlinear and depend on the excitation amplitude. Although a linear assumption can provide an acceptable result at low-amplitude sloshing, when the excitation amplitude becomes high or the liquid level in the tank becomes shallow, significant nonlinearity is observed during liquid sloshing. This makes it difficult to estimate the supplemental damping being provided by the tank damper to the primary structure.

1.4 TUNED LIQUID COLUMN DAMPER

The TLCD is a type of tuned liquid damper constituting of a partially filled liquid container with a pair of opposite upstanding end portions wherein the liquid-free surface is formed (Balendra et al., 1995; Xu et al., 1992). U-shaped containers are most commonly used for TLCDs (see Figure 1.16). When the damper is laterally excited, oscillatory motion is induced in the liquid column and the frequency of oscillation of the liquid column is tuned to the structural frequency. Energy dissipation in the TLCD is caused by the passage of the liquid through orifice(s) which has intrinsic head-loss properties. This provides a definitive amount of damping that is lacking in the tank sloshing damper. Moreover, the oscillating liquid in the TLCD leads to higher volumetric efficiency in comparison to the tank sloshing damper.

The earliest application of such U-shaped partially-filled liquid containers in motion control of the host system is found in the anti-rolling tank introduced by Frahm (1911) (see Figure 1.17a). Fraham's anti-rolling tank consists of two water tanks connected near the base through a horizontal link which allows water to flow. The tanks are also connected at the top through an air valve to regulate the flow of water between the tanks. Here, the frequency of oscillatory motion of the contained water is tuned to (i.e., designed to be equal to) the fundamental rolling frequency of a ship so that this component of motion could be effectively reduced. One of the initial successful installations of Fraham's anti-rolling tank was made in S.S.

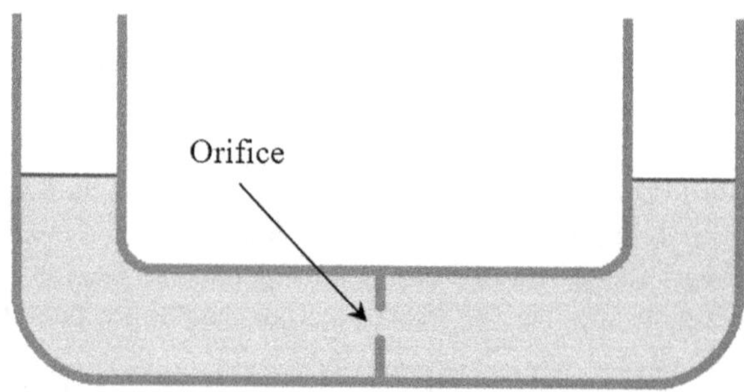

FIGURE 1.16 Tuned liquid column damper (TLCD).

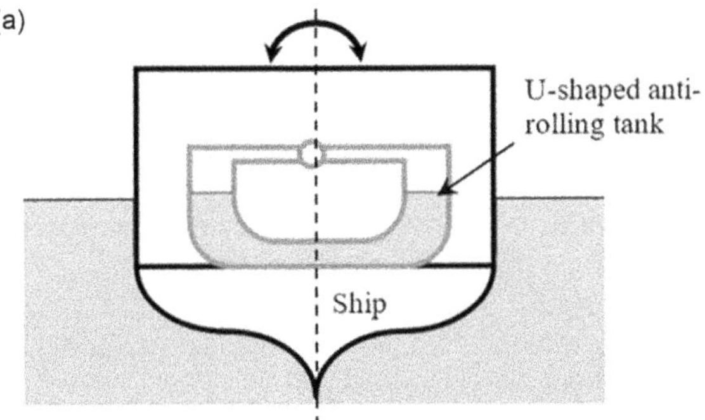

FIGURE 1.17   (a) Schematic diagram of U-shaped anti-rolling tank (Konar and Ghosh, 2023a) and (b) Official Hamburg America Line (HAPAG) postcard for SS Ypiranga (https://en.wikipedia.org/wiki/SS_Ypiranga#/media/File:Ypiranga_Postkarte.jpg).

Ypiranga (see Figure 1.17b), a German-registered cargo ship, which was reported to be highly unsteady at sea prior to the installation of the anti-rolling device (Frahm, 1911b). Subsequently, Fraham's anti-rolling tank and its improved varieties have been installed in numerous marine liners (Alujević et al., 2019; Den Hartog, 1947). However, the concept of the modern-day TLCD for structural vibration control was initiated only in the late 1980s (Saoka et al., 1988). The first US patent in this regard was obtained by Sakai et al. (1991) under the title "Damping device for tower-like structure." Since then, the TLCD is fast emerging as one of the most promising vibration control devices. This is because, apart from the well-defined damping mechanism that readily allows the quantification of the supplemental damping value, and the fact that for a given volume of damper liquid, the TLCD is more effective as compared to other varieties of liquid dampers, it also produces a consistent performance across a broad range of excitation levels and can be easily tuned to different frequencies by simply varying the length of the liquid column.

It is pertinent to note that although active, semi-active and hybrid forms of the TLCD have been studied and developed, most of the research efforts on the TLCD have been directed towards its passive variety. This is because passive dampers are, in general, economical and easy to install and maintain.

In the upcoming chapters, a detailed account of the TLCD, including its working mechanism, configurational variations, design and performance under different loads and implementational issues is presented.

REFERENCES

Al-Hajj, S., Dhaini, H.R., Mondello, S., Kaafarani, H., Kobeissy, F., DePalma, R.G., 2021. Beirut Ammonium nitrate blast: analysis, review, and recommendations. *Front. Public Heal.* 9, 657996. https://doi.org/10.3389/fpubh.2021.657996

Alujević, N., Ćatipović, I., Malenica, Š., Senjanović, I., Vladimir, N., 2019. Ship roll control and power absorption using a U-tube anti-roll tank. *Ocean Eng.* 172, 857–870. https://doi.org/10.1016/j.oceaneng.2018.12.007

Antuono, M., Bardazzi, A., Lugni, C., Brocchini, M., 2014. A shallow-water sloshing model for wave breaking in rectangular tanks. *J. Fluid Mech.* 746, 437–465. https://doi.org/10.1017/jfm.2014.127

Ayache, K., Lynch, R., 1969. Analyses of the performance of liquid dampers for nutation in spacecraft. *J. Spacecr. Rockets* 6, 1385–1389. https://doi.org/10.2514/3.29835

Baker, C.J., 2007. Wind engineering: past, present and future. *J. Wind Eng. Ind. Aerodyn.* 95, 843–870. https://doi.org/10.1016/j.jweia.2007.01.011

Balendra, T., Wang, C.M., Cheong, H.F., 1995. Effectiveness of tuned liquid column dampers for vibration control of towers. *Eng. Struct.* 17, 668–675. https://doi.org/10.1016/0141-0296(95)00036-7

Bauer, H.F., 1984. Oscillations of immiscible liquids in a rectangular container: a new damper for excited structures. *J. Sound Vib.* 93, 117–133. https://doi.org/10.1016/0022-460X(84)90354-7

Bayraktar, A., Keypour, H., Naderzadeh, A., 2012. Application of ancient earthquake resistant method in modern construction technology, in: *15th World Conference on Earthquake Engineering (15WCEE)*. Lisbon, Portugal, p. 9.

Carpani, B., 2017. Base Isolation drom a historical perspective, in: *16WCEE*, Santiago, Chile, Paper No. 4934

Cartwright, W.F., Massingill, E.C., Trueblood, R.D., 1963. Circular constraint nutation damper. *AIAA J.* 1, 1375–1380. https://doi.org/10.2514/3.1797

Chen, J., Zhan, G., Zhao, Y., 2016. Application of spherical tuned liquid damper in vibration control of wind turbine due to earthquake excitations. *Struct. Des. Tall Spec. Build.* 24, 421–439. https://doi.org/10.1002/tal.1266

Christenson, R.E., 2001. *Semi-Active Control of Civil Structures for Natural Hazard Mitigation: Analytical and Experimental Studies.* University of Notre Dame, Notre Dame, IN

Cooper, A.F., 1870. Improved foundation for buildings. US100262A.

CRED, 2019. *The Human Cost of Disasters: An Overview of the Last 20 years 2000–2019.* CRED, Brussels.

CRED, 2022. *Disasters Year in Review 2021.* CRED, Brussels.

Daka, T., Udatha, L., Pasupuleti, V.D.K., Kalapatapu, P., Rajaram, B., 2018. Ancient sandbox technique: an experimental study using piezoelectric sensors, Lecture Notes in Computer Science. Springer International Publishing. https://doi.org/10.1007/978-3-030-01765-1_20

Dallard, P., Fitzpatrick, T., Flint, A., Low, A., Smith, R.R., Willford, M., Roche, M., 2001. London millennium bridge: pedestrian-induced lateral vibration. *J. Bridg. Eng.* 6, 412–417.

Das, A., Konar, T., Maity, D., Bhattacharyya, S.K., 2023. Revisiting equivalent tuned mass damper analogy for tuned liquid damper utilizing CFD-FEA framework. *Soil Dyn. Earthq. Eng.* 172, 108051. https://doi.org/10.1016/j.soildyn.2023.108051

Dean, R.G., Dalrymple, R.A., 1991. *Water Wave Mechanics for Engineers and Scientists.* World Scientific Publishing Company, Singapoore. https://doi.org/10.1142/9789812385512

Den Hartog, J.P., 1947. Mechanical Vibrations, third ed. McGraw-Hill, New York.

Eguchi, R.T., Goltz, J.D., Taylor, C.E., Chang, S.E., Flores, P.J., Johnson, L.A., Seligson, H.A., Blais, N.C., 1998. Direct economic losses in the northridge earthquake: a three-year post-event perspective. *Earthq. Spectra* 14. https://doi.org/10.1193/1.1585998

Elias, S., Matsagar, V., 2017. Research developments in vibration control of structures using passive tuned mass dampers. *Annu. Rev. Control* 44, 129–156. https://doi.org/10.1016/j.arcontrol.2017.09.015

El-Sherif, F.S., Farghaly, S.H., Farghaly, M.H., El-Mandy, T.H., 1985. Analysis of fluid sloshing in tanks for vibration control, in: *ASAT Conferance Proceedings.* pp. 347–356. https://doi.org/10.21608/ASAT.1985.26514

Frahm, H., 1911a. Device for damping vibrations of bodies. US989958A.

Frahm, H., 1911b. Results of trials of the anti-rolling tanks at sea. *Trans. Inst. Nav. Archit.* 53, 183–201.

Frandsen, J.B., 2005. Numerical predictions of tuned liquid tank structural systems. *J. Fluids Struct.* 20, 309–329. https://doi.org/10.1016/j.jfluidstructs.2004.10.003

Froude, W., 1861. On the rolling of ships. *Trans. Inst. Nav. Archit.* 2, 180–227.

Fujii, K., Tamura, Y., Sato, T., Wakahara, T., 1990. Wind induced vibration of tower and practical applications of tuned sloshing damper. *J. Wind Eng. Ind. Aerodyn.* 33, 263–272.

Fujino, Y., Fujii, K., 1989. Tuned liquid damper and its application to tower-like structures, in: *Structural Design, Analysis and Testing, ASCE Structures Congress'89,* San Francisco, USA, pp. 490–499.

Fujino, Y., Sun, L., Pacheco, B.M., Member, A., Chaiseri, P., Limin, S., Pacheco, B.M., Chaiseri, P., Sun, L., Pacheco, B.M., Chaiseri, P., 1992. Tuned liquid damper (TLD) for supressing horizontal motion of structures. *J. Eng. Mech.* 118, 2017–2030.

Gardarsson, S., Yeh, H., Reed, D., 2001. Behavior of sloped-bottom tuned liquid dampers. *J. Eng. Mech.* 127, 266–271.

Georgakis, C.T., 2011. United States Patent No. US 7971397B2.

Ghosh, A.D., Basu, B., 2007. A closed-form optimal tuning criterion for TMD in damped structures. *Struct. Control Heal. Monit.* 14, 681–692. https://doi.org/10.1002/stc.176

Guevara-Perez, L.T., 2008. Seismic regulations versus modern architectural and urban configurations, in: *14th World Conference on Earthquake Engineering.* Beijing, China.

Holmes, J.D., 1995. Listing of installations. *Eng. Struct.* 17, 676–678. https://doi.org/10.1016/0141-0296(95)90027-6

https://commons.wikimedia.org/wiki/File:Tacoma-narrows-bridge-collapse.jpg, 2010.

Hussain, E., Kalaycıoğlu, S., Milliner, C.W.D., Çakir, Z., 2023. Preconditioning the 2023 Kahramanmaraş (Türkiye) earthquake disaster. *Nat. Rev. Earth Environ.* 4, 287–289. https://doi.org/10.1038/s43017-023-00411-2

Ibrahim, R.A., 2005. *Liquid Sloshing Dynamics: Theory and Applications.* Cambridge University Press, London, UK.

Kareem, A., 1993. Liquid tuned mass damper: past, present and future, in: *7th US National Conference on Wind Engineering.* Los Angeles, CA.

Kareem, A., Sun, W.J., 1987. Stochastic response of structures with fluid-containing appendages. *J. Sound Vib.* 119, 389–408. https://doi.org/10.1016/0022-460X(87)90405-6

Konar, T., 2023. Design of intermediate - depth tuned liquid damper with horizontal baffles for seismic control and carbon footprint reduction of buildings. *J. Vib. Eng. Technol.* https://doi.org/10.1007/s42417-023-01005-4

Konar, T., 2024. Seismic vibration control of a building by overhead water tank designed as slender tuned sloshing damper. *Pract. Period. Struct. Des. Constr.* 29, 1–11. https://doi.org/10.1061/ppscfx.sceng-1393

Konar, T., Ghosh, A.D., 2021. Flow damping devices in tuned liquid damper for structural vibration control: a review. *Arch. Comput. Methods Eng.* 28, 2195–2207. https://doi.org/10.1007/s11831-020-09450-0

Konar, T., Ghosh, A.D., 2022a. Use of deep liquid-containing tanks as dynamic vibration absorbers for lateral vibration control of structures: a review. *Iran. J. Sci. Technol. Trans. Civ. Eng.* 46, 753–769. https://doi.org/10.1007/s40996-021-00679-8

Konar, T., Ghosh, A.D., 2022b. Enhanced damping in a TLD by slat screens and horizontal baffles: a comparative study, in: Sudarshan, T.S., Pandey, K.M., Misra, R.D., Patowari, P.K., Bhaumik, S. (Eds.), *Recent Advancements in Mechanical Engineering-Select Proceedings of ICRAME 2021.* Springer Nature, Singapore, pp. 287–295. https://doi.org/10.1007/978-981-19-3266-3_22

Konar, T., Ghosh, A.D., 2022c. A study on the economic perspective of utilizing liquid tanks as dynamic vibration absorbers in building structures, in: Babu, S. A. (Ed.), *5th World Congress on Disaster Management*, Vol. I. Routledge, London, UK, pp. 115–122. https://doi.org/10.4324/9781003341956

Konar, T., Ghosh, A.D., 2023a. A review on various configurations of the passive tuned liquid damper. *J. Vib. Control* 29, 1945–1980. https://doi.org/10.1177/10775463221074077

Konar, T., Ghosh, A.D., 2023b. Adaptive design of an overhead water tank as dynamic vibration absorber for buildings by use of a stiffness-varying support arrangement. *J. Vib. Eng. Technol.* 11, 827–843. https://doi.org/10.1007/s42417-022-00611-y

Lago, A., Trabucco, D., Wood, A., 2019. *Damping Technologies for Tall Buildings.* Butterworth-Heinemann, Oxford, UK. https://doi.org/10.1016/b978-0-12-815963-7.00008-7

Lamb, S., Kwok, K.C.S., 2019. The effects of motion sickness and sopite syndrome on office workers in an 18-month field study of tall buildings. *J. Wind Eng. Ind. Aerodyn.* 186, 105–122. https://doi.org/10.1016/j.jweia.2019.01.004

Lee, S.C., Reddy, D. V, 1982. Frequency tuning of offshore platforms by liquid sloshing. *Appl. Ocean Res.* 4, 226–231. https://doi.org/10.1016/S0141-1187(82)80029-1

Lepelletier, F., Raichlen, T.G., 1988. Nonlinear oscillations in rectangular tanks. *J. Eng. Mech.* 114, 1–23.

Lewis, S., 2017. The 10 largest base-isolated buildings in the world. *Eng. News-Record.* https://www.enr.com/articles/42366-the-10-largest-base-isolated-buildings-in-the-world

Li, H.N., Yi, T.H., Jing, Q.Y., Huo, L.S., Wang, G.X., 2012. Wind-induced vibration control of Dalian international trade mansion by tuned liquid dampers. *Math. Probl. Eng.* 2012. https://doi.org/10.1155/2012/848031

Love, J.S., Tait, M.J., 2015. The response of structures equipped with tuned liquid dampers of complex geometry. *J. Vib. Control* 21, 1171–1187. https://doi.org/10.1177/1077546313495074

Luca, A. De, Guidi, L.G., 2019. State of art in the worldwide evolution of base isolation design. *Soil Dyn. Earthq. Eng.* 125, 105722. https://doi.org/10.1016/j.soildyn.2019.105722

Makris, N., 2019. Seismic isolation: early history. *Earthq. Eng. Struct. Dyn.* 48, 269–283. https://doi.org/10.1002/eqe.3124

McNamara, R.J., 1977. Tuned mass dampers for towers and buildings. *J. Struct. Div.* 103, 1785–1798. https://doi.org/10.1061/JSDEAG.0004721

Mikishev, G.N., Dorozhkin, N.Y., 1961. An experimental investigation of free oscillations of a liquid in containers. *Izv. Akad. Nauk SSSR, Otd. Tekhnicheskikh Nauk. Mekhanika i Mashinostr.* 4, 48–83.

Miles, J.W., 1967. Surface-wave damping in closed basins. *Proc. R. Soc. London. Ser. A. Math. Phys. Sci.* 297, 459–475. https://doi.org/10.1098/rspa.1967.0081

Modi, V.J., Welt, F., 1988. Damping of wind induced oscillations through liquid sloshing. *J. Wind Eng. Ind. Aerodyn.* 30, 85–94. https://doi.org/10.1016/0167-6105(88)90074-8

Moon, K.S., 2010. Vertically distributed multiple tuned mass dampers in tall buildings: performance analysis and preliminary design. *Struct. Des. Tall Spec. Build.* 19, 347–366.

Munich RE, 2018. *NatCatSERVICE: Relevant Natural Loss Events Worldwide 1980–2018.* https://natcatservice.munichre.com

Noji, T., Yoshida, H., Tatsumi, E., Kosaka, H., Hagiuda, H., 1988. Study on vibration control damper utilizing sloshing of water. *J. Wind Eng.* 37, 557–566.

Ormondroyd, J., Den Hartog, J.P., 1928. *The Theory of the Dynamic Vibration Absorber.* ASME APM-50-7, New York, USA.

Pryor, F., 2003. *Britain BC: Life in Britain and Ireland before the Romans.* Harper Collins, London. https://doi.org/10.1080/00665983.2004.11020579

Rana, R., Soong, T.T., 1998. Parametric study and simplified design of tuned mass dampers. *Eng. Struct.* 20, 193–204. https://doi.org/10.1016/S0141-0296(97)00078-3

Renlund, M., 2008. https://www.flickr.com/photos/deltamike/2423266991/

Saaed, T.E., Nikolakopoulos, G., Jonasson, J.E., Hedlund, H., 2015. A state-of-the-art review of structural control systems. *J. Vib. Control* 21, 919–937. https://doi.org/10.1177/1077546313478294

Sakai, F., Takaeda, S., Tamaki, T., 1991. Damping device for tower-like structure. US5070663.

Saoka, Y., Sakai, F., Takaeda, S., Tamaki, T., 1988. On the suppression of vibrations by tuned liquid column dampers, in: *Annual Meeting of JSCE.* JSCE, Tokyo, Japan.

Sayar, B.A., Baumgarten, J.R., 1981. Pendulum analogy for nonlinear fluid oscillations in spherical containers. *J. Appl. Mech.* 48, 769–772. https://doi.org/10.1115/1.3157731

Sayar, B.A., Baumgarten, J.R., 1982. Linear and nonlinear analysis of a fluid slosh dampers. *AIAA J.* 20, 1534–1538.

Shu, Z., Li, S., Sun, X., He, M., 2019. Performance-based seismic design of a pendulum tuned mass damper system. *J. Earthq. Eng.* 23, 334–355. https://doi.org/10.1080/13632469.2017.1323042

Sinha, R., Brzev, S., Kharel, G., 2004. Indigenous earthquake-resistant technologies: an overview, in: *13th World Conference on Earthquake Engineering.* Vancouver, Canada, p. 5053.

Soong, T.T., Dargush, G.F., 1997. *Passive Energy Dissipation Systems in Structural Engineering.* Wiley, Chichester, UK.

Soto, M.G., Adeli, H., 2013. Tuned mass dampers. *Arch. Comput. Methods Eng.* 19, 419–431. https://doi.org/10.1007/978-3-319-72541-3_18

Stevenson, D., 1868. Notice of aseismatic arrangements, adapted to structures in countries subject to earthquake shocks. *Trans. R. Scottish Soc. Arts* 7, 557–566.

Stiros, S.C., 2020. Monumental articulated ancient Greek and Roman columns and temples and earthquakes : archaeological, historical, and engineering approaches. *J. Sesimol.* 4, 853–881. https://doi.org/10.1007/s10950-019-09902-6

Sun, L.M., Fujino, Y., Pacheco, B.M., Isobe, M., 1989. Nonlinear waves and dynamic pressures in rectangular TLD - simulation and experimental verification. *Struct. Eng. Earthq. Eng.* 6, 81–92.

Tait, M.J., El Damatty, A.A., Iayumov, N., 2004. Testing of tuned liquid damper with screens and development of equivalent TMD model. *Wind Struct. An Int. J.* 7, 215–234. https://doi.org/10.12989/was.2004.7.4.215

Tamura, Y., Fujii, K., Ohtsuki, T., Wakahara, T., Kohsaka, R., 1995. Effectiveness of tuned liquid dampers under wind excitations. *Eng. Struct.* 17, 609–621. https://doi.org/10.101 6/0141-0296(95)00031-2

Tamura, Y., Kousaka, R., Modi, V.J., 1992. Practical application of nutation damper for suppressing wind-induced vibrations of airport towers. *J. Wind Eng. Ind. Aerodyn.* 41–44, 1919–1930.

Touaillon, J., 1870. Improvement in Buildings. US99973A.

Vandiver, J.K., Mitome, S., 1978. The effect of liquid storage tanks on the dynamic response of offshore platforms, in: *10th Annual Offshore Technology Conference*. Houston, USA, pp. 993–1000. https://doi.org/10.4043/3162-MS

Watts, P., 1883. On a method of reducing the rolling of ships at sea. *Trans. Inst. Nav. Archit.* 24, 165–191.

Watts, P., 1885. The use of water chambers for reducing the rolling of ships at sea. *Trans. Inst. Nav. Archit.* 26, 30–49.

Welch, S., 2021. https://commons.wikimedia.org/wiki/File:13th_century_Ramappa_temple,_ Rudresvara,_Palampet_Telangana_India_-_02.jpg

Wu, H., 2011. Analysis of vibration control due to strong earthquakes. *Adv. Mater. Res.* 163– 167, 4179–4184. https://doi.org/10.4028/www.scientific.net/AMR.163-167.4179

Xu, Y.L., Kwok, K.C.S., Samali, B., 1992. The effect of tuned mass dampers and liquid dampers on cross-wind response of tall/slender structures. *J. Wind Eng. Ind. Aerodyn.* 40, 33–54. https://doi.org/10.1016/0167-6105(92)90519-G

Yamada, M., Goto, T., 1975. The criteria to motions in tall buildings, in: *Pan-Pacific Tall Buildings Conference*. Hawaii, pp. 233–244.

Yu, E., Xu, G., Han, Y., Hu, P., Townsend, J.F., Li, Y., 2022. Bridge vibration under complex wind field and corresponding measurements: a review. *J. Traffic Transp. Eng.* (English Ed. In Press, 1–24. https://doi.org/10.1016/j.jtte.2021.12.001

Zhang, Z., 2020. Numerical and experimental investigations of the sloshing modal properties of sloped-bottom tuned liquid dampers for structural vibration control. *Eng. Struct.* 204, 110042. https://doi.org/10.1016/j.engstruct.2019.110042

MATLAB® CODE 1.1

```
% Control of an SDOF structure with TMD under harmonic base excitation
clear all
clc
close all
% Definition of general constants
g=9.81; % gravitational acceleration, unit m/s^2
pi=3.141593;

% Input structural parameters
Ms=1000000; % mass of the structure, unit kg
ZETAs=0.005; % damping ratio of the structure
Ts=1.5; % natural period of the structure, unit s

% Input damper parameters
mu=0.01; % mass ratio
ZETAd=0.01; % damper damping ratio
tunr=0.95; % tuning ratio
OMEGAs=2*pi/Ts;
Md=Ms*mu;
OMEGAd=tunr*OMEGAs;
ks=Ms*OMEGAs*OMEGAs;
```

```matlab
cs=2*ZETAs*sqrt(Ms*ks);
kd=Md*OMEGAd*OMEGAd;
cd=2*ZETAd*sqrt(Md*kd);

% Input harmonic base excitation in the form of a sine wave
dt=0.02; % time interval of base excitation, unit s
t=0:dt:300; % duration of excitation taken as 300 s
A=0.5; % amplitude of base acceleration, unit m/s^2
beta=0.98; % forced frequency ratio
omg=OMEGAs*beta;
abc=A*sin(omg*t);
[n,m]=size(abc);

Gabc=abc;

plot(t,(Gabc/9.81))
set(gca,{'FontName','FontSize'},{'Times New Roman',25})
set(gca,{'XMinorTick','YMinorTick'},{'on','on'})
xlabel('Time (s)')
ylabel('Base acceleration (g)')

% Determination of response of structure with TMD
yy=zeros(4,1);
tt=0;
Gabcd=0;
Yin=[0 0 0 0];
time=[];
dispS=[];
velS=[];
dispD=[];
velD=[];
stateSD=[];
accelS=[];

for i=1:(m*n)

   ttf=tt+dt;

[T, Y]=ode45(@(tt,yy) rk_ma_damp(tt,yy,Gabcd,Ms,ks,cs,Md,kd,cd),[tt
ttf],Yin); % Solve ODE

   time=[time; T(:,1)];
   dispS=[dispS; Y(:,1)];
   velS=[velS; Y(:,2)];
   dispD=[dispD; Y(:,3)];
   velD=[velD; Y(:,4)];
   stateSD=[stateSD; Y(:,:)];

   xdd=-(ks/Ms).*Y(:,1)...
     -(cs/Ms).*Y(:,2)...
     +(kd/Ms).*Y(:,3)...
     +(cd/Ms).*Y(:,4)...
     -Gabcd;
```

```matlab
accelS=[accelS; xdd];

    tt=ttf;
    Gabcd=Gabc(1,i);

    Yin=Y(end,:);
end

% Determination of response of uncontrolled structure
tts=0;
Gabcd=0;
Yin=[0 0];
times=[];
yys=zeros(2,1);
disp=[];
vel=[];
state=[];
accel=[];
for i=1:(m*n)
   ttsf=tts+dt;
   [T, Y]=ode45(@(tts,yys) rk_str(tts,yys,Gabcd,OMEGAs,ZETAs),[tts
ttsf],Yin); % Solve ODE

   times=[times; T(:,1)];
   disp=[disp; Y(:,1)];
   vel=[vel; Y(:,2)];
   state=[state; Y(:,:)];

   xdd=-(OMEGAs^2).*Y(:,1)-(2*ZETAs*OMEGAs).*Y(:,2)-Gabcd;
   accel=[accel; xdd];

   tts=ttsf;
   Gabcd=Gabc(1,i);
   Yin=Y(end,:);
end

% Determination of control effectiveness of TMD
        rms=sqrt(mean(disp.^2));
        rmsd=sqrt(mean(dispS.^2));
        redrms_disp=(rms-rmsd)*100/rms

        peak=max(abs(disp));
        peakd=max(abs(dispS));
        redpeak_disp=(peak-peakd)*100/peak

        rms=sqrt(mean(vel.^2));
        rmsd=sqrt(mean(velS.^2));
        redrms_vel=(rms-rmsd)*100/rms

        peak=max(abs(vel));
        peakd=max(abs(velS));
        redpeak_vel=(peak-peakd)*100/peak
```

```
        rms=sqrt(mean(accel.^2));
        rmsd=sqrt(mean(accelS.^2));
        redrms_accel=(rms-rmsd)*100/rms

        peak=max(abs(accel));
        peakd=max(abs(accelS));
        redpeak_accel=(peak-peakd)*100/peak

% Plots of time histories

        figure

plot(times,disp,'r-',time,dispS)
set(gca,{'FontName','FontSize'},{'Times New Roman',35})
set(gca,{'XMinorTick','YMinorTick'},{'on','on'})
legend(' without damper',' with damper')
xlabel('Time (s)')
ylabel('Structural displacement (m)')
grid on

figure
subplot(3,1,1); plot(times,disp,'r-',time,dispS)
legend('without damper','with damper')
xlabel('Time (s)')
ylabel('Displacement (m)')
grid on
subplot(3,1,2); plot(times,vel,'r-',time,velS)
legend('without damper','with damper')
xlabel('Time (s)')
ylabel('Velocity (m/s)')
grid on
subplot(3,1,3); plot(times,accel,'r-',time,accelS)
legend('only structure','structure with damper')
xlabel('Time (s)')
ylabel('Acceleration (m/s^2)')
hold on
grid on

function[dydtt]= rk_ma_damp(tt,yy,Gabcd,Ms,ks,cs,Md,kd,cd)
dydtt=zeros(4,1);  % a column vector

% Structural displacement=yy(1), velocity=dydtt(1)=yy(2)

dydtt(1)=yy(2);

% Structural acceleration
dydtt(2)=-(ks/Ms)*yy(1)...
    -(cs/Ms)*yy(2)...
    +(kd/Ms)*yy(3)...
    +(cd/Ms)*yy(4)...
    -Gabcd;

% TMD displacement=yy(3), velocity=dydtt(3)=yy(4)
dydtt(3)=yy(4);
```

```matlab
% TMD acceleration
dydtt(4)=-(kd/Md)*yy(3)...
    -(cd/Md)*yy(4)...
    +(cs/Ms)*yy(2)...
    +(ks/Ms)*yy(1)...
    -(kd/Ms)*yy(3)...
    -(cd/Ms)*yy(4);

end
function[dydtt] = rk_str(tts,yys,Gabcd,OMEGAs,ZETAs)
dydtt=zeros(2,1);   % a column vector

% Structural displacement=yys(1), velocity=dydtt(1)=yys(2)

dydtt(1)=yys(2);

% Structural acceleration

dydtt(2)=-(OMEGAs^2)*yys(1)-(2*ZETAs*OMEGAs)*yys(2)-Gabcd;

end
```

MATLAB® CODE 1.2

```matlab
% Control of an SDOF structure with TLD under harmonic base excitation
% TLD is considered to be a circular tank damper
% Tank damper is modelled as an equivalent mechanical system
clear all
clc
close all
% Definition of general constants
g=9.81; % gravitational acceleration, unit m/s^2
pi=3.141593;

% Input structural parameters
Ms=1000000; % mass of the structure, unit kg
ZETAs=0.005; % equivalent viscous damping ratio of the structure
Ts=1.5; % natural period of the structure, unit s

% Input damper parameters
d=1; % diameter of tank, unit m
ro=1000; % density of damper liquid, unit kg/m^3
nu=0.000001; % kinematic viscosity of damper liquid, unit m^2/s
ZETA_fdd=0.01; % additional damping due to flow damping device
% in case of absence of any flow damping device set ZETA_fdd=0
tunr=1.0; % tuning ratio
mu=0.01; % mass ratio (in terms of total liquid mass)
OMEGAs=2*pi/Ts;
OMEGAd=tunr*OMEGAs;
h=d/3.682*atanh(OMEGAd*OMEGAd*d/3.682/g);
liquid_height_in_damper=h % unit m
N=round(Ms*mu/(pi*d*d*h/4)/ro);
nunber_of_damper_tanks_required=N
mu=N*(pi*d*d*h/4)*ro/Ms;
```

```matlab
Md=Ms*mu*d/4.4/h*tanh(3.684*h/d);
ZETAd0=4.86/pi*sqrt(nu/d^1.5/g^0.5)*(1+(0.318/sinh(3.68*h/d))...
    *(1+(1-(2*h/d))/cosh(3.68*h/d)));
ZETAd=ZETAd0+ZETA_fdd;

ks=Ms*OMEGAs*OMEGAs;
cs=2*ZETAs*sqrt(Ms*ks);
kd=Md*OMEGAd*OMEGAd;
cd=2*ZETAd*sqrt(Md*kd);

% Input harmonic base excitation in the form of a sine wave
dt=0.02; % time interval of base excitation, unit s
t=0:dt:300; % duration of excitation taken as 300 s
A=0.5; % amplitude of base acceleration, unit m/s^2
beta=0.98; % forced frequency ratio
omg=OMEGAs*beta;
abc=A*sin(omg*t);
[n,m]=size(abc);
Gabc=abc;

plot(t,(Gabc/9.81))
set(gca,{'FontName','FontSize'},{'Times New Roman',25})
set(gca,{'XMinorTick','YMinorTick'},{'on','on'})
xlabel('Time (s)')
ylabel('Base acceleration (g)')

% Determination of response of structure with TLD
yy=zeros(4,1);
tt=0;
Gabcd=0;
Yin=[0 0 0 0];
time=[];
dispS=[];
velS=[];
dispD=[];
velD=[];
stateSD=[];
accelS=[];

for i=1:(m*n)

  ttf=tt+dt;

[T, Y]=ode45(@(tt,yy) rk_ma_damp(tt,yy,Gabcd,Ms,ks,cs,Md,kd,cd),[tt
ttf],Yin); % Solve ODE

    time=[time; T(:,1)];
    dispS=[dispS; Y(:,1)];
    velS=[velS; Y(:,2)];
    dispD=[dispD; Y(:,3)];
    velD=[velD; Y(:,4)];
    stateSD=[stateSD; Y(:,:)];
```

```matlab
    xdd=-(ks/Ms).*Y(:,1)...
      -(cs/Ms).*Y(:,2)...
      +(kd/Ms).*Y(:,3)...
      +(cd/Ms).*Y(:,4)...
      -Gabcd;
    accelS=[accelS; xdd];
    tt=ttf;
    Gabcd=Gabc(1,i);
    Yin=Y(end,:);
end

% Determination of response of uncontrolled structure
tts=0;
Gabcd=0;
Yin=[0 0];
times=[];
yys=zeros(2,1);
disp=[];
vel=[];
state=[];
accel=[];
for i=1:(m*n)
  ttsf=tts+dt;

   [T, Y]=ode45(@(tts,yys) rk_str(tts,yys,Gabcd,OMEGAs,ZETAs),[tts
ttsf],Yin); % Solve ODE

    times=[times; T(:,1)];
    disp=[disp; Y(:,1)];
    vel=[vel; Y(:,2)];
    state=[state; Y(:,:)];

    xdd=-(OMEGAs^2).*Y(:,1)-(2*ZETAs*OMEGAs).*Y(:,2)-Gabcd;
    accel=[accel; xdd];

    tts=ttsf;
    Gabcd=Gabc(1,i);
    Yin=Y(end,:);
end

% Determination of control effectiveness of TLD
        rms=sqrt(mean(disp.^2));
        rmsd=sqrt(mean(dispS.^2));
        redrms_disp=(rms-rmsd)*100/rms

        peak=max(abs(disp));
        peakd=max(abs(dispS));
        redpeak_disp=(peak-peakd)*100/peak

        rms=sqrt(mean(vel.^2));
        rmsd=sqrt(mean(velS.^2));
        redrms_vel=(rms-rmsd)*100/rms
```

```
        peak=max(abs(vel));
        peakd=max(abs(velS));
        redpeak_vel=(peak-peakd)*100/peak

        rms=sqrt(mean(accel.^2));
        rmsd=sqrt(mean(accelS.^2));
        redrms_accel=(rms-rmsd)*100/rms

        peak=max(abs(accel));
        peakd=max(abs(accelS));
        redpeak_accel=(peak-peakd)*100/peak

% Plots of time histories

        figure

plot(times,disp,'r-',time,dispS)
set(gca,{'FontName','FontSize'},{'Times New Roman',35})
set(gca,{'XMinorTick','YMinorTick'},{'on','on'})
legend(' without damper',' with damper')
xlabel('Time (s)')
ylabel('Structural displacement (m)')
grid on

figure
subplot(3,1,1); plot(times,disp,'r-',time,dispS)
legend('without damper','with damper')
xlabel('Time (s)')
ylabel('Displacement (m)')
grid on
subplot(3,1,2); plot(times,vel,'r-',time,velS)
legend('without damper','with damper')
xlabel('Time (s)')
ylabel('Velocity (m/s)')
grid on
subplot(3,1,3); plot(times,accel,'r-',time,accelS)
legend('only structure','structure with damper')
xlabel('Time (s)')
ylabel('Acceleration (m/s^2)')
hold on
grid on

function[dydtt]= rk_ma_damp(tt,yy,Gabcd,Ms,ks,cs,Md,kd,cd)
dydtt=zeros(4,1);  % a column vector

% Structural displacement=yy(1), velocity=dydtt(1)=yy(2)

dydtt(1)=yy(2);

% Structural acceleration
dydtt(2)=-(ks/Ms)*yy(1)...
    -(cs/Ms)*yy(2)...
    +(kd/Ms)*yy(3)...
    +(cd/Ms)*yy(4)...
```

```
    -Gabcd;

% equivalent mass displacement=yy(3), velocity=dydtt(3)=yy(4)
dydtt(3)=yy(4);

% equivalent mass acceleration
dydtt(4)=-(kd/Md)*yy(3)...
    -(cd/Md)*yy(4)...
    +(cs/Ms)*yy(2)...
    +(ks/Ms)*yy(1)...
    -(kd/Ms)*yy(3)...
    -(cd/Ms)*yy(4);

end
function[dydtt] = rk_str(tts,yys,Gabcd,OMEGAs,ZETAs)
dydtt=zeros(2,1);   % a column vector

% Structural displacement=yys(1), velocity=dydtt(1)=yys(2)

dydtt(1)=yys(2);

% Structural acceleration

dydtt(2)=-(OMEGAs^2)*yys(1)-(2*ZETAs*OMEGAs)*yys(2)-Gabcd;

end
```

2 Tuned Liquid Column Dampers

Working Mechanism and Design Considerations

2.1 BACKGROUND

As discussed in Sections 1.1 and 1.4 of Chapter 1, tuned liquid column damper (TLCD) is an inertia-based vibration control device, the forerunner of which was Fraham's anti-rolling tank, introduced in the early 1900s (Den Hartog, 1947; Frahm, 1911). Subsequently, Frahm's anti-rolling tanks were installed on many large German liners, such as Bremen and Europa (Den Hartog, 1947). An estimate suggests that Frahm's anti-rolling tanks were installed in over 1 million tons of German shipping before World War II (Gawad et al., 2001). Subsequently, several passive, semi-active and active devices were developed, with the aim of controlling the rolling motion of ships by extending the concept of Frahm's anti-rolling tank (Alujević et al., 2020, 2019). By 1975, about 2000 ships were fitted with different types of tank-based rolling stabilization systems (Field and Martin, 1976). More than a decade later, Saoka et al. (1988) developed the idea of the modern-day TLCD for vibration control of civil engineering structures, and Sakai et al. (1989) studied the feasibility of TLCD as a vibration control device considering real-life example structures. In the first US patent on the TLCD (Sakai et al., 1991, Patent No, US5070663), it was claimed that the device would be effective in the vibration suppression of tall tower-like structures during their service life as well as in the construction period (see Figure 2.1).

In the initial studies on TLCD, Samali (1990), Won (1994) and Haroun et al. (1996) demonstrated the effectiveness of the device in seismically excited structures, while Kwok et al. (1991), Xu et al. (1992a,b) examined its control performance under wind-induced excitations. Investigations related to control of wind- and seismic-induced vibration of structures using TLCD are discussed in Chapters 4 and 5, respectively. In other pioneering research, Balendra et al. (1995) and Xue et al. (1998) experimentally identified the damping characteristics of the TLCD. The first instance of a real-world installation of a TLCD system took place in the Higashi-Kobe Bridge, Japan (Kitazawa et al., 1992; Sakai et al., 1991a). In this, TLCDs were used as vibration control devices in the free-standing pylons of the cable stayed bridge during the construction stage. Some of the prominent real-world installations of TLCD, including that in the Higashi-Kobe Bridge, are examined in Chapter 8. In this chapter, the TLCD working mechanism, mathematical modelling and design considerations are discussed.

DOI: 10.1201/9781003377894-2

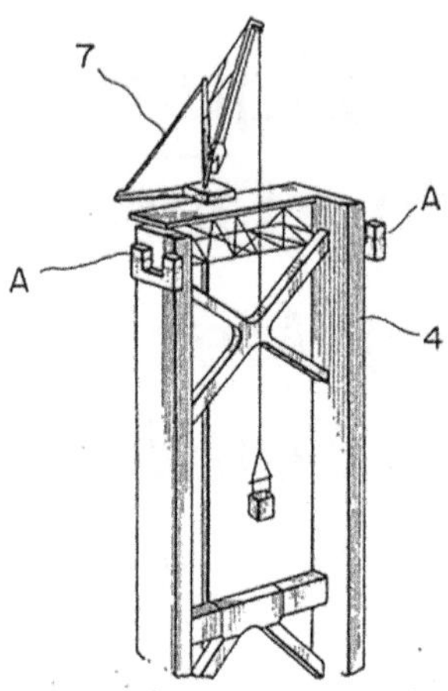

FIGURE 2.1 Sketch showing tuned liquid column dampers (TLCDs) in an under-construction tower-like structure (Sakai et al., 1991b); 4 = under-construction tower-like structure, 7 = crane, A = TLCD.

2.2 WORKING MECHANISM AND MATHEMATICAL MODELLING

2.2.1 WORKING MECHANISM

In a TLCD attached to a laterally excited structure, vibration control of the structure is realized through the oscillatory motion of liquid mass in the TLCD container. The latter is composed of a pair of opposite upstanding end portions, in which the liquid free surfaces are formed. In most cases, the TLCD containers are of a U-shape (Chang, 1999; Hemmati et al., 2019; Konar and Ghosh, 2010; Roy et al., 2023), although containers of other shapes, such as V-shape (Gao et al., 1997), semi-circular-shape (Sakai et al., 1991b) circular-shape (Basu et al., 2016) and other irregular shapes (Xu et al., 1992a, 1992b) have also been studied. In this book, the TLCDs are assumed to be U-shaped, unless mentioned otherwise. The restoring force in the TLCD is developed due to gravity acting upon the liquid, and the damping effect is generated as a result of hydraulic head-loss due to the orifice(s) installed inside the container (Balendra et al., 1995; Xu et al., 1992b). The frequency of oscillation of the liquid column is tuned, that is, maintained close to the frequency of the structural mode required to be controlled. The working mechanism of the TLCD is based on the condition that the upstanding limbs of the damper would have some positive liquid height throughout the duration of the oscillation of the liquid column. Further, it is assumed that during vibration, the sloshing of liquid in the upstanding limbs of the TLCD is negligible. The liquid in the damper is considered to be incompressible, and the damper container is taken to be rigid. In the mathematical model, the TLCD is generally considered as a single-degree-of-freedom (SDOF) damped oscillator, which is rigidly attached to the vibrating primary structure. Here, the displacement of the liquid free surface in the upstanding limbs of the TLCD along the vertical direction is considered as the degree-of-freedom (DOF) of the damper.

2.2.2 EQUATIONS OF MOTION OF THE STRUCTURE-TLCD SYSTEM

The schematic of a conventional U-shaped TLCD is shown in Figure 2.2. The overall length, horizontal length and cross-sectional area of the liquid column are denoted by L, b and A, respectively. The density of the contained liquid is represented by ρ. Figure 2.3 presents a simplified TLCD-structure system. Here, the structure is modelled as an SDOF system having mass, stiffness and damping equal to M_S, K_S and C_S, respectively. The mass of the damper container is assumed to be lumped with the mass of the structure. Kareem (1983) demonstrated that structures with well-separated natural frequencies, which is valid for most civil engineering structures, can be adequately represented by an equivalent SDOF primary system for the purpose of damper design. The structure is considered to be excited by an external force $F(t)$. The combined structure-damper system, thus, has two DOFs; first, the lateral displacement of the structural system relative to its base, which is denoted by $x(t)$, and second, the change in elevation of the free surface of the liquid column, which is represented by $y(t)$.

Using energy principles, the incremental total energy of the system should equal the work done by the external force. At equilibrium, with $x = y = 0$, the potential energy is zero. At time t, the potential energy, V and the kinetic energy, T of the structure-damper system are expressed as

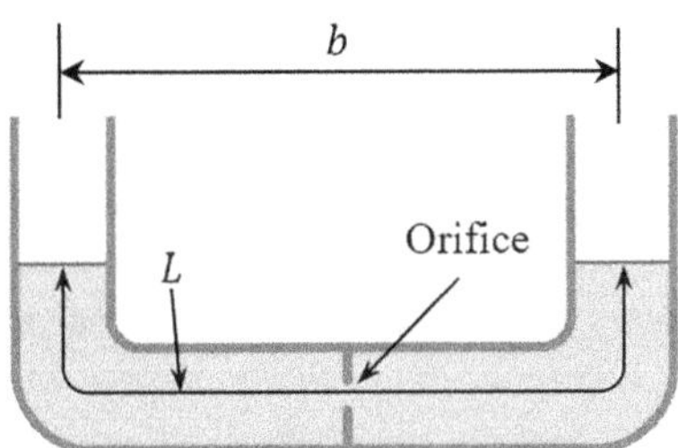

FIGURE 2.2 Schematic of tuned liquid column damper (TLCD).

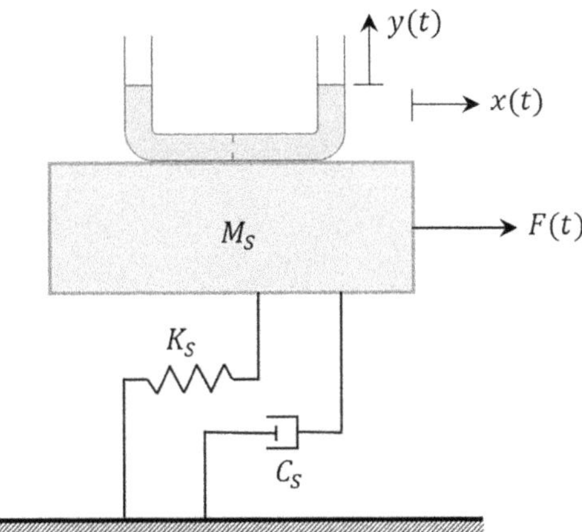

FIGURE 2.3 Schematic of the structure–tuned liquid column damper system.

$$T = \frac{1}{2} M_S \, \dot{x}^2 + \frac{1}{2} \rho A b (\dot{x} + \dot{y})^2 + \frac{1}{2} \rho A \left(\frac{L-b}{2} - y \right)(\dot{y}^2 + \dot{x}^2)$$

$$+ \frac{1}{2} \rho A \left(\frac{L-b}{2} + y \right)(\dot{y}^2 + \dot{x}^2) \tag{2.1}$$

and

$$V = \frac{1}{2} K_S x^2 + \rho A g y^2. \tag{2.2}$$

In Eq. (2.2), g is the acceleration due to gravity.

Simplification of Eq. (2.1) leads to

$$T = \frac{1}{2}(M_S + \rho A L)\dot{x}^2 + \frac{1}{2}\rho A L \dot{y}^2 + \rho A b \dot{x} \dot{y}. \tag{2.3}$$

Hence, the Lagrangian of the system is expressed as

$$\mathcal{L} = \frac{1}{2}(M_S + \rho A L)\dot{x}^2 + \frac{1}{2}\rho A L \dot{y}^2 + \rho A b \dot{x} \dot{y} - \frac{1}{2}K_S x^2 - \rho A g y^2. \tag{2.4}$$

For a non-conservative system, such as the present one, Lagrange's equation may be written as

$$\frac{d}{dt}\left(\frac{\partial T}{\partial \dot{q}_i} \right) - \frac{\partial \mathcal{L}}{\partial q_i} = Q_i, \tag{2.5}$$

where q_i and Q_i, respectively, denote the generalized coordinate and the non-conservative generalized force corresponding to the ith degree-of-freedom.

For the structure-damper system, the generalized coordinates are

$$q_1 = x \text{ and } q_2 = y. \tag{2.6}$$

The external force and the damping forces are the non-conservative forces in this system. By using the principle of virtual work, the non-conservative generalized forces corresponding to the first and second DOF may be written as

$$Q_1 = F(t) - C_S \dot{x} \tag{2.7}$$

and

$$Q_2 = -\frac{1}{2}\rho A \xi |\dot{y}| \dot{y}. \tag{2.8}$$

In Eq. (2.8), ξ is a dimensionless term that represents the coefficient of head-loss of the orifice(s).

The equations of motion of the structure-TLCD system may be obtained from

$$\frac{d}{dt}\left(\frac{\partial T}{\partial \dot{x}} \right) - \frac{\partial \mathcal{L}}{\partial x} = Q_1 \tag{2.9}$$

and

$$\frac{d}{dt}\left(\frac{\partial T}{\partial \dot{y}}\right) - \frac{\partial \mathcal{L}}{\partial y} = Q_2.$$ (2.10)

The substitution of Eqs. (2.3), (2.4) and (2.7) in Eq. (2.9) leads to

$$\left(M_S + \rho AL\right)\ddot{x} + \rho Ab\ddot{y} + K_S x = F(t) - C_S \dot{x}.$$ (2.11)

Let, $\alpha\left[= b/L\right]$ be the length ratio, defined by the ratio of the horizontal length to the total length of the liquid column and $\mu\left[= \rho AL/M_S\right]$ the mass ratio, which is the ratio of the mass of the liquid in the damper to the mass of the structure.

The normalization of Eq. (2.11) with respect to $\left(1+\mu\right)M_S$ leads to

$$\ddot{x} + \frac{2\zeta_s\omega_s}{\left(1+\mu\right)}\dot{x} + \frac{\omega_s^2}{\left(1+\mu\right)}x = \frac{F(t)}{\left(1+\mu\right)M_S} - \frac{\alpha\mu}{\left(1+\mu\right)}\ddot{y}$$ (2.12)

in which, $\zeta_s\left[= \sqrt{C_S/M_S}\right]$ is the equivalent viscous damping ratio and $\omega_s\left[= \sqrt{K_S/M_S}\right]$ is the natural frequency of the structure.

On substitution of Eqs. (2.3), (2.4) and (2.8) in Eq. (2.10),

$$\rho Ab\ddot{x} + \rho AL\ddot{y} + 2\rho Agy = -\frac{1}{2}\rho A\xi|\dot{y}|\dot{y}$$ (2.13)

is obtained. The rearrangement of terms in Eq. (2.13) leads to

$$\ddot{y} + \frac{1}{2}\frac{\xi}{L}|\dot{y}|\dot{y} + \omega_{\text{TLCD}}^2 y = -\alpha\ddot{x}.$$ (2.14)

In Eq. (2.14), ω_{TLCD} is the natural frequency of oscillation of the liquid column in the TLCD to be tuned to the frequency of the structural mode to be controlled. Equations (2.12) and (2.14) are coupled differential equations that represent the motion of the structure-damper system.

The expressions of ω_{TLCD} and that of the corresponding natural time period of the TLCD, T_{TLCD}, are given as

$$\omega_{\text{TLCD}} = \sqrt{\frac{2g}{L}}$$ (2.15)

and

$$T_{\text{TLCD}} = 2\pi\sqrt{\frac{L}{2g}}.$$ (2.16)

As can be observed from Eq. (2.16), T_{TLCD} depends only on a single TLCD parameter, L. The variation of T_{TLCD} with L is presented in Figure 2.4, which indicates a parabolic profile. This implies that in order to have a low value of T_{TLCD}, say less than 1 s,

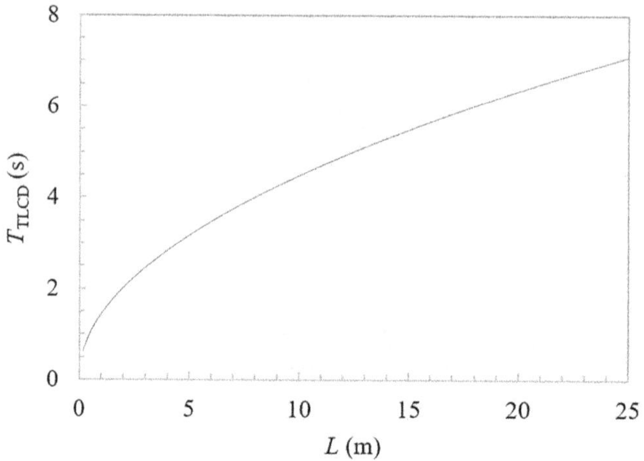

FIGURE 2.4 Variation in T_{TLCD} with L.

the value of L should be less than 0.5 m, which is too small to be implemented for a TLCD intended for application in civil engineering structure. That is why, TLCD, in its conventional form, is considered feasible only for long-period structures.

Instead of an external force $F(t)$, acting on the structure, for the case of an input base acceleration, $\ddot{Z}_0(t)$, the equations of motion of the structure-TLCD system are modified to

$$\ddot{x} + \frac{2\zeta_s\omega_s}{(1+\mu)}\,\dot{x} + \frac{\omega_s^2}{(1+\mu)}\,x = -\ddot{Z}_0 - \frac{\alpha\mu}{(1+\mu)}\,\ddot{y} \tag{2.17}$$

and

$$\ddot{y} + \frac{1}{2}\frac{\xi}{L}|\dot{y}|\dot{y} + \omega_{\text{TLCD}}^2 y = -\alpha\left(\ddot{x} + \ddot{Z}_0\right). \tag{2.18}$$

In Eqs. (2.17) and (2.18), as mentioned earlier, it should be borne in mind that the equations of motion are valid for the condition $|y| > (L-b)/2$.

It may be noted that Eqs. (2.14) and (2.18) are nonlinear equations due to the presence of the quadratic liquid damping term. Equivalent linear equations corresponding to Eqs. (2.14) and (2.18) are

$$\ddot{y} + 2\frac{C_p}{L}\dot{y} + \omega_{\text{TLCD}}^2 y = -\alpha\ddot{x} \tag{2.19}$$

and

$$\ddot{y} + 2\frac{C_p}{L}\dot{y} + \omega_{\text{TLCD}}^2 y = -\alpha\left(\ddot{x} + \ddot{Z}_0\right) \tag{2.20}$$

wherein, C_p represents the equivalent linearized damping coefficient, which has a unit of velocity and is not identical to the equivalent viscous damping ratio, which is dimensionless.

The error ϵ, incorporated due to this linearization is given by

$$\epsilon = \frac{1}{2}\frac{\xi}{L}|\dot{y}|\dot{y} - 2\frac{C_p}{L}\dot{y}. \tag{2.21}$$

The value of C_p can be obtained by minimizing the mean square value of the error ϵ, that is,

$$\frac{\partial}{\partial C_p}\left\langle\epsilon^2\right\rangle = 0 \tag{2.22}$$

In Eq. (2.22), $\langle\cdot\rangle$ is the operator of expectation. The value of C_p may then be evaluated as

$$C_p = \frac{\xi\left\langle|\dot{y}|\dot{y}^2\right\rangle}{4\left\langle\dot{y}^2\right\rangle}. \tag{2.23}$$

If $\dot{y}$ is assumed to be a zero-mean stationary Gaussian process, then

$$\left\langle|\dot{y}|\dot{y}^2\right\rangle = 2\sqrt{\frac{2}{\pi}}\sigma_{\dot{y}}^3 \tag{2.24}$$

and

$$\left\langle\dot{y}^2\right\rangle = \sigma_{\dot{y}}^2. \tag{2.25}$$

In Eqs. (2.24) and (2.25), $\sigma_{\dot{y}}$ is the standard deviation of the velocity of the liquid elevation in the vertical limb of the damper. The expression of C_p may then be rewritten as

$$C_p = \frac{\xi\sigma_{\dot{y}}}{\sqrt{2\pi}}. \tag{2.26}$$

As C_p and $\sigma_{\dot{y}}$ are interdependent, the responses are evaluated by the use of iterative methods with an initial value of C_p, till convergence is achieved. To assess the impact of this linearization on the prediction of the TLCD behaviour, Balendra et al. (1995) compared the root-mean-square (rms) velocity of the liquid free surface determined from Eqs. (2.14) and (2.19), when the TLCD was subjected to sinusoidal harmonic excitation. Three cases, with the amplitude of the harmonic excitation as 0.05, 0.2 and $0.4\,\mathrm{m/s^2}$, were considered. The coefficient of head-loss of the orifice was assumed in the range of 2–10. It was reported that the error due to the assumption of linearization increases with an increase in the head-loss coefficient of the orifice, as well as with an increase in the amplitude of excitation. However, for a practical value of length ratio equal to 0.7, and for the considered range of orifice head-loss coefficients and excitation amplitude, the maximum error due to linearization of the damping term was found to be within 9%.

2.2.3 FREQUENCY DOMAIN REPRESENTATION OF THE EQUATIONS OF MOTION OF THE STRUCTURE-TLCD SYSTEM

Very often, frequency domain studies are conducted for a more fundamental understanding of the behaviour of the structure-damper system to the external excitation and for evaluating parametric variations. Usually, frequency domain analysis has the advantage of simplicity, ease of computation and fast execution in a linear system (Konar and Ghosh, 2023). The key approximation used in a frequency domain analysis of a nonlinear system is the linearization of any nonlinearity present in the equations of motion of the structure-damper system (Roberts and Spanos, 1990). The input–output relations of the system in frequency domain are obtained through Fourier transformation of the equations of motion derived in the time domain, leading to a set of algebraic equations.

Let $X(\omega)$, $Y(\omega)$ and $\mathcal{F}(\omega)$ be the Fourier transformation of the time dependent variables $x(t)$, $y(t)$ and $F(t)$, respectively.

The Fourier transformation of Eq. (2.12) leads to

$$X(\omega) = H_1(\omega)\left[\frac{\mathcal{F}(\omega)}{M_S} + \omega^2 \alpha \mu Y(\omega)\right]. \tag{2.27}$$

In Eq. (2.27),

$$H_1(\omega) = \frac{1}{\omega_s^2 - \omega^2(1+\mu) + 2\zeta_s \omega_s i\omega}, \tag{2.28}$$

where $i = \sqrt{-1}$. Fourier transforming Eq. (2.19) yields

$$Y(\omega) = H_2(\omega)\omega^2 X(\omega). \tag{2.29}$$

In Eq. (2.29),

$$H_2(\omega) = \frac{\alpha}{\omega_{\text{TLCD}}^2 - \omega^2 + 2\dfrac{C_p}{L}i\omega}. \tag{2.30}$$

The solution of Eqs. (2.27) and (2.29) leads to

$$X(\omega) = H_X(\omega)\mathcal{F}(\omega) \tag{2.31}$$

and

$$Y(\omega) = H_Y(\omega)\mathcal{F}(\omega). \tag{2.32}$$

In Eqs. (2.31) and (2.32), $H_X(\omega)$ and $H_Y(\omega)$ denote the transfer functions relating the displacement of the structure to the external excitation force and the displacement of the liquid in the vertical limb of the TLCD to the external excitation force, respectively. These are expressed as

$$H_X(\omega) = \frac{H_1(\omega)}{M_S\left[1 - \omega^4 \alpha\mu H_1(\omega) H_2(\omega)\right]} \tag{2.33}$$

and

$$H_Y(\omega) = \frac{H_1(\omega) H_2(\omega)\omega^2}{M_S\left[1 - \omega^4 \alpha\mu H_1(\omega) H_2(\omega)\right]}. \tag{2.34}$$

For the specific case where the vibration is caused by a base acceleration $\ddot{Z}_0(t)$ acting as an excitation to the structure, the transfer functions in Eqs. (2.33) and (2.34) represented by $\bar{H}_X$ and $\bar{H}_Y$, respectively, take the form

$$\bar{H}_X = H_1(\omega)\frac{1 + \mu + \omega^2 \alpha\mu H_2(\omega)}{\omega^4 \alpha\mu H_1(\omega) H_2(\omega) - 1} \tag{2.35}$$

and

$$\bar{H}_Y = H_2(\omega)\left[\omega^2 H_1(\omega)\frac{1 + \mu + \omega^2 \alpha\mu H_2(\omega)}{\omega^4 \alpha\mu H_1(\omega) H_2(\omega) - 1} - 1\right]. \tag{2.36}$$

Now, one can compute the response quantities of interest using random vibration analysis in the frequency domain. In particular, there are two quantities of interest: first, the rms displacement of the structure σ_x, which is used to assess the effectiveness of the damper; second, the standard deviation of the velocity of the liquid elevation in the vertical limbs of the TLCD $\sigma_{\dot{y}}$, which is required to compute the optimum head-loss coefficient for the orifice. These response quantities are obtained using Eqs. (2.37) and (2.38) as

$$\sigma_x^2 = \int_{-\infty}^{\infty} |H_X(\omega)|^2 S_f(\omega) d\omega \tag{2.37}$$

and

$$\sigma_{\dot{y}}^2 = \int_{-\infty}^{\infty} \omega^2 |H_Y(\omega)|^2 S_f(\omega) d\omega; \tag{2.38}$$

or using Eqs. (2.39) and (2.40) as

$$\sigma_x^2 = \int_{-\infty}^{\infty} |\bar{H}_X(\omega)|^2 S_z(\omega) d\omega \tag{2.39}$$

and

$$\sigma_{\dot{y}}^2 = \int_{-\infty}^{\infty} \omega^2 |\bar{H}_Y(\omega)|^2 S_z(\omega) d\omega. \tag{2.40}$$

In Eqs. (2.37)–(2.40), $S_f(\omega)$ and $S_z(\omega)$ are the power spectral density (PSD) of the forcing function acting on the structure and of the base displacement applied to the structure, respectively. For further details, see Roberts and Spanos (1990), and Newland (1993).

2.3 DESIGN CONSIDERATIONS

From Section 2.2, it is clear that the independent parameters required to determine the response of a structure with an attached TLCD include the mass of the structure M_S, the damping ratio of the structure ζ_s, the natural frequency of the structural mode to be controlled ω_s, the mass ratio μ, the length ratio α, the coefficient of head-loss of the orifice ξ, the tuning ratio $\gamma = \omega_{\text{TLCD}}/\omega_s$ and the input excitation. The mass of the structure M_S is required only when the structure vibrates under an external force acting on the structure and not for the case when the structure is base-excited. It may be noted that the parameters related to the structure, such as M_S, ζ_s and ω_s, as well as the input excitation, are defined by the problem specification itself. Thus, the design of a TLCD aims to achieve the best possible performance of the damper under the expected loading by providing the optimum values of the remaining four parameters, namely μ, α, ξ and γ, within some feasible bounds.

2.3.1 MASS RATIO

In case of civil engineering structures, it is practically impossible to accommodate a TLCD with a high mass ratio due to limited availability of space and restriction on the addition of dead load on the structure. Rather, during the design of TLCDs, the mass ratio is determined based on the trade-off between the desired reduction in the structural response, the cost and the practical feasibility of providing the damper on the structure. For the design of TLCDs for real-world installations, the mass ratio is normally restricted to 1.5% (Cammelli et al., 2016; Hitchcock et al., 1999; Kitazawa et al., 1992; Konar et al., 2024; Konar and Ghosh, 2013; Lago et al., 2019; Teramura and Yoshida, 1996). However, even for such small values of mass ratio, the performance of a TLCD improves significantly with the increase in the mass ratio (Gao et al., 1997; Min et al., 2005; Sadek et al., 1998). For instance, Sadek et al. (1998) showed that the effectiveness of a TLCD in terms of displacement response reduction is doubled due to an increase in mass ratio from 1% to 4%.

2.3.2 LENGTH RATIO

The length ratio of a TLCD has a significant influence on the performance of the damper. When other parameters are constant, the structural response reduction by the damper increases with the increase in the length ratio. The maximum possible value of the length ratio is unity. However, for $\alpha = 1$, that is, when the liquid is present only in the horizontal portion of the damper container, the system would cease to function as a TLCD (Yalla and Kareem, 2000). Further, as the governing equations of the structure-TLCD system presented in Section 2.2

are valid only when the vertical limbs of the damper have some positive liquid height throughout the duration of the vibration, the threshold value of α may be determined as (Min et al., 2005)

$$\alpha \leq 1 - 2\frac{|y_{max}|}{L}. \tag{2.41}$$

In Eq. (2.41), y_{max} denotes the maximum vertical displacement of the liquid column during vibration. Since y_{max} depends on the excitation, an iterative procedure is needed for determining the optimum value of α.

In some early studies on TLCD, especially in the investigations where frequency domain studies were conducted, α was taken as 0.9 (Xu et al., 1992a, 1992b). In these studies, since only rms displacements were determined, it was not possible to check the condition given in Eq. (2.41). However, in the practical implementations of TLCDs (Cammelli et al., 2016; Hitchcock et al., 1999; Kitazawa et al., 1992; Lago et al., 2019; Teramura and Yoshida, 1996) and in experimental studies (Balendra et al., 1995; Colwell and Basu, 2008; Das et al., 2020), the value of α is within 0.7. Thus, while determining the optimum value of α using an iterative approach for a given design excitation, $\alpha = 0.7$ could be a good initial guess. Further, during practical design, there may be a restriction on the horizontal length of the TLCD due to space constraints. In such cases, the designer has to consider a value of α much lower than the optimum value determined from Eq. (2.41).

It may be noted that, while designing a TLCD system, the freeboard in the vertical limbs of the TLCD should be greater than y_{max}, else, the damper liquid would spill out of the damper container, leading to an alteration in the mass ratio and the frequency of the damper.

2.3.3 Coefficient of Head-Loss of the Orifice(s)

The flow of liquid through an orifice, as shown in Figure 2.5, involves the formation of ring-shaped vortices in the upstream and downstream of the orifice plate due to a sudden change in cross section. These vortices cause a partial loss of pressure head across the orifice, and the pressure drop is not recovered on the downstream side (Blevins, 1984). The supplemental damping afforded by a TLCD is a result of this irreversible head-loss characteristic of the orifice(s). The energy dissipation in an orifice is represented by the coefficient of head-loss ξ that is mainly dependent upon the opening ratio $\varnothing$, defined by the ratio of the cross-sectional area

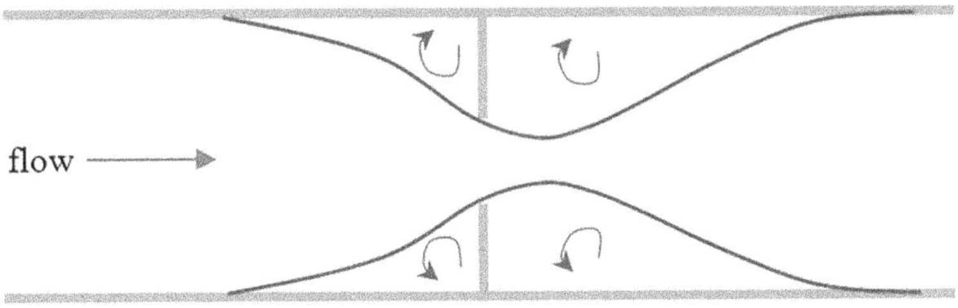

FIGURE 2.5 Flow of liquid through the orifice of a TLCD.

of the orifice to the cross-sectional area of the damper container (Jianhua et al., 2010; Wu et al., 2005). The smaller the opening ratio, the larger is the coefficient of head-loss. The empirical formula

$$\xi = \left(-0.6\psi + 2.1\psi^{0.1}\right)^{1.6} \left(1-\psi\right)^{-2} \tag{2.42}$$

may be used for the determination of the coefficient of head-loss of the orifice(s) for a particular opening ratio (Wu, 2005; Wu et al., 2005). In Eq. (2.42), ψ is the blocking ratio, expressed as $\left(1-\varnothing\right)$.

In addition to $\varnothing$, the Reynolds number, Re, of the flow approaching the orifice plate is another factor that can influence the coefficient of head-loss of an orifice (Jianhua et al., 2010). For flow through a circular section, $Re = \dot{y}D\rho/\tau$, where D and τ, respectively, denote the diameter of the TLCD container and the dynamic viscosity of the damper liquid. Depending upon $\varnothing$, there exists some threshold Reynolds number Re_T, beyond which the influence of Re on ξ is insignificant (Alvi et al., 1978; Ntamba and Fester, 2012). Table 2.1 shows the values of Re_T for different $\varnothing$.

In practical situations, when a TLCD experiences the design load, it is expected that Re would be greater than Re_T. To illustrate this, a case of a TLCD having a circular cross section with a diameter of 1 m and water as the damper liquid is considered as an example. The values of ρ and τ of water are $1,000\,kg/m^3$ and $0.001\ N.s/m^2$, respectively. The orifice opening ratio, $\varnothing$ is considered to have a value between 0.2 and 0.7 depending upon the optimality condition of the damper. In such a case, $Re > Re_T$ for any value of the velocity of the liquid column $\dot{y}$ greater than 0.008 m/s, a value much less than the expected velocity of the liquid column in a TLCD under typical design loading.

Jianhua et al. (2010) showed that the thickness of the orifice plate can also influence ξ, especially for the cases with lower value of $\varnothing$. The following empirical formula

$$\xi = \frac{0.7418}{\delta^{0.1142}} \left(\frac{3.196}{\varphi^4} - \frac{5.646}{\varphi^2} + 2.45 \right) \tag{2.43}$$

was proposed to include the effect of orifice plate thickness on ξ. In Eq. (2.43), δ is the thickness ratio, given by the ratio of the thickness of the orifice plate to the diameter of the cross-sectional area of the TLCD, and φ is the diameter ratio given by $\sqrt{\varnothing}$.

A comparison between the values of ξ obtained from the equations given by Wu et al. (2005) and Jianhua et al. (2010) for a range of $\varnothing$ is shown in Figure 2.6. Two different values of δ, namely 0.01 and 0.05, are considered in the determination of the values of ξ as per Jianhua et al. (2010). It can be observed that the variation of ξ with $\varnothing$ has a similar trend in all three cases. However, the equation given by Wu et al. (2005)

TABLE 2.1

Re_T for Different ϕ (Ntamba and Fester, 2012)

ϕ	0.2	0.3	0.57	0.7
Re_T	1,000	2,000	6,000	8,000

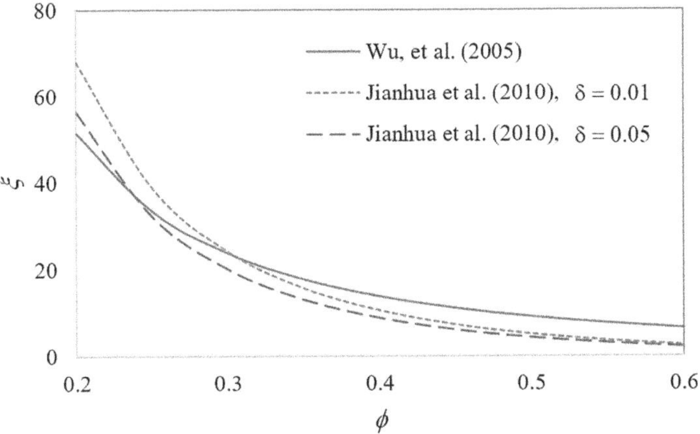

FIGURE 2.6 Comparison between the value of ξ obtained from the equations given by Wu et al. (2005) and Jianhua et al. (2010).

overestimates the value of ξ as compared to the equation given by Jianhua et al. (2010) for higher values of $\varnothing$. For lower values of $\varnothing$, it is the reverse. The equation given by Jianhua et al. (2010) suggests that ξ increases with the decrease in δ, and this effect is more prominent for lower values of $\varnothing$. The difference between the results obtained from Eqs. (2.42) and (2.43) may be attributed to the fact that Wu et al. (2005) conducted their study on TLCDs having rectangular cross section, while Jianhua et al. (2010) considered a circular cross section of the liquid column. A more detailed study in this regard is required to arrive at a more generalized expression for ξ.

To maximize the effectiveness of the TLCD, the coefficient of head-less of the orifice(s) is required to be optimized. The optimum coefficient of head-loss of the orifice(s), ξ_{opt}, mainly depends upon the damping ratio of the host structure, the excitation causing the vibration as well as on the damper mass ratio and length ratio. Although some existing literature provide mathematical expressions for ξ_{opt} for some specific cases (Wu et al., 2005; Yalla and Kareem, 2000), ξ_{opt} is generally determined numerically.

In some instances, especially when large TLCDs are used, the orifice is replaced with other flow damping devices (FDDs), such as baffles, nets and screens, to have an improved damper performance (Konar and Ghosh, 2022, 2021). For example, the 307-m tall Comcast Center, Philadelphia, USA, has a TLCD which contains 1179 ton of water as damper liquid for wind-induced vibration control. Both ends of the horizontal limb of the TLCD have vertical steel vanes to form a screen-like FDD for energy dissipation (Konar et al., 2024; Stephens, 2009). As sufficient research on the use of those FDDs in the TLCD is not available yet, results from experimental studies are relied upon to determine the energy dissipation in such cases.

2.3.4 Tuning Ratio

The tuning ratio, defined by the ratio of the frequency of the damper to the frequency of the structural mode to be controlled, is one of the most important design parameters of the TLCD. Although the degree of sensitivity varies from case to case, the performance of the TLCD is fairly sensitive to tuning ratio. The structural damping ratio, the excitation, the mass ratio and the damper length ratio all influence the optimum value of the tuning ratio, γ_{opt}. For a TLCD with low μ and a primary structure with low ζ_s, which are fairly common for civil engineering structures, γ_{opt} remains close to unity. However, γ_{opt} moves towards the lower side of unity with the increase in μ as well as in ζ_s. Similar to ξ_{opt}, mathematical expressions for γ_{opt} for some special cases are available in the literature (Ghosh and Basu, 2007; Wu et al., 2005; Yalla and Kareem, 2000), else γ_{opt} is determined numerically for the considered loading. Different values of ξ_{opt} and γ_{opt} would be obtained if the optimization procedure is carried out with different objectives, such as the maximum reduction in rms structural displacement, the maximum reduction in peak structural displacement, the maximum reduction in rms structural acceleration or the maximum reduction in peak structural acceleration.

In a civil engineering structural system, uncertainties in structural parameters as well as in the loading are expected. The efficiency of dampers may drastically reduce in detuned conditions resulting from such uncertainties (Debbarma et al., 2010a,b). In such cases, a structural system may be provided with multiple TLCDs (MTLCDs) which would have their natural frequencies distributed over a range of frequencies spanning either side of the optimum tuning frequency. Such a distributed system is more robust against uncertainties in structural parameters and effective against excitation having frequencies distributed over a wide band. It is reported that the performance of an MTLCD system is comparable to an equivalent single TLCD (Gao et al., 1999; Sadek et al., 1998). Usually, in an MTLCD system, the natural frequencies of the individual TLCDs are separated by a constant interval (Gao et al., 1999; Ghosh et al., 2011; Sadek et al., 1998; Shum and Xu, 2002). The tuning criteria are described by two parameters, namely the tuning bandwidth $\Delta\gamma$ and the central tuning ratio γ_0 defined as

$$\gamma_0 = \frac{\gamma_N + \gamma_1}{2} \tag{2.44}$$

and

$$\Delta\gamma = \frac{\gamma_N - \gamma_1}{\gamma_0}. \tag{2.45}$$

In Eqs. (2.44) and (2.45), γ_1 and γ_N denote the tuning ratio of the TLCDs having the lowest and the highest frequencies, respectively, and N is the total number of TLCDs in the MTLCD system.

In some studies, γ_0 has been considered equal to unity (Sadek et al., 1998), whereas other researchers have taken a value of γ_0 equal to γ_{opt} of an equivalent TLCD (Shum and Xu, 2002). Sadek et al. (1998) had numerically derived an optimum value of N for the most effective performance of the MTLCD system for an example structure. It was also demonstrated that $\Delta\gamma$ depends upon μ and that its value increases with the increase in μ.

2.3.5 Other Design Considerations

Water is normally considered to be the residing liquid in the TLCD. There are limited studies on the use of other liquids in the TLCD. Colwell and Basu (2008) showed that the use of glycol instead of water does not transmit into a significant difference in the passive damping characteristic of the TLCD system, despite glycol having a much higher viscosity than water. They suggested the use of glycol or glycol/water mixtures for TLCDs where the primary structure is located in cold regions. The use of denser liquid in TLCD would reduce the overall size of the damper for a particular mass ratio. In case of building structures, the overhead water tank for fire fighting can be designed to serve the dual function of the TLCD. This would reduce the cost of the damper significantly.

Researchers have also considered the use of smart materials, such as magneto-rheological (MR) fluids as the damper liquid in the TLCD (Colwell and Basu, 2008; Hokmabady et al., 2019; Mohammadyzadeh et al., 2022; Vázquez-Greciano et al., 2023; Wang et al., 2005). MR fluids are composed of micron-sized magnetizable particles and additives dispersed in a base fluid. MR fluid particle sizes typically range from 0.1 to 10 µm (Carlson and Jolly, 2000). MR fluids are capable of reversibly transforming from a fluid state with low viscosity to a semi-solid state with controllable yield strength in a matter of milliseconds when subjected to a magnetic field. Thus, using MR fluids, semi-active TLCDs with alterable fluid viscosity are devised. Such TLCDs are often designated as MR-TLCDs. Due to the rapidly alterable fluid viscosity, it is possible to adjust the control force generated by the damper based on the feedback of the semi-active system. A more detailed discussion on MR-TLCDs is presented in Chapter 7.

REFERENCES

AlujAlujević, N., Ćatipović, I., Malenica, Š., Senjanovi, I., Vladimir, N., 2020. Stability, performance and power flow of active U-tube anti-roll tank. *Eng. Struct.* 211, 110267. https://doi.org/10.1016/j.engstruct.2020.110267

Alujević, N., Ćatipović, I., Malenica, Š., Senjanović, I., Vladimir, N., 2019. Ship roll control and power absorption using a U-tube anti-roll tank. *Ocean Eng.* 172, 857–870. https://doi.org/10.1016/j.oceaneng.2018.12.007

Alvi, S.H., Sridharan, K., Rao, N.S.L., 1978. Loss claracteristics of orifices and nozzles. *J. Fluids Eng.* 100, 299–307.

Balendra, T., Wang, C.M., Cheong, H.F., 1995. Effectiveness of tuned liquid column dampers for vibration control of towers. *Eng. Struct.* 17, 668–675. https://doi.org/10.1016/0141-0296(95)00036-7

Basu, B., Zhang, Z., Nielsen, S.R.K., 2016. Damping of edgewise vibration in wind turbine blades by means of circular liquid dampers. *Wind Energy* 19, 213–226. https://doi.org/10.1002/we.1827

Blevins, R.D., 1984. *Applied Fluid Dynamics Handbook*. Van Nostrand Reinhold Co., New York.

Cammelli, S., Li, Y.F., Mijorski, S., 2016. Mitigation of wind-induced accelerations using tuned liquid column dampers : experimental and numerical studies. *Jnl. Wind Eng. Ind. Aerodyn.* 155, 174–181. https://doi.org/10.1016/j.jweia.2016.06.002

Carlson, J.D., Jolly, M.R., 2000. MR fluid, foam and elastomer devices. *Mechatronics* 10, 555–569. https://doi.org/10.1016/S0957-4158(99)00064-1

Chang, C.C., 1999. Mass dampers and their optimal designs for building vibration control. *Eng. Struct.* 21, 454–463. https://doi.org/10.1016/S0141-0296(97)00213-7

Colwell, S., Basu, B., 2008. Experimental and theoretical investigations of equivalent viscous damping of structures with TLCD for different fluids. *J. Struct. Eng.* 134, 154–163.

Das, A., Maity, D., Bhattacharyya, S.K., 2020. Characterization of liquid sloshing in U-shaped containers as dampers in high-rise buildings. *Ocean Eng.* 210, 1–22. https://doi.org/10.1016/j.oceaneng.2020.107462

Debbarma, R., Chakraborty, S., Ghosh, S., 2010a. Optimum design of tuned liquid column dampers under stochastic earthquake load considering uncertain bounded system parameters. *Int. J. Mech. Sci.* 52, 1385–1393. https://doi.org/10.1016/j.ijmecsci.2010.07.004

Debbarma, R., Chakraborty, S., Ghosh, S., 2010b. Unconditional reliability-based design of tuned liquid column dampers under stochastic earthquake load considering system parameters uncertainties. *J. Earthq. Eng.* 14, 970–988. https://doi.org/10.1080/13632461003611103

Den Hartog, J.P., 1947. *Mechanical Vibrations*, third ed. McGraw-Hill, New York.

Field, S.B., Martin, J.P., 1976. Comparative effects of U-tube and free surface type passive roll stabilisation systems. *Nav. Archit.* 73–92.

Frahm, H., 1911. Results of trials of the anti-rolling tanks at sea. *Trans. Inst. Nav. Archit.* 53, 183–201.

Gao, H., Kwok, K.C.S., Samali, B., 1997. Optimization of tuned liquid column dampers. *Eng. Struct.* 19, 476–486. https://doi.org/10.1016/S0141-0296(96)00099-5

Gao, H., Kwok, K.S.C., Samali, B., 1999. Characteristics of multiple tuned liquid column dampers in suppressing structural vibration. *Eng. Struct.* 21, 316–331.

Gawad, A.F.A., Ragab, S.A., Nayfeh, A.H., Mook, D.T., 2001. Roll stabilization by anti-roll passive tanks. *Ocean Eng.* 28, 457–469. https://doi.org/10.1016/S0029-8018(00)00015-9

Ghosh, A., Basu, B., 2007. Alternative approach to optimal tuning parameter of liquid column damper for seismic applications. *J. Struct. Eng.* 133, 1848–1852.

Ghosh, A., Gangopadhyay, A., Basu, B., 2011. Performance investigation of multiple compliant liquid column dampers for control of seismic vibrations, in: *Proceedings of the 8th International Conference on Structural Dynamics*, Lueven, Belgium, EURODYN 2011. pp. 1671–1677.

Haroun, M.A., Pires, J.A., Won, A.Y.J., 1996. Suppression of environmentally-induced vibrations in tall buildings by hybrid liquid column dampers. *Struct. Des. Tall Build.* 5, 45–54.

Hemmati, A., Oterkus, E., Barltrop, N., 2019. Fragility reduction of offshore wind turbines using tuned liquid column dampers. *Soil Dyn. Earthq. Eng.* 125, 105705. https://doi.org/10.1016/j.soildyn.2019.105705

Hitchcock, P.A., Glanville, M.J., Kwok, K.C.S., Watkins, R.D., Samali, B., 1999. Damping properties and wind-induced response of a steel frame tower fitted with liquid column vibration absorbers. *J. Wind Eng. Ind. Aerodyn.* 83, 183–196. https://doi.org/10.1016/S0167-6105(99)00071-9

Hokmabady, H., Mohammadyzadeh, S., Mojtahedi, A., 2019. Suppressing structural vibration of a jacket-type platform employing a novel magneto-rheological tuned liquid column gas damper (MR-TLCGD). *Ocean Eng.* 180, 60–70. https://doi.org/10.1016/j.oceaneng.2019.03.055

Jianhua, W., Wanzheng, A., Qi, Z., 2010. Head loss coefficient of orifice plate energy dissipator. *J. Hydraul. Res.* 48, 526–530. https://doi.org/10.1080/00221686.2010.507347

Kareem, A., 1983. Mitigation of wind induced motion of tall buildings. *J. Wind Eng. Ind. Aerodyn.* 11, 273–284. https://doi.org/10.1016/0167-6105(83)90106-X

Kitazawa, M., Nishimori, K., Noguchi, J., Shimoda, I., 1992. Earthquake resistant design of a long-period cable-stayed bridge, in: *10th World Conferance on Earthquake Engineering*, Madrid, Spain. pp. 4797–4802.

Konar, T., Ghosh, A.D., 2023. A review on various configurations of the passive tuned liquid damper. *J. Vib. Control* 29, 1945–1980. https://doi.org/10.1177/10775463221074077

Konar, T., Ghosh, A.D., 2022. Enhanced damping in a TLD by slat screens and horizontal baffles: a comparative study, in: Sudarshan, T.S., Pandey, K.M., Misra, R.D., Patowari, P.K., Bhaumik, S. (Eds.), *Recent Advancements in Mechanical Engineering-Select Proceedings of ICRAME 2021*. Springer Nature, Singapore, pp. 287–295. https://doi.org/10.1007/978-981-19-3266-3_22

Konar, T., Ghosh, A.D., 2021. Flow damping devices in tuned liquid damper for structural vibration control: a review. *Arch. Comput. Methods Eng.* 28, 2195–2207. https://doi.org/10.1007/s11831-020-09450-0

Konar, T., Ghosh, A.D., 2013. Bimodal vibration control of seismically excited structures by the liquid column vibration absorber. *J. Vib. Control* 19, 385–394. https://doi.org/10.1177/1077546311430718

Konar, T., Ghosh, A.D., 2010. Passive control of seismically excited structures by the liquid column vibration absorber. *Struct. Eng. Mech.* 36, 561–573. https://doi.org/10.12989/sem.2010.36.5.561

Konar, T., Ghosh, A.D., Basu, B., 2024. Real-world installations of tuned liquid column dampers for wind- induced vibration control of buildings: some important case studies. *Struct. Infrastruct. Eng.* in press. https://doi.org/10.1080/15732479.2024.2420174

Kwok, K.C.S., Xu, Y.L., Samali, B., 1991. Control of wind-induced vibrations of tall structures by optimized tuned liquid column dampers, in: *Proceedings of the Asian Pacific Conference on Computational Mechanics*. Hong Kong, pp. 249–254.

Lago, A., Trabucco, D., Wood, A., 2019. *Damping Technologies for Tall Buildings*. Butterworth-Heinemann, Oxford, UK. https://doi.org/10.1016/b978-0-12-815963-7.00008-7

Min, K., Kim, Hyoung-seop, Lee, S., Kim, Hongjin, 2005. Performance evaluation of tuned liquid column dampers for response control of a 76-story benchmark building. *Eng. Struct.* 27, 1101–1112. https://doi.org/10.1016/j.engstruct.2005.02.008

Mohammadyzadeh, S., Mojtahedi, A., Hokmabady, H., Farajpour, I., 2022. Performance of a magneto-rheological tuned liquid column damper (MR-TLCD) in mitigating vibration of an offshore structure. *J. Vib. Eng. Technol.* 10, 2999–3010. https://doi.org/10.1007/s42417-022-00532-w

Newland, D.E., 1993. *An Introduction to Random Vibrations, Spectral and Wavelet Analysis*. Longman, New York and London.

Ntamba, B.N., Fester, V., 2012. Pressure losses and limiting Reynolds numbers for non-newtonian fluids in short square-edged orifice plates. *J. Fluids Eng.* 134, 091204. https://doi.org/10.1115/1.4007156

Roberts, J.B., Spanos, P.D., 1990. Random vibration and statistical linearization. John Wiley & Sons, Inc., Chichester, UK. https://doi.org/10.1016/0022-460x(90)90647-i

Roy, A.K., Konar, T., Ghosh, A., 2023. Mitigation of structural vibrations due to pulse-type-near-fault earthquake by the compliant liquid column damper. *J. Earthq. Tsunami* 17, 2350004. https://doi.org/10.1142/S1793431123500045

Sadek, F., Mohraz, B., Lew, H.S., 1998. Single- and multiple-tuned liquid column dampers for seismic applications. *Earthq. Eng. Struct. Dyn.* 27, 439–463.

Sakai, F., Takaeda, S., Tamaki, T., 1991a. Tuned liquid column dampers (TLCD) for cable-stayed bridges, in: *Proceeding of Specialty Conference on Innovation in Cable-Stayed Bridges*. Fukuoka, Japan, pp. 197–205.

Sakai, F., Takaeda, S., Tamaki, T., 1991b. Damping device for tower-like structure. US5070663.

Sakai, F., Takaeda, S., Tamaki, T., Takeda, S., Tamaki, T., 1989. Tuned liquid column dampers: new type device for suppression of building vibrations, in: *Proceedings of International Conference on High Rise Buildings*. Nanjing, China, pp. 926–931.

Samali, B., 1990. Dynamic response of structures equipped with tuned liquid column dampers, in: *Australian Vibration and Noise Conference 1990: Vibration and Noise-Measurement Prediction and Control 18–20*. Melbourne, Australia, pp. 138–143.

Saoka, Y., Sakai, F., Takaeda, S., Tamaki, T., 1988. On the suppression of vibrations by tuned liquid column dampers, in: *Annual Meeting of JSCE*. JSCE, Tokyo, Japan.

Shum, K.M., Xu, Y.L., 2002. Multiple-tuned liquid column dampers for torsional vibration control of structures: experimental investigation. *Earthq. Engng Struct. Dyn.* 31, 977–991. https://doi.org/10.1002/eqe.133

Stephens, S., 2009. Robert A.M. Stern Architects raises the bar with Philadelphia's Comcast Center. *Archit. Rec.* 9, 168–174.

Teramura, A., Yoshida, O., 1996. Development of vibration control system using U-shaped water tank, in: *11th World Conferance on Earthquake Engineering*. Acapulco, Mexico, p. Paper No. 1343.

Vázquez-Greciano, A., Aznar López, A., Buratti, N., Ortiz Herrera, J.M., 2023. Magnetic fields to enhance tuned liquid damper performance for vibration control: a review. *Arch. Comput. Methods Eng.* 31, 25–45. https://doi.org/10.1007/s11831-023-09971-4

Wang, J.Y., Ni, Y.Q., Ko, J.M., Spencer, B.F., 2005. Magneto-rheological tuned liquid column dampers (MR-TLCDs) for vibration mitigation of tall buildings: modelling and analysis of open-loop control. *Comput. Struct.* 83, 2023–2034. https://doi.org/10.1016/j.compstruc.2005.03.011

Won, A.Y.J., 1994. *Tuned Liquid Column Dampers and Hybrid Liquid Column Dampers to Suppress Earthquake-Induced Motions in Flexible Structures*. University of California, California.

Wu, J., 2005. Experimental calibration and head loss prediction of tuned liquid column damper. *Tamkang J. Sci. Eng.* 8, 319–325. https://doi.org/10.6180/jase.2005.8.4.08

Wu, J., Shih, M., Lin, Y., Shen, Y., 2005. Design guidelines for tuned liquid column damper for structures responding to wind. *Eng. Struct.* 27, 1893–1905. https://doi.org/10.1016/j.engstruct.2005.05.009

Xu, Y.L., Kwok, K.C.S., Samali, B., 1992a. The effect of tuned mass dampers and liquid dampers on cross-wind response of tall/slender structures. *J. Wind Eng. Ind. Aerodyn.* 40, 33–54. https://doi.org/10.1016/0167-6105(92)90519-G

Xu, Y.L., Samali, B., Kwok, K.C.S., 1992b. Control of along: wind response of structures by mass and liquid dampers. *J. Eng. Mech.* 118, 20–39.

Xue, S.D., Ko, J.M., Xu, Y.L., 1998. Experimental investigation on structural pitching motion control by tuned liquid column damper, in: *5th Annual International Symposium on Smart Structures and Materials*. San Diego, IL, USA, pp. 474–483.

Yalla, S.K., Kareem, A., 2000. Optimum absorber parameters for tuned liquid column dampers. *J. Struct. Eng.* 126, 906–915.

MATLAB® CODE 2.1

```
% Response of an SDOF structure with and without TLCD under harmonic
base excitation
clear all
clc
close all
% Definition of general constants
g=9.81; % gravitational acceleration, unit m/s^2
pi=3.141593;

% Input structural parameters
Ms=1000000; % mass of the structure, unit kg
ZETAs=0.01; % damping ratio of the structure
Ts=2; % natural period of the structure, unit s
```

```matlab
% Input damper parameters
tunr=1.0; % tuning ratio
mu=0.05; % mass ratio
alpha=0.7; % length ratio
zi=20; % head-loss coefficient of the orifice(s)

OMEGAs=2*pi/Ts;
OMEGAd=tunr*OMEGAs;
L=2*g/OMEGAd/OMEGAd

% Input harmonic base excitation in the form of a sine wave
dt=0.02; % time interval of base excitation, unit s
t=0:dt:300; % duration of excitation taken as 300 s
A=0.2; % amplitude of base acceleration, unit m/s^2
beta=0.99; % forced frequency ratio
omg=OMEGAs*beta;
abc=A*sin(omg*t);
[n,m]=size(abc);

Gabc=abc;

plot(t,(Gabc/9.81))
set(gca,{'FontName','FontSize'},{'Times New Roman',25})
set(gca,{'XMinorTick','YMinorTick'},{'on','on'})
xlabel('Time (s)')
ylabel('Base acceleration (g)')

% Determination of response of structure with TLCD
yy=zeros(4,1);
tt=0;
Gabcd=0;
Yin=[0 0 0 0];
time=[];
dispS=[];
velS=[];
dispD=[];
velD=[];
stateSD=[];
accelS=[];

for i=1:(m*n)

  ttf=tt+dt;

[T, Y]=ode45(@(tt,yy) New_rk_TLCD(tt,yy,Gabcd,OMEGAs,OMEGAd,ZETAs,L
,zi,...
    mu,alpha),[tt ttf],Yin); % Solve ODE

  time=[time; T(:,1)];
  dispS=[dispS; Y(:,1)];
  velS=[velS; Y(:,2)];
  dispD=[dispD; Y(:,3)];
```

```matlab
  velD=[velD; Y(:,4)];
  stateSD=[stateSD; Y(:,:)];

xdd=(1/(1+mu-alpha*alpha*mu))*(-(OMEGAs^2).*Y(:,1)...
    -(2*ZETAs*OMEGAs).*Y(:,2)...
    +(mu*alpha)*((OMEGAd^2).*Y(:,3)...
    +(0.5*zi/L).*abs(Y(:,4)).*Y(:,4)))...
    -Gabcd;

  accelS=[accelS; xdd];

  tt=ttf;
  Gabcd=Gabc(1,i);

  Yin=Y(end,:);

end

% Determination of response of uncontrolled structure
tts=0;
Gabcd=0;
Yin=[0 0];
times=[];
yys=zeros(2,1);
disp=[];
vel=[];
state=[];
accel=[];
for i=1:(m*n)
  ttsf=tts+dt;

  [T, Y]=ode45(@(tts,yys) rk_str(tts,yys,Gabcd,OMEGAs,ZETAs),[tts
ttsf],Yin); % Solve ODE

  times=[times; T(:,1)];
  disp=[disp; Y(:,1)];
  vel=[vel; Y(:,2)];
  state=[state; Y(:,:)];

  xdd=-(OMEGAs^2).*Y(:,1)-(2*ZETAs*OMEGAs).*Y(:,2)-Gabcd;
  accel=[accel; xdd];
  tts=ttsf;
  Gabcd=Gabc(1,i);
  Yin=Y(end,:);
end

% Determination of control effectiveness of TLCD

        rms=sqrt(mean(disp.^2));
        rmsd=sqrt(mean(dispS.^2));
        redrms_disp=(rms-rmsd)*100/rms

        peak=max(abs(disp));
```

```matlab
        peakd=max(abs(dispS));
        redpeak_disp=(peak-peakd)*100/peak

        rms=sqrt(mean(vel.^2));
        rmsd=sqrt(mean(velS.^2));
        redrms_vel=(rms-rmsd)*100/rms

        peak=max(abs(vel));
        peakd=max(abs(velS));
        redpeak_vel=(peak-peakd)*100/peak

        rms=sqrt(mean(accel.^2));
        rmsd=sqrt(mean(accelS.^2));
        redrms_accel=(rms-rmsd)*100/rms

        peak=max(abs(accel));
        peakd=max(abs(accelS));
        redpeak_accel=(peak-peakd)*100/peak

        figure
plot(times,disp,'r-',time,dispS)
set(gca,{'FontName','FontSize'},{'Times New Roman',35})
set(gca,{'XMinorTick','YMinorTick'},{'on','on'})
legend(' without damper',' with damper')
xlabel('Time (s)')
ylabel('Structural displacement (m)')
grid on

figure
subplot(3,1,1); plot(times,disp,'r-',time,dispS)
legend('without damper','with damper')
xlabel('Time (s)')
ylabel('Displacement (m)')
grid on

subplot(3,1,2); plot(times,vel,'r-',time,velS)
legend('without damper','with damper')
xlabel('Time (s)')
ylabel('Velocity (m/s)')
grid on

subplot(3,1,3); plot(times,accel,'r-',time,accelS)
legend('only structure','structure with damper')
xlabel('Time (s)')
ylabel('Acceleration (m/s^2)')
hold on
grid on

function[dydtt]= New_rk_TLCD(tt,yy,Gabcd,OMEGAs,OMEGAd,ZETAs,L,zi,...
    mu,alpha)

dydtt=zeros(4,1);  % a column vector

% structural displacement=yy(1), velocity=dydtt(1)=yy(2)
```

```
dydtt(1)=yy(2);
%structural acceleration
dydtt(2)=(1/(1+mu-alpha*alpha*mu))*(-(OMEGAs^2)*yy(1)...
    -(2*ZETAs*OMEGAs)*yy(2)...
    +(mu*alpha)*((OMEGAd^2)*yy(3)...
    +0.5*zi/L*abs(yy(4))*yy(4)))...
    -Gabcd;

% displacement of liquid column=yy(3), velocity of liquid
column=dydtt(3)=yy(4)

dydtt(3)=yy(4);

% acceleration of liquid column
dydtt(4)=(1/(1+mu-alpha*alpha*mu))*((-(OMEGAd^2)*yy(3)...
    -0.5*zi/L*abs(yy(4))*yy(4))*(1+mu)...
    +2*ZETAs*OMEGAs*alpha*yy(2)...
    +(OMEGAs^2)*alpha*yy(1));
end

function[dydtt] = rk_str(tts,yys,Gabcd,OMEGAs,ZETAs)
dydtt=zeros(2,1);   % a column vector

% Structural displacement=yys(1), velocity=dydtt(1)=yys(2)

dydtt(1)=yys(2);

% Structural acceleration
dydtt(2)=-(OMEGAs^2)*yys(1)-(2*ZETAs*OMEGAs)*yys(2)-Gabcd;
 end
```

3 Different Configurations of the Tuned Liquid Column Damper

3.1 INTRODUCTION

Although the conventional tuned liquid column damper (TLCD) has been proven to be effective as a vibration control device, it has two major drawbacks. First, it only acts unidirectionally in the horizontal plane, which is a major deterrent in its applicability to structures requiring either control of lateral vibrations along two orthogonal directions or control of vertical vibrations. Second, being an inherently long-period system, the conventional TLCD becomes ineffective for short-period structures due to difficulty in tuning (Corte et al., 2007; Ghosh and Basu, 2004). In order to address these shortcomings of the conventional TLCD, significant research has been carried out that has led to the development of some special configurations of the conventional TLCD. Additionally, certain modifications to the original TLCD have been proposed to further improve the effectiveness of the damper and to allow for greater design flexibility and architectural adaptability. The bi-directional TLCD (BTLCD), the toroidal TLCD (TTLCD), the omnidirectional TLCD (OTLCD), the liquid column vibration absorber (LCVA), the tuned liquid column ball damper (TLCBD), the tuned liquid multi-column damper (TLMCD), the circular TLCD, the S-shaped TLCD, the sealed TLCD, the tuned liquid column and sloshing damper (TLCSD), the tank-pipe damper (TPD), the compliant TLCD (CLCD), the pendulum TLCD (PLCD), the tuned liquid column damper inerter (TLCDI), the vertical sealed TLCD (VSTLCD), the adaptive TLCD (Ad-TLCD) and the vertical TLCD (VTLCD) are the notable modified forms of the TLCD. The configurations can be subdivided into four major groups, as shown in Table 3.1, based on the primary advantage offered by them. The first group offers multi-directional vibration control with a single device and thereby removes the shortcoming of unidirectional applicability of the conventional TLCD. The second group provides enhanced effectiveness as compared to the conventional TLCD. The enhancement in effectiveness may be in terms of an increase in response control, multi-modal control, torsional response control, design flexibility and architectural adaptability. The third group includes the TLCD configurations with improved tunability. These TLCD configurations can be tuned to short-period structures and thereby expand the applicability of TLCD beyond the long-period structural systems. The dampers in the fourth group are specifically designed for vertical vibration control. In the following sections, these TLCD configurations are discussed in detail. Various combinations of the TLCD configurations have also been developed, which are covered in the appropriate sections.

DOI: 10.1201/9781003377894-3

TABLE 3.1

Classification of the Configurations of the TLCD Based on the Advantage Offered

Advantage	Configurations
Multi-directional vibration control	• Bi-directional TLCD (BTLCD) • Toroidal TLCD (TTLCD) • Omnidirectional TLCD (OTLCD)
Enhanced effectiveness	• Liquid column vibration absorber (LCVA) • Tuned liquid column ball damper (TLCBD) • Tuned liquid multi-column damper (TLMCD) • Circular TLCD • S-shaped TLCD • Tuned liquid column and sloshing damper (TLCSD) • Tank-pipe damper (TPD)
Improved tunability	• Sealed TLCD • Compliant TLCD (CLCD) • Pendulum TLCD (PLCD) • Tuned liquid column damper inerter (TLCDI)
Vertical vibration control	• Adaptive TLCD (Ad-TLCD) • Vertical sealed TLCD (VSTLCD) • Vertical TLCD (VTLCD)

3.2 TLCD CONFIGURATIONS FOR MULTI-DIRECTIONAL VIBRATION CONTROL

3.2.1 BI-DIRECTIONAL TLCD (BTLCD)

As the name implies, in a BTLCD, the damper liquid can oscillate along two orthogonal directions in the horizontal plane, and thus, the damper can suppress lateral vibrations along the two principal axes of the primary structure. Sakai et al. (1991) coined the idea of the BTLCD when they proposed combining two conventional TLCDs in a crossed formation for use as a single device (see Figure 3.1). They also patented the concept alongside their patent for the conventional TLCD (Patent No. US5070663).

The expressions for the frequency of oscillation of the liquid column in a BTLCD along each of its principal directions X and Y are (Konar and Ghosh, 2023; Rozas et al., 2016)

$$\omega_{\text{TLCD},X} = \sqrt{\frac{2g}{L_X}} \tag{3.1}$$

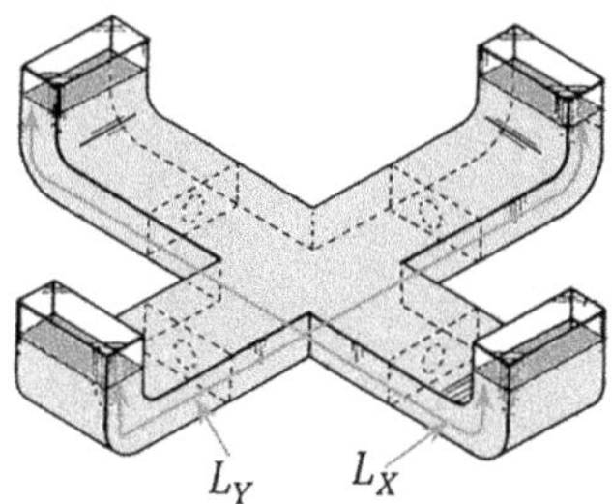

FIGURE 3.1 Bi-directional tuned liquid column damper configuration as proposed by Sakai et al. (1991).

and

$$\omega_{\text{TLCD},Y} = \sqrt{\frac{2g}{L_Y}} \tag{3.2}$$

respectively.

In Eqs. (3.1) and (3.2), L_X and L_Y are the lengths of the liquid column along the principal directions X and Y, respectively, of the BTLCD (see Figure 3.1), and g is the acceleration due to gravity. If the primary structure has different dominant frequencies along its orthogonal directions, the BTLCD would have different lengths of the horizontal limbs along the orthogonal directions.

Although the functioning of a BTLCD is similar to that of a set of two independent conventional TLCDs arranged perpendicular to each other, in this device, a portion of the liquid participates in the control of vibration along both directions. Thus, a BTLCD is more cost-effective as it requires a smaller volume of liquid as compared to the case when a set of two independent conventional TLCDs is used. Similar to the conventional TLCD, the BTLCD is also provided with orifice(s) in the horizontal limbs to introduce damping. In the BTLCD with crossed formation, under bi-directional excitation, the displacements of the free surface of the liquid columns are uncoupled. Thus, based on the derivation in Section 2.2.2 of Chapter 2, the equations of motion of a BTLCD rigidly attached to a base-excited 2-degree-of-freedom (DOF) primary structure (see Figure 3.2) may be written as

$$\ddot{x}_X + \frac{2\zeta_{sX}\omega_{sX}}{(1+\mu_X)}\dot{x}_X + \frac{\omega_{sX}^2}{(1+\mu_X)}x_X = -\ddot{Z}_X - \frac{\alpha_X\mu_X}{(1+\mu_X)}\ddot{y}_X, \tag{3.3}$$

$$\ddot{x}_Y + \frac{2\zeta_{sY}\omega_{sY}}{(1+\mu_Y)}\dot{x}_Y + \frac{\omega_{sY}^2}{(1+\mu_Y)}x_Y = -\ddot{Z}_Y - \frac{\alpha_Y\mu_Y}{(1+\mu_Y)}\ddot{y}_Y, \tag{3.4}$$

$$\ddot{y}_X + \frac{1}{2}\frac{\xi_X}{L_X}|\dot{y}_X|\dot{y}_X + \omega_{\text{TLCD},X}^2 y_X = -\alpha_X\left(\ddot{x}_X + \ddot{Z}_X\right), \tag{3.5}$$

and

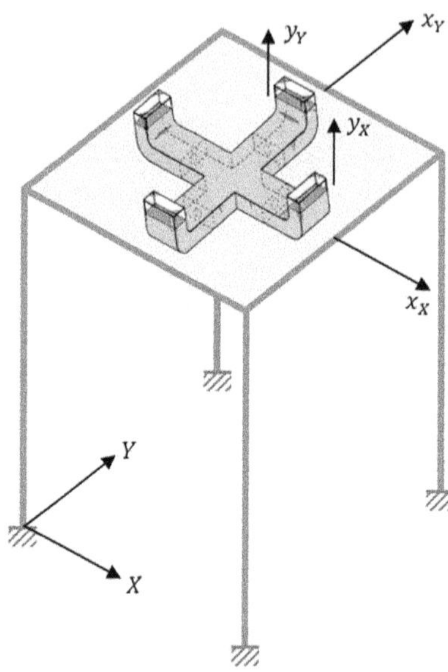

FIGURE 3.2 A 2-degrees of freedom structure with bi-directional tuned liquid column damper.

$$\ddot{y}_Y + \frac{1}{2}\frac{\xi_Y}{L_Y}|\dot{y}_Y|\dot{y}_Y + \omega_{\text{TLCD},Y}{}^2 y_Y = -\alpha_Y\left(\ddot{x}_Y + \ddot{Z}_Y\right). \tag{3.6}$$

In Eqs. (3.3)–(3.6), x_X and x_Y denote the structural displacements along the X and Y axes, respectively, and the overdot denotes the time derivative. The displacement of the free surface of the liquid columns parallel to the X and Y axes is y_X and y_Y, respectively. The length ratio of the liquid columns parallel to the X and Y axes is represented by α_X and α_Y, respectively. The natural frequency and the damping ratio of the structure along the X direction are denoted by ω_{sX} and ζ_{sX}, respectively, whereas the said parameters along the Y direction are denoted by ω_{sY} and ζ_{sY}. The head-loss coefficients of the orifice(s) in the liquid columns parallel to the X and Y axes are ξ_X and ξ_Y, respectively. The mass ratio along the X direction, defined by the ratio of the mass of the liquid column parallel to the X axis to the structural mass, is μ_X while μ_Y represents the mass ratio along the Y direction. The components of the base excitation along the X and Y axes are $\ddot{Z}_X$ and $\ddot{Z}_Y$, respectively.

The equivalent linear damping coefficient of a BTLCD along the X and Y directions may be obtained by following the procedure discussed in Section 2.2.2 of Chapter 2. There is no detailed research work available that deals with the parametric optimization of the BTLCD. However, as a BTLCD is essentially a combination of two conventional TLCDs fused in a crossed formation, the optimum parameters of the conventional TLCD may be used for the purpose of preliminary design of a BTLCD.

Teramura and Yoshida (1996) designed a different form of the BTLCD that may be viewed as four separate TLCDs integrated into a single unit and, in plan, has the shape of an annular rectangle (see Figure 3.3). This configuration of the BTLCD is

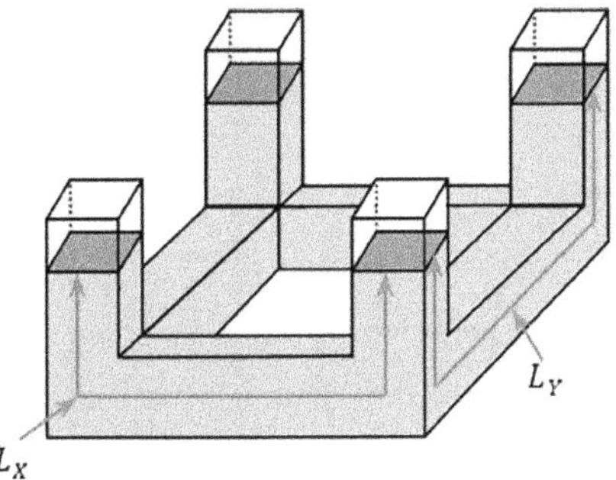

FIGURE 3.3 Bi-directional tuned liquid column damper having the shape of an annular rectangle in plan (BTLCD-AR).

represented by the acronym BTLCD-AR, to differentiate it from the one proposed by Sakai et al. (1991). Much later, Rozas et al. (2016) presented a detailed analytical, experimental and parametric study on the BTLCD-AR. To represent the dynamics of a base excited 2-DOF primary structure with a BTLCD-AR, Eqs. (3.3)–(3.6) can be utilized with two modifications. First, the mass ratio along a particular axis would be the ratio of the total mass of both the liquid columns parallel to that axis to the structural mass. Second, y_X and y_Y, respectively, would be the liquid displacement in the horizontal limbs of the damper that are parallel to the X and Y axes. The latter is required because, in the BTLCD-AR, under bi-directional excitation, the displacement of the free surface of the liquid column in any vertical limb of the damper is dependent on the excitation along both the directions. For example, under the action of external loading, if the liquid in two adjacent horizontal limbs moves in the direction of a vertical limb, the displacement of the free surface of the liquid columns in that vertical limb would be $y_X + y_Y$.

Here, it is pertinent to mention that a BTLCD-AR was successfully installed in the 106 m tall Hotel Cosima (later known as Hotel Sofitel) in Tokyo, Japan, in the year 1994 (Kareem et al., 1999; Konar et al., 2024). This particular BTLCD-AR contains 58 kilolitres of water and is provided with a period adjustment arrangement for tuning the damper frequency with that of the building. It has been estimated that due to the installation of the damper, the effective damping ratio of the building along both its principal axis has been enhanced by around ten times (Teramura and Yoshida, 1996).

3.2.2 TOROIDAL TLCD (TTLCD)

Recently, Ding et al. (2020) have introduced the concept of the toroidal TLCD (TTLCD) which can be utilized for multi-directional vibrational control if the primary structure has identical frequencies along different directions (Ding et al., 2023a). A TTLCD consists of two vertical concentric cylindrical shells with top open and bottom closed (see Figure 3.4a and b). The tops of both the cylindrical shells are at the same level, while the bottom of the inner shell is kept above that of the outer shell. The annular portion between the cylindrical shells is divided into compartments by an even number of equally spaced radial plates that are solid from the top

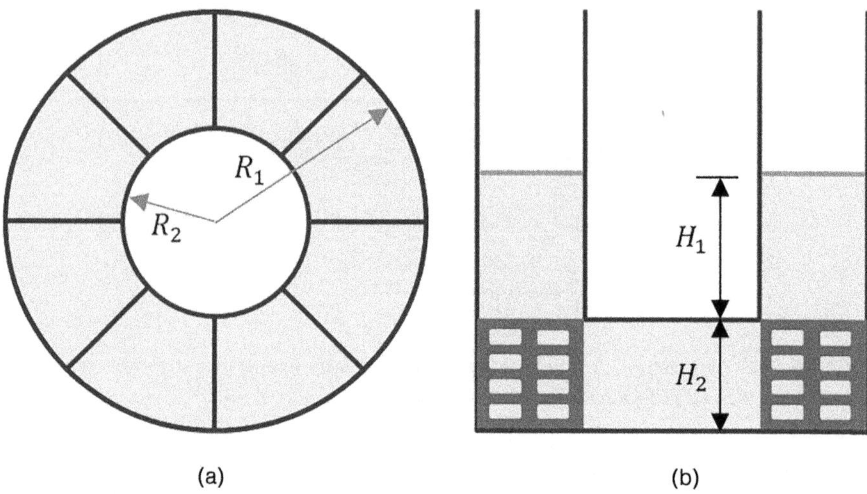

(a)   (b)

FIGURE 3.4 Toroidal tuned liquid column damper (TTLCD) (a) top view and (b) vertical cross-sectional view.

to the bottom of the inner shell. The remaining portions of the plates are perforated, which acts as a set of orifices (see Figure 3.4b). Further, a cylindrical perforated shell may be provided from the bottom of the inner shell up to the bottom of the outer shell to enhance damping. When the portion between the inner and outer shells is partially filled with liquid, in-between two adjacent radial plates, a vertical liquid column is formed. As all the vertical liquid columns are linked at their bottom, any two different columns may be considered to constitute a TLCD. An experimental study with lateral excitation on the TTLCD has shown that all the liquid columns of the damper have an identical frequency of oscillation and the displacements of the liquid free surface are different for different liquid columns (Ding et al., 2021b, 2022).

The key dimensions of the TTLCD are marked in Figure 3.4a and b. The radii of the external and internal cylinder shells are denoted by R_1 and R_2, respectively. The gap between the bottom of the external and the bottom of the internal cylinder shells is H_2, and H_1 represents the height of the internal cylinder shell, which remains filled with liquid when the damper is at rest.

If a TTLCD has $2N$ numbers of radial plates, the damper may be considered as a combination of n_c number of conventional TLCDs, where, $n_c = (N+1)/2$ if N is odd, and $n_c = N/2$ if N is even (see Figure 3.5a and b). The cross-sectional areas of the vertical and horizontal limbs of the equivalent conventional TLCDs are designated as A_{vt} and A_{ht}, respectively, and their expressions are (Ding et al., 2020)

$$A_{vt} = \frac{\pi}{2N}\left(R_1^{\,2} - R_2^{\,2}\right)$$ (3.7)

and

$$A_{ht} = \frac{2R_1 H_2}{2n_c - 1}.$$ (3.8)

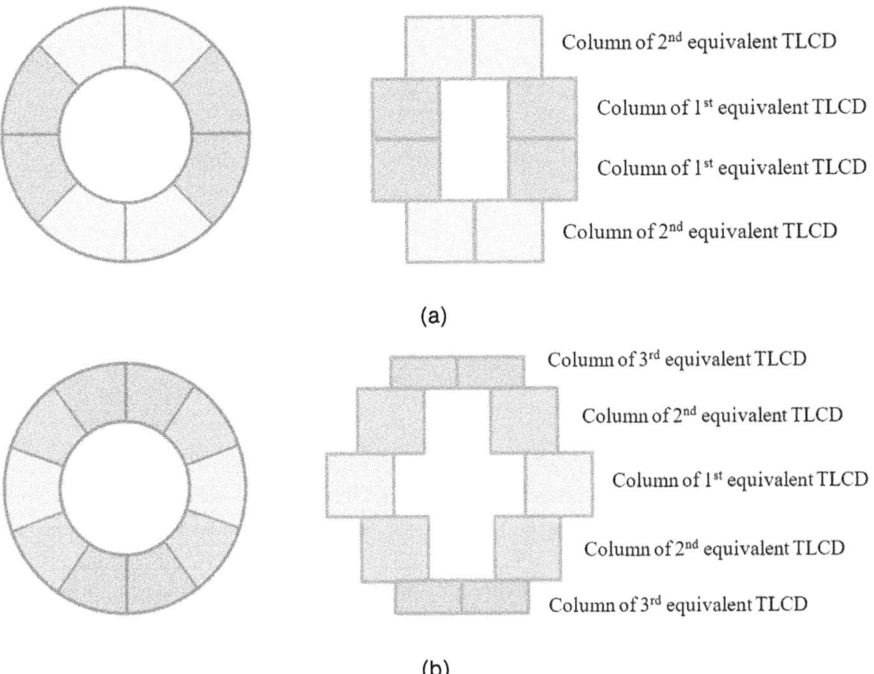

FIGURE 3.5 Top view of toroidal tuned liquid column damper (TTLCD) and equivalent conventional TLCDs (a) TTLCD where N is even and (b) TTLCD where N is odd.

As stated earlier, the frequency of oscillation of all the liquid columns in a TTLCD is identical, which may be determined from

$$\omega_{\text{TTLCD}} = \sqrt{\frac{2g}{L_{E-T}}}. \tag{3.9}$$

The experimental validation of the expression in Eq. (3.9) is available in Ding et al. (2020). In Eq. (3.9), L_{E-T} is the effective length of the liquid column of the TTLCD and is expressed by

$$L_{E-T} = 2H_1 + H_2 + r_i W b_1. \tag{3.10}$$

In Eq. (3.10), r_i denotes the area ratio defined by the ratio of the cross-sectional area of vertical limbs to the cross-sectional area of the horizontal limb, and W is the correction factor expressed by

$$W = \frac{\sum_{i=1}^{n_c} W_i^3}{\sum_{i=1}^{n_c} W_i^2} \tag{3.11a}$$

for excitation along one of the radial plates of the damper, and

$$W = \frac{1 + 2\sum_{i=2}^{n_c} W_i^3}{1 + 2\sum_{i=2}^{n_c} W_i^2} \tag{3.11b}$$

for excitation along a symmetrical axis of one of the liquid columns with,

$$W_i = \frac{b_i}{b_1}. \tag{3.12}$$

In Eqs. (3.10) and (3.12), b_1 denotes the length of the horizontal limb of the first equivalent TLCD and is given by

$$b_1 = R_1 + R_2. \tag{3.13}$$

In Eq. (3.12), b_i denotes the length of the horizontal limb of the ith equivalent TLCD.

Now, consider a TTLCD attached to a single DOF (SDOF) structural system having mass, stiffness and damping equal to M_S, K_S and C_S, respectively (see Figure 3.6). The mass of the damper container is included in M_S. The structure is subjected to a base acceleration, $\ddot{Z}_0$. The lateral displacement of the structure is denoted by $x(t)$. The displacement of the free surface of the liquid column in the vertical direction of the ith equivalent TLCD is denoted by $y_i(t)$, and ρ is the mass density of the damper liquid. Now, at any time instant t, the potential energy, V and the kinetic energy, T of the structure-damper system may be expressed as

$$V = \frac{1}{2}K_S x^2 + \rho A_{vt} g y_1^2 + 2\sum_{i=2}^{n_c} \rho A_{vt} g y_i^2 \tag{3.14}$$

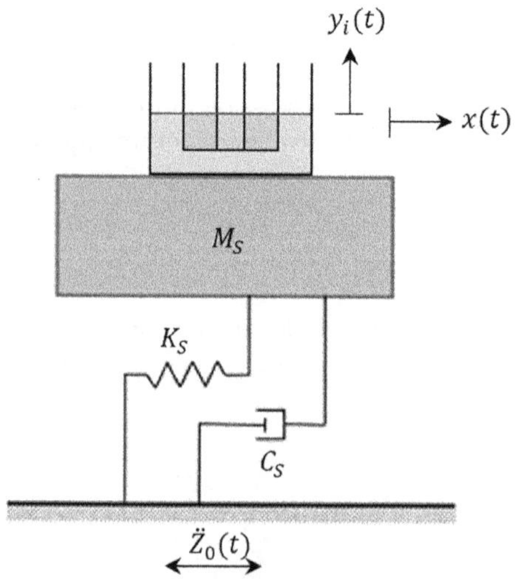

FIGURE 3.6 Schematic of structure–toroidal tuned liquid column damper system.

and

$$T = \frac{1}{2} M_S (\dot{x} + \dot{Z}_0)^2 + \frac{1}{2} \rho A_{ht} b_1 (\dot{x} + \dot{Z}_0 + r_t \dot{y}_1)^2$$

$$+ \sum_{i=2}^{n_c} \rho A_{ht} b_i (\dot{x} + \dot{Z}_0 + r_t \dot{y}_i)^2 + \frac{1}{2} \rho A_{vt} \left(H_1 + \frac{H_2}{2} - y_1 \right)$$

$$\left[\dot{y}_1^2 + (\dot{x} + \dot{Z}_0)^2 \right] + \frac{1}{2} \rho A_{vt} \left(H_1 + \frac{H_2}{2} + y_1 \right) \left[\dot{y}_1^2 + (\dot{x} + \dot{Z}_0)^2 \right]$$

$$+ \sum_{i=2}^{n_c} \rho A_{vt} \left(H_1 + \frac{H_2}{2} - y_i \right) \left[\dot{y}_i^2 + (\dot{x} + \dot{Z}_0)^2 \right] + \sum_{i=2}^{n_c} \rho A_{vt} \left(H_1 + \frac{H_2}{2} + y_i \right) \left[\dot{y}_i^2 + (\dot{x} + \dot{Z}_0)^2 \right].$$

$$(3.15)$$

The non-conservative damping forces in the structure and the damper may be, respectively, expressed as

$$Q_x = -C_S \dot{x} \tag{3.16}$$

and

$$Q_y = -\frac{1}{2} \rho r_t A_{vt} \xi_1 |\dot{y}_1| \dot{y}_1 - \sum_{i=2}^{n_c} \rho r_t A_{vt} \xi_i |\dot{y}_i| \dot{y}_i \tag{3.17}$$

where ξ_i is the coefficient of flow resistance of the ith equivalent TLCD.

Using Eqs. (3.14) and (3.15), the Lagrangian of the system may be formed and the same may be substituted in the Lagrange's equation given in Eq. (2.5). After subsequent simplifications, governing equations of motion of the structure-TTLCD system are obtained as

$$\begin{bmatrix} M_s + m_f & m_1 \\ m_1 & m_2 \end{bmatrix} \begin{Bmatrix} \ddot{x} \\ \ddot{y}_1 \end{Bmatrix} + \begin{bmatrix} C_s & 0 \\ 0 & c_f \end{bmatrix} \begin{Bmatrix} \dot{x} \\ \dot{y}_1 \end{Bmatrix} + \begin{bmatrix} K_s & 0 \\ 0 & k_f \end{bmatrix} \begin{Bmatrix} x \\ y_1 \end{Bmatrix}$$

$$= - \begin{Bmatrix} M_s + m_f \\ m_1 \end{Bmatrix} \ddot{Z}_0 \tag{3.18}$$

where m_1, m_2, m_f, c_f and k_f are given by

$$m_1 = \rho A_{vt} \left(b_1 + 2 \sum_{i=2}^{n_c} \frac{b_i^2}{b_1} \right), \tag{3.19}$$

$$m_2 = \rho A_{vt} \left[2 \left(H_1 + \frac{H_2}{2} \right) + 4 \left(H_1 + \frac{H_2}{2} \right) \sum_{i=2}^{n_c} \frac{b_i^2}{b_1^2} + b_1 r_t + 2 r_t \sum_{i=2}^{n_c} \frac{b_i^3}{b_1^2} \right], \tag{3.20}$$

$$m_f = \rho\left[\left(4n_c - 2\right)A_{vt}\left(H_1 + \frac{H_2}{2}\right) + A_{ht}b_1 + 2A_{ht}\sum_{i=2}^{n_c}b_i\right], \qquad (3.21)$$

$$c_f = \frac{1}{2}\rho r_t A_{vt}\xi_1\left|\dot{y}_1\right| + \rho r_t A_{vt}\sum_{i=2}^{n_c}\frac{b_i^{\,2}}{b_1^{\,2}}\xi_i\left|\dot{y}_1\right| \qquad (3.22)$$

and

$$k_f = 2\rho A_{vt}g\left(1 + 2\sum_{i=2}^{n_c}\frac{b_i^{\,2}}{b_1^{\,2}}\right). \qquad (3.23)$$

The effectiveness of the TTLCD in controlling multi-directional vibration of structures with identical frequencies along different directions, such as a monopole, has been demonstrated by Ding et al. (2023a,b). However, when vibration control is required only along a specific direction, the performance of the TTLCD is found to be slightly inferior to that of a conventional TLCD having the same amount of liquid mass (Ding et al., 2021b). In case of multi-modal vibration control, a multi-layer TTLCD that consists of more than two concentric cylindrical shells, in comparison to the simple TTLCD with only two concentric cylindrical shells, would prove more effective (Ding et al., 2021a).

3.2.3 OMNIDIRECTIONAL TLCD (OTLCD)

An omnidirectional TLCD (OTLCD), as the name suggests, too can control structural vibration induced by excitations acting along multiple directions in the horizontal plane, provided the structure possesses identical frequency along these degrees of freedom. Mehrkian and Altay (2020) were the first to propose the concept of the OTLCD and experimentally validate its working principles as well. The OTLCD is constituted by the circular distribution of N (with N an integer having value of 3 or more) L-shaped limbs of equal size about a common centre point, through which all the L-shaped limbs are connected to each other. The L-shaped limbs are equally spaced at an angle of $2\pi/N$, and the ith L-shaped limb makes an angle, θ_i, equal to $2(i-1)\pi/N$, with the first limb. The horizontal portions of all the L-shaped limbs are provided with orifices. An OTLCD with three L-shaped limbs is shown in Figure 3.7a and b. The height of liquid in the vertical limb is denoted by H_o, and the liquid length in the horizontal portion of the L-shaped limb is $b_o/2$.

   The frequency of liquid oscillation along the different limbs of the OTLCD is the same and is expressed as

$$\omega_{\mathrm{OTLCD}} = \sqrt{\frac{2g}{L_{E-O}}} \qquad (3.24)$$

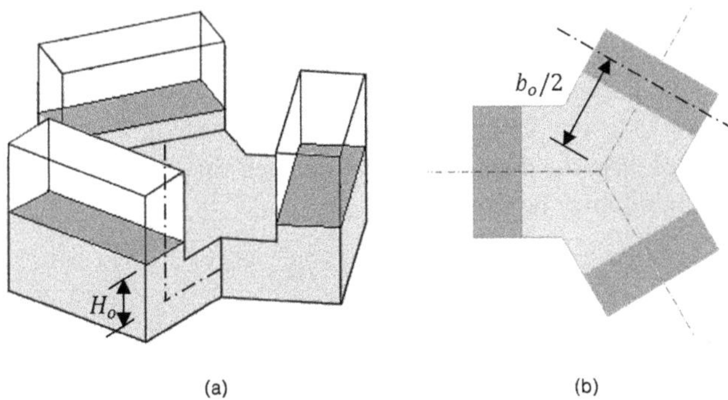

(a)   (b)

FIGURE 3.7 Omnidirectional tuned liquid column damper with three L-shaped limbs: (a) three-dimensional view and (b) top view.

where L_{E-O} represents the length of the liquid column of an equivalent conventional TLCD representing the OTLCD and is expressed by

$$L_{E-O} = 2H_o + b_o. \tag{3.25}$$

An OTLCD mounted on a structural system is shown in Figure 3.8. The structure is assumed to be an SDOF system where $x(t)$ is the lateral displacement along the X axis. It is subjected to a base acceleration, $\ddot{Z}_0$, along the X axis and the resulting displacement of the structure is $x(t)$. The other parameters of the structure are the same as those of the SDOF structural system described in Section 3.2.2. The dead

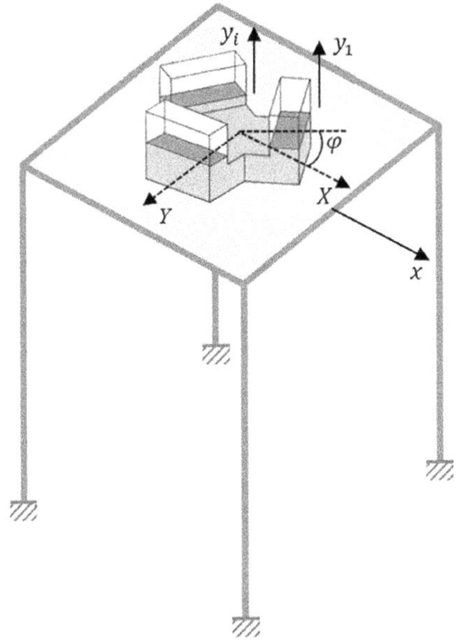

FIGURE 3.8 A single-degree-of-freedom structure with omnidirectional tuned liquid column damper.

mass of the OTLCD container is assumed to be lumped with the structural mass. The first L-shaped limb of the OTLCD makes an angle φ with the X axis, that is, the direction of the excitation. Let us assume that the displacement of the free surface of the liquid column in an equivalent conventional TLCD along X axis representing the OTLCD is $y(t)$. Hence, the displacement of the free surface of the liquid column in the ith L-shaped limb may be expressed as

$$y_i = y\cos(\varphi - \theta_i).\tag{3.26}$$

The potential energy, V and the kinetic energy, T of the structure-OTLCD system at a time instant t, may be expressed as

$$V = \frac{1}{2}K_S x^2 + \frac{1}{2}\rho g A_o \sum_{i=1}^{N} y_i^2 \tag{3.27}$$

and

$$T = \frac{1}{2}M_S(\dot{x}+\dot{Z}_0)^2 + \frac{1}{2}\rho A_o \sum_{i=1}^{N}\left(\frac{b_o}{2}|\dot{y}_{h,i}|^2 + H_o|\dot{y}_{v,i}|^2\right).\tag{3.28}$$

In Eqs. (3.27) and (3.28), A_o is the cross-sectional area of the L-shaped limb of the OTLCD. Further in Eq. (3.28), $\dot{y}_{h,i}$ and $\dot{y}_{v,i}$ represent the liquid velocity in the horizontal and vertical portion of the ith L-shaped limb, respectively. The magnitude of their components along X and Y axes, and the vertical direction, respectively, are expressed by

$$\dot{y}_{v,i} = \left[\begin{array}{ccc}(\dot{x}+\dot{Z}_0) & 0 & \dot{y}_i\end{array}\right]\tag{3.29}$$

and

$$\dot{y}_{h,i} = \left[\begin{array}{ccc}\dot{y}_i\cos(\theta_i - \varphi)+(\dot{x}+\dot{Z}_0) & \dot{y}_i\sin(\theta_i - \varphi) & 0\end{array}\right].\tag{3.30}$$

The non-conservative damping forces in the structure and the damper are, respectively, given by

$$Q_x = -C_S\dot{x}\tag{3.31}$$

and

$$Q_y = -\frac{1}{4}A_o\rho\xi_o\sum_{i=1}^{N}|\dot{y}_i|\dot{y}_i.\tag{3.32}$$

In Eq. (3.32), ξ_o is the coefficient of head-loss of the conventional TLCD that is equivalent to the OTLCD.

Now, the Lagrangian of the system may be obtained from Eqs. (3.26)–(3.30). Further, by the use of Eqs. (2.5), (3.31), (3.32) and the Lagrangian of the system, the equations of motion of the structure-OTLCD system are derived as

$$\ddot{x} + \frac{2\zeta_s \omega_s}{\left(1+\mu_o\right)}\,\dot{x} + \frac{\omega_s^{\,2}}{\left(1+\mu_o\right)}\,x = -\ddot{Z}_0 - \frac{1}{2}\frac{\alpha_o \mu_o}{\left(1+\mu_o\right)}\,\ddot{y} \tag{3.33}$$

and

$$\ddot{y} + \frac{1}{2}\frac{\xi_o}{L_{E-O}}\,|\dot{y}|\,\dot{y} + \omega_{\mathrm{OTLCD}}{}^2 y = -\alpha_o\left(\ddot{x} + \ddot{Z}_0\right) \tag{3.34}$$

In Eqs. (3.33)–(3.34), α_o is the length ratio of the OTLCD given by b_o/L_{E-O}, and μ_o is the mass ratio defined as the ratio of the total mass of the liquid in the OTLCD to the mass of the SDOF structural system.

On comparing Eqs. (3.33)–(3.34) with the equations of motion of an SDOF structure with a conventional TLCD as presented in Section 2.2.2 of Chapter 2, it may be observed that the equations are identical in both cases, except for the last term in Eq. (3.33).

Mehrkian and Altay (2020) conducted detailed experimental studies on an OTLCD attached to a base-excited SDOF system. Effective performance of the OTLCD independent of the excitation direction was reported. In a subsequent work, Mehrkian and Altay (2022) extended the earlier study to the control of multi-DOF structural systems by the OTLCD under base excitation as well as under external forces acting on the structure.

3.3 TLCD CONFIGURATIONS TO ENHANCE CONTROL PERFORMANCE

3.3.1 Liquid Column Vibration Absorber

The liquid column vibration absorber (LCVA) is a modified version of the conventional TLCD and was first proposed by Hitchcock et al. (1997a,b). Similar to the TLCD, the LCVA consists of a liquid column in a partly filled rigid U-shaped container. However, unlike the TLCD, the LCVA has different cross-sectional areas in the vertical and horizontal limbs of the U-shaped damper container (Hitchcock et al., 1997a; Konar and Ghosh, 2010) (see Figure 3.9).

The oscillatory frequency of the liquid column in an LCVA is expressed by (Hitchcock et al., 1997a)

$$\omega_{\mathrm{LCVA}} = \sqrt{\frac{2g}{L_{E-L}}} \tag{3.35}$$

where L_{E-L} is the effective length of the liquid column given by

$$L_{E-L} = 2h + b_L r. \tag{3.36}$$

In Eq. (3.36), h and b_L denote the liquid height in the vertical limbs and the length of the horizontal limb, respectively, while r is the area ratio defined by

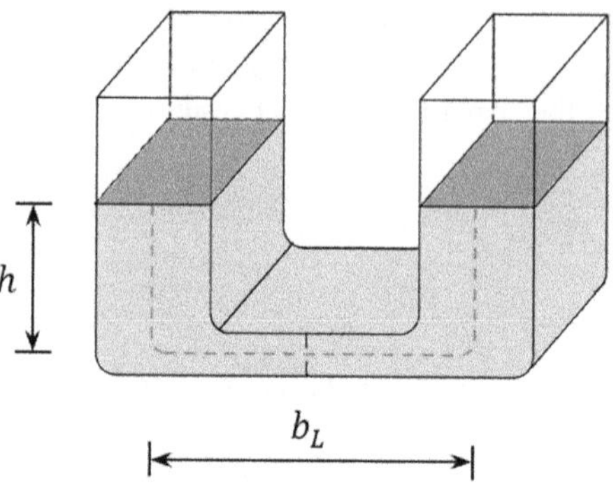

FIGURE 3.9   Liquid column vibration absorber.

$$r = \frac{A_v}{A_h}. \tag{3.37}$$

In Eq. (3.37), A_v and A_h represent the cross-sectional area of the vertical limbs and the horizontal limb, respectively.

It may be observed from Eq. (3.36) that the natural frequency of an LCVA depends not only on the liquid column length, $(2h + b_L)$, but also on the area ratio. Thus, an LCVA with a required frequency can be designed with different lengths of the liquid column. This enhances the architectural adaptability and versatility of the LCVA. Further, in addition to the energy dissipation due to the presence of the orifice(s), the change in the cross-sectional area between the horizontal and vertical limbs of the LCVA induces a transitional effect due to the sudden variation in the velocity of the moving liquid, which provides additional damping.

In order to derive the governing equations of motion of a structure equipped with an LCVA, the simplified model of a structure-LCVA system shown in Figure 3.10 is considered. Here, the structure is an SDOF system with mass, stiffness and damping as described in Section 3.2.2. Similar to the other cases discussed in this chapter, here too the structure is subjected to a base excitation represented by the acceleration $\ddot{Z}_0(t)$. The vertical displacement of the free surface of the liquid column is represented by $y(t)$.

At an instant of time t, the potential energy, V and the kinetic energy, T of the structure-LCVA system may be expressed as

$$V = \frac{1}{2} K_S x^2 + \rho A_v g y^2 \tag{3.38}$$

and

$$
\begin{aligned}
T = {} &\frac{1}{2} M_S (\dot{x} + \dot{Z}_0)^2 + \frac{1}{2} \rho A_h b_L \left(\dot{x} + \dot{Z}_0 + r\dot{y} \right)^2 \\
&+ \frac{1}{2} \rho A_v (h - y) \left[\dot{y}^2 + \left(\dot{x} + \dot{Z}_0 \right)^2 \right] + \frac{1}{2} \rho A_v (h + y) \left[\dot{y}^2 + \left(\dot{x} + \dot{Z}_0 \right)^2 \right].
\end{aligned}
\tag{3.39}
$$

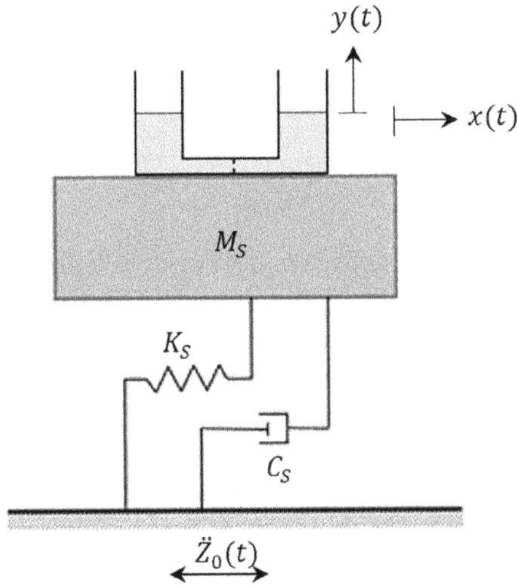

FIGURE 3.10 A single-degree-of-freedom structure with liquid column vibration absorber.

Further, the non-conservative forces in the structure and the damper are, respectively, given by

$$Q_x = -C_S \dot{x} \tag{3.40}$$

and

$$Q_y = -\frac{1}{2} \rho A_h \xi_{\text{LCVA}} |r\dot{y}| r\dot{y}, \tag{3.41}$$

where ξ_{LCVA} is the coefficient of flow resistance or head-loss due to orifice(s) and due to the change in the cross-sectional area between the horizontal and vertical limbs of the damper.

The procedure described in Section 2.2.2 of Chapter 2 for the conventional TLCD can be applied here as well in order to obtain the governing equations of motion of the structure-LCVA system yielding

$$\ddot{x} + \frac{2\zeta_s \omega_s}{(1+\mu_L)} \dot{x} + \frac{\omega_s^{\,2}}{(1+\mu_L)} x = -\ddot{Z}_0 - \frac{\rho b_L r A_h}{M_S(1+\mu_L)} \ddot{y} \tag{3.42}$$

and

$$\ddot{y} + \frac{1}{2}\frac{\xi_{\text{LCVA}}}{L_{E-L}} r|\dot{y}|\dot{y} + \omega_{\text{LCVA}}^{\,2} y = -\alpha_L \left(\ddot{x} + \ddot{Z}_0\right). \tag{3.43}$$

In Eqs. (3.42)–(3.43), α_L is the length ratio of the LCVA defined by the ratio (b_L / L_{E-L}), and μ_L is the mass ratio defined by the ratio of the mass of the damper liquid to the mass of the SDOF structural system.

Similar to the conventional TLCD, the equivalent damping coefficient, ζ_{LCVA} of an LCVA is often used in the design of the damper (Chang and Hus, 1998; Konar and Ghosh, 2013, 2010) and may be expressed as (Chang and Hus, 1998)

$$\zeta_{\mathrm{LCVA}} = \frac{r}{\sqrt{2\pi}}\xi_{\mathrm{LCVA}}\sigma_{\dot{y}}. \tag{3.44}$$

The efficacy of the LCVA as a passive control device is found to be comparable and in some instances even superior to the conventional TLCD (Chang and Hus, 1998; Konar and Ghosh, 2010). Similar to the bi-directional TLCD, a bi-directional LCVA has also been studied by some researchers (Hitchcock et al., 1999). Further, in line with the concept of the OTLCD, an omnidirectional LCVA (OLCVA) has also been proposed by Mehrkian and Altay (2022) for multi-directional vibration control of building structures. Unlike the OTLCD, the OLCVA has different cross-sectional areas in the vertical and the horizontal portions of the L-shaped limbs constituting the damper. It is important to note that the cross section of the vertical limb of the LCVA is generally designed to be higher than that of the horizontal limb, in order to allow for architectural adaptability. Hence, the sloshing action in the vertical limbs of the LCVA can also be considered, along with liquid column oscillation, for bimodal control of structural vibration (Das et al., 2020, 2023a; Konar, 2008; Konar and Ghosh, 2013).

3.3.2 TUNED LIQUID COLUMN BALL DAMPER

The tuned liquid column ball damper (TLCBD) is a configuration which significantly enhances the performance of the conventional TLCD. The study on the TLCBD was initiated by Al-saif et al. (2011). The cross section of the TLCBD is circular and accommodates a ball in the horizontal limb of the TLCD, instead of orifice(s) (see Figure 3.11). The ball is normally made of metal, such as steel. When laterally excited, the ball within the damper liquid rolls horizontally, functioning as a moving orifice. Thus, the TLCBD is also sometimes referred to as a moving orifice TLCD (Pandey and Mishra, 2018).

Except for the ball functioning as an orifice, the other parameters of a TLCBD, such as the overall length of the liquid column L, the horizontal length of the liquid column b, cross-sectional area A and length ratio α, are the same as in a conventional

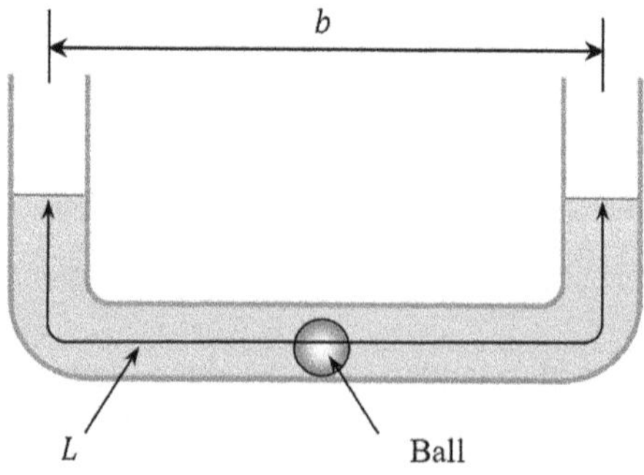

FIGURE 3.11 Tuned liquid column ball damper.

TLCD. The frequency of oscillation of the liquid column in the TLCBD is also the same as in the case of a conventional TLCD (see Eq. (2.15)).

A schematic of a structure-TLCBD system, excited at the base by acceleration $\ddot{Z}_0(t)$ is shown in Figure 3.12. The structure is modelled as an SDOF system with properties as described in Section 3.2.2. The vertical displacement of the free surface of the liquid column and the lateral displacement of the ball with respect to the damper container are denoted by $y(t)$ and $x_b(t)$, respectively.

At an instant of time t, the potential energy, V and the kinetic energy, T of the structure-damper system are given by

$$V = \frac{1}{2}K_S x^2 + \rho A g y^2 \tag{3.45}$$

and

$$
\begin{aligned}
T &= \frac{1}{2}M_S(\dot{x}+\dot{Z}_0)^2 + \frac{1}{2}\rho Ab\left(\dot{x}+\dot{Z}_0+\dot{y}\right)^2 \\
&\quad + \frac{1}{2}\rho A\left(\frac{L-b}{2}-y\right)\left[\dot{y}^2+\left(\dot{x}+\dot{Z}_0\right)^2\right] \\
&\quad + \frac{1}{2}\rho A\left(\frac{L-b}{2}+y\right)\left[\dot{y}^2+\left(\dot{x}+\dot{Z}_0\right)^2\right] \\
&\quad + \frac{1}{2}m_b(\dot{x}_b+\dot{x}+\dot{Z}_0)^2 + \frac{1}{2}J_b\left(\frac{\dot{x}_b}{R_b}\right)^2.
\end{aligned}
\tag{3.46}
$$

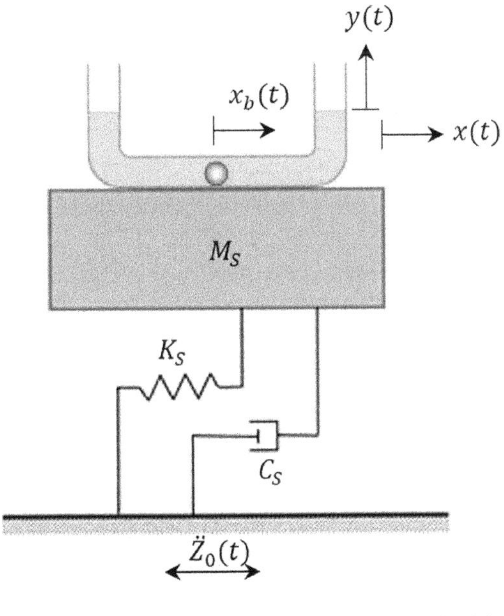

FIGURE 3.12 A single-degree-of-freedom structure with tuned liquid column ball damper.

In Eq. (3.46), m_b and R_b represent the mass and radius of the ball, respectively, and the mass moment of inertia of the ball is denoted by $J_b (= 2m_b R_b^2 /5)$.

The non-conservative forces in the structure, TLCBD and the ball may be expressed as

$$Q_x = -C_S \dot{x}, \tag{3.47}$$

$$Q_y = -C_d \dot{y} \tag{3.48}$$

and

$$Q_{xb} = -d_{eq} \dot{x}_b + 2\rho A_b yg, \tag{3.49}$$

respectively.

In Eqs. (3.47)–(3.49), C_d represents the equivalent damping coefficient of the TLCBD, while $d_{eq} (= 6\pi \rho v R_b)$ denotes the equivalent viscous damping coefficient due to the liquid drag acting on the ball (Nakayama and Boucher, 1999). A_b stands for the cross-sectional area of the ball, and the kinematic viscosity of the damper liquid is denoted by v. It may be noted that the damping term in the case of the TLCBD is assumed to be linear.

Using Eqs. (3.45)–(3.49) and following the Lagrange principle, the equations of motion of the structure-TLCBD system may be derived as

$$(M_S + \rho AL + m_b)\ddot{x} + C_S \dot{x} + K_S x = -(M_S + \rho AL + m_b)\ddot{Z}_0 - \alpha \rho AL\ddot{y} + m_b \ddot{x}_b, \tag{3.50}$$

$$\ddot{y} + 2\zeta_d \omega_{\text{TLCBD}} \dot{y} + \omega_{\text{TLCBD}}^2 y = -\alpha(\ddot{x} + \ddot{Z}_0) \tag{3.51}$$

and

$$\left(m_b + \frac{J_b}{R_b^2}\right)\ddot{x}_b + d_{eq} \dot{x}_b = m_b(\ddot{x} + \ddot{Z}_0) + 2\rho A_b yg. \tag{3.52}$$

In Eqs. (3.50)–(3.52), the equivalent damping ratio of the TLCBD is denoted by ζ_d, which is primarily dependent on the ratio of the ball diameter to the diameter of the damper container tube, r_d. Consequently, the efficacy of the TLCBD is significantly influenced by r_d. Through experimental studies on a small-scale model of TLCBD, Al-saif et al. (2011) determined ζ_d for six different values of r_d. A plot of the experimental results is presented in Figure 3.13, wherein it is observed that ζ_d is practically insensitive to r_d over a range of values of r_d, until r_d is equal to 0.8, whereas for r_d equal to 0.9, the value of ζ_d is almost doubled from its value at r_d equal to 0.8. However, existing research on TLCBD does not yet provide a standardized expression to link the ball-to-limb diameter ratio to the supplemental damping afforded by the device as evidenced from experiments. As a result, a universally applicable analytical or empirical formula to determine the equivalent linear damping coefficient of a TLCBD remains elusive.

An optimally designed TLCBD has demonstrated superior effectiveness as a vibration absorber as compared to the conventional TLCD, when incorporated into an SDOF primary structure under the action of a harmonic force (Al-saif et al., 2011),

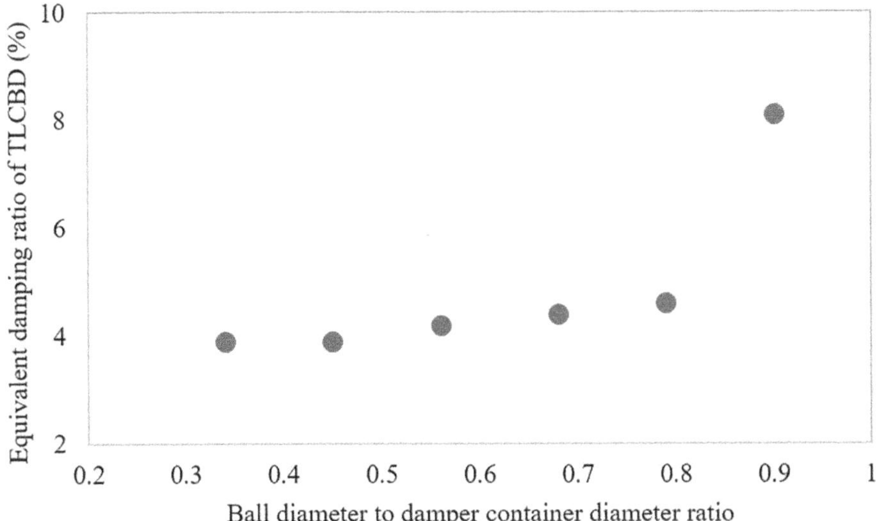

FIGURE 3.13 Equivalent damping ratio of the tuned liquid column ball damper vs. ball diameter to the damper container diameter ratio (plotted based on experimental data available in Al-saif et al. (2011)).

as well as under the action of recorded ground motions (Gur et al., 2015). The effectiveness of the TLCBD under a seismically excited MDOF system has also been demonstrated (Tanveer et al., 2020).

A recently proposed variation from the original TLCBD (Shah et al., 2022) has a spring-controlled tuned liquid column ball damper, in which the motion of the ball in the horizontal limb of the damper is controlled through a spring to further improve the performance of the damper. In another variation, the ball of the TLCBD is replaced by a circular cylinder, and the cross section of the damper container is made rectangular (Das, 2022). This development is significant as during the real-life implementation of passive TLCDs, a damper container with a rectangular cross section is normally preferred as it is more adaptable from a construction point of view.

3.3.3 Tuned Liquid Multi-Column Damper

The tuned liquid multi-column damper (TLMCD), conceptualized by Wu et al. (2018), is a recently developed configurational variation of the conventional TLCD. It is constructed with several vertical liquid columns interconnected by a horizontal liquid column, as illustrated in Figure 3.14. This system, characterized by multiple DOF, offers the unique capability of tuning to several frequencies simultaneously. In a symmetrical TLMCD with N identical vertical limbs, $N/2$ distinct frequencies can be achieved, where N is an even number. However, it is important to note that the TLMCD is a complex nonlinear system featuring interlinked liquid motion within its multiple vertical limbs, and as of now, there are no generalized expressions available for determining the frequencies of the various modes of liquid oscillation in the TLMCD. Nevertheless, the equations of motion of a TLMCD may be derived

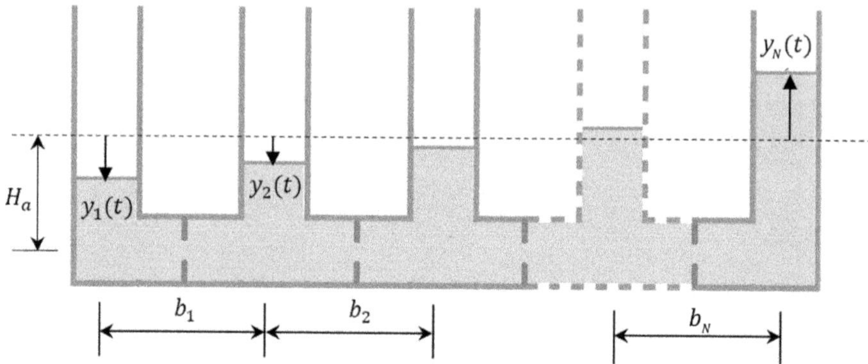

FIGURE 3.14 Tuned liquid multi-column damper.

using the Lagrange equations. The potential energy, V and the kinetic energy, T of the TLMCD may be written as

$$V = \frac{1}{2}\rho A g \sum_{i=1}^{N} y_i^2 \tag{3.53}$$

and

$$T = \frac{1}{2}\rho A \sum_{i=1}^{N} \dot{y}_i^{\,2}\left(H_a + y_i\right) + \frac{1}{2}\rho A \sum_{i=1}^{N-1} b_i \left(\dot{x} + \sum_{j=1}^{i}\dot{y}_i\right)^2. \tag{3.54}$$

In Eqs. (3.53)–(3.54), H_a, b_i and A denote the height of liquid in the vertical limbs of the damper at rest condition, the horizontal centre-to-centre distance between the ith and $(i + 1)$th columns of the damper and the cross-sectional area of the damper tube, respectively. It is assumed that the cross-sectional area of all the vertical limbs and the horizontal limb of the damper are equal. The vertical displacement of the liquid free surface in the ith column and the lateral displacement of the structure on which the TLMCD is mounted are represented by y_i and x, respectively.

The non-conservative orifice damping forces acting along the ith DOF of the TLMCD may be written as

$$Q_{yi} = -\frac{1}{2}\rho A \sum_{k=i}^{N-1}\left[\xi_k \left|\sum_{j=1}^{k}\dot{y}_j\right|\left|\sum_{j=1}^{k}\dot{y}_j\right|\right]. \tag{3.55}$$

Using Eqs. (3.53)–(3.55) and the Lagrange's equations, a set of nonlinear equations of motion of the TLMCD may be written in the form

$$\mathrm{M}(y)\ddot{y} + \mathrm{K}y + \mathrm{F}(y,\dot{y}) = 0. \tag{3.56}$$

In Eq. (3.56), $\mathrm{M}(y)$ and K, respectively, denote the state-dependent mass matrix and the linear stiffness matrix, both having dimensions $(N-1)\times(N-1)$. $\mathrm{F}(y,\dot{y})$ is the restoring force matrix containing all the nonlinear terms and has a dimension of $(N-1)\times 1$. The derivation of the analytical expressions for the frequencies of the damper from Eq. (3.56) is difficult due to the nonlinear nature of the equation. However, the damper frequencies may be determined numerically, considering undamped free vibration of the TLMCD system and using the invariant points approach. The detailed procedure in this regard is available in Cao et al. (2020).

Let us consider a case where a TLMCD is attached to an SDOF structure with properties as described in Section 3.2.2 (see Figure 3.15). Using the energy principle, the equations of motion of the structure-damper system may be expressed as (Wang et al., 2021)

$$M\ddot{u} + C\dot{u} + Ku = E_d F_d + E_e F_e.$$
(3.57)

In Eq. (3.57), $\mathbf{M}$, $\mathbf{C}$ and K denote the mass, damping and stiffness matrices, respectively, and are given by

$$M = \begin{bmatrix} M_S + m_d & 2\rho A\left(\sum_{k=1}^{N/2} b_k - b_{N/2}\right) & 2\rho A\left(\sum_{k=2}^{N/2} b_k - b_{N/2}\right) & \cdots & \rho A b_{N/2} \\ 2\rho A\left(\sum_{k=1}^{N/2} b_k - b_{N/2}\right) & 2\rho A\left\{\left(\sum_{k=1}^{N/2} b_k - b_{N/2}\right) + h\right\} & 2\rho A\left(\sum_{k=2}^{N/2} b_k - b_{N/2}\right) & \cdots & \rho A b_{N/2} \\ 2\rho A\left(\sum_{k=2}^{N/2} b_k - b_{N/2}\right) & 2\rho A\left(\sum_{k=2}^{N/2} b_k - b_{N/2}\right) & 2\rho A\left\{\left(\sum_{k=2}^{N/2} b_k - b_{N/2}\right) + h\right\} & \cdots & \rho A b_{N/2} \\ \vdots & \vdots & \vdots & \ddots & \vdots \\ \rho A b_{N/2} & \rho A b_{N/2} & \rho A b_{N/2} & \cdots & \rho A\{b_{N/2} + 2h\} \end{bmatrix},$$
(3.58)

$$C = \begin{bmatrix} C_S & 0 & 0 & \cdots & 0 \\ 0 & 0 & 0 & \cdots & 0 \\ 0 & 0 & 0 & \cdots & 0 \\ \vdots & \vdots & \vdots & \ddots & \vdots \\ 0 & 0 & 0 & \cdots & 0 \end{bmatrix}$$
(3.59)

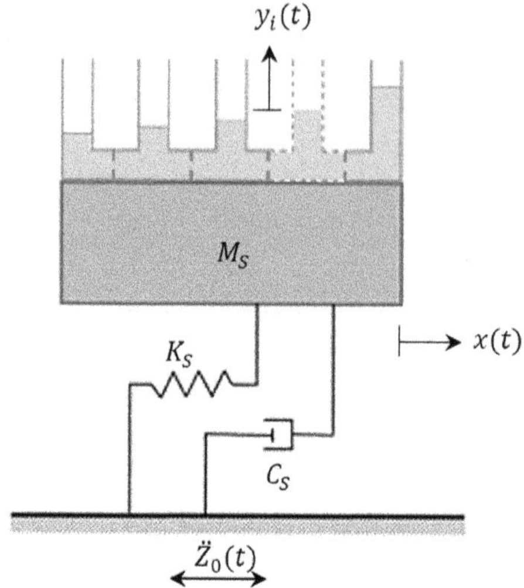

FIGURE 3.15   A single-degree-of-freedom structure with tuned liquid multi-column damper.

and

$$
K = \begin{bmatrix}
K_S & 0 & 0 & \cdots & 0 \\
0 & 2\rho Ag & 0 & \cdots & 0 \\
0 & 0 & 2\rho Ag & \cdots & 0 \\
\vdots & \vdots & \vdots & \ddots & \vdots \\
0 & 0 & 0 & \cdots & 2\rho Ag
\end{bmatrix}
\tag{3.60}
$$

Further in Eq. (3.57), u is the displacement vector as given by

$$
\{u\} = \begin{bmatrix} x & y_1 & y_2 & \cdots & y_{N/2} \end{bmatrix},
\tag{3.61}
$$

and F_d and F_e represent the damping force and the excitation force vectors, respectively, while E_d and E_e are the force location matrices containing only 0s and 1s as elements.

In Eq. (3.58), m_d is the total mass of the liquid in the TLMCD.

To evaluate the performance of the TLMCD, the coupled equations in Eq. (3.57) may be solved under different loading conditions. However, there is limited work on this front. The efficacy of the TLMCD in controlling the vibration of SDOF structural systems under harmonic excitation (Cao et al., 2020) and wind-induced excitation (Wang et al., 2021) has been studied. The possible multi-modal vibration control capability of the TLMCD has not yet been explored.

3.3.4   CIRCULAR TLCD

There are two types of circular TLCDs. The first variety was developed to effectively mitigate the torsional response of structures, while the second variety was developed for the vibration control of wind turbine blades.

The first variant of the circular TLCD features a circular horizontal limb and two vertical limbs in an extended plane perpendicular to the plane of the horizontal limb, as illustrated in Figure 3.16. The radius of the circular limb, the total length of the liquid column and the height of the liquid column in the vertical limbs in the at-rest condition are denoted by R, L_c and h, respectively. Due to its specific applicability in controlling torsional vibration, these circular TLCDs are often denoted as torsional TLCDs.

The natural frequency of a circular TLCD may be determined using the same expression as that for a conventional TLCD, by the use of Eq. (2.15) in Chapter 2. Again, similar to conventional TLCDs, the energy dissipation in circular TLCDs occurs through orifice damping. Consequently, the equivalent linear damping coefficient for a circular TLCD may be obtained by using the formulation provided in Eq. (2.26) of Chapter 2.

When a circular TLCD is attached to an SDOF structural system (see Figure 3.17), the potential energy, V and the kinetic energy, T of the structure-damper system at a time instant t, may be written as

$$V = \frac{1}{2}K_{St}\theta^2 + \rho Agy^2 \tag{3.62}$$

and

$$T = \frac{1}{2}M_S\Re^2\left(\dot{\theta}+\dot{Z}_\theta\right)^2 + \frac{1}{2}\rho A(L_c - 2h)\left[R\left(\dot{\theta}+\dot{Z}_\theta\right)+\dot{y}\right]^2$$
$$+ \frac{1}{2}\rho A(h-y)\left[\dot{y}^2 + R^2\left(\dot{\theta}+\dot{Z}_\theta\right)^2\right] + \frac{1}{2}\rho A(h+y)\left[\dot{y}^2 + R^2\left(\dot{\theta}+\dot{Z}_\theta\right)^2\right]. \tag{3.63}$$

In Eqs. (3.62) and (3.63), θ and y denote the rotational displacement of the structure and the vertical displacement of the free liquid surface, respectively. Further, M_S, K_{St} and $\Re$ represent the mass of the structure, the torsional stiffness of the structure and the radius of gyration of the structure about the vertical axis through the centre of mass, respectively. The torsional acceleration of ground motion is denoted by $\ddot{Z}_\theta$.

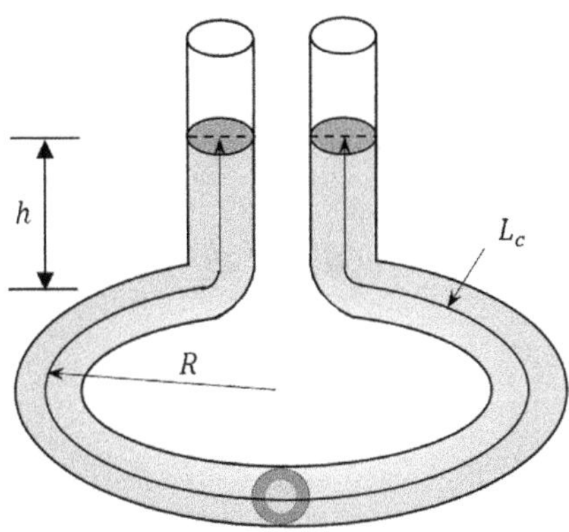

FIGURE 3.16 Circular tuned liquid column damper.

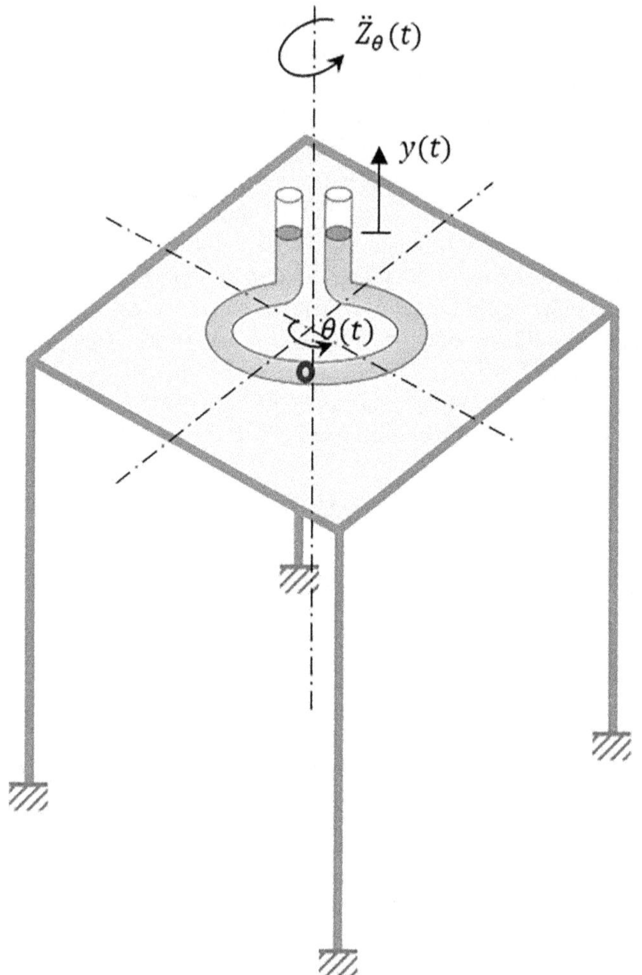

FIGURE 3.17   A single-degree-of-freedom structural system with circular tuned liquid column damper.

From the principle of virtual work, the generalized non-conservative forces corresponding to the DOF of the structure and the damper system may be, respectively, written as

$$Q_\theta = -C_{St}\dot{\theta} \tag{3.64}$$

and

$$Q_y = -\frac{1}{2}\rho A \xi |\dot{y}|\dot{y}. \tag{3.65}$$

In Eq. (3.64), C_{St} is the coefficient of viscous damping of the structure for rotational motion.

Now, following the procedure described in Section 2.2.2 of Chapter 2 for the conventional TLCD, the governing equations of motion of the structure with circular TLCD may be formulated as

$$\left(M_S\Re^2 + \rho AL_c R^2\right)\ddot{\theta} + C_{St}\dot{\theta} + K_{St}\theta = -\left(M_S\Re^2 + \rho AL_c R^2\right)\ddot{Z}_\theta - \rho AL_c R\alpha_c\ddot{y} \quad (3.66)$$

and

$$\ddot{y} + \frac{1}{2}\frac{\xi}{L_c}|\dot{y}|\dot{y} + \omega_{\text{C-TLCD}}^2 y = -\alpha_c R\left(\ddot{\theta} + \ddot{Z}_\theta\right). \quad (3.67)$$

In Eq. (3.66)–(3.67), α_c denotes the length ratio of the circular TLCD given by $\left(L_c - 2h\right)/L_c$, while $\omega_{\text{C-TLCD}}$ denotes the natural frequency of the damper.

The torsional vibration control effectiveness of the circular TLCD was first studied by Liang (1996), who conducted some preliminary experimental works. Subsequently, a detailed analytical and parametric study on the circular TLCD was carried out by Huo and Li (2004a, 2004b, 2005), who reported significant positive performance of the circular TLCD in controlling the torsional vibration of base-excited structures. However, circular TLCDs have not yet been examined under wind-induced vibration.

Some configurational variations of the circular TLCD have also been studied. One such example involves the incorporation of a ball instead of the traditional orifice in the damper (Pandey and Mishra, 2018). This adaptation thus extends the concept of the TLCBD to the circular TLCD. Another modified version of the circular TLCD is the torsional tuned liquid column gas damper (Fu, 2011), which is essentially a circular TLCD with the ends of the damper container sealed to produce a gas-spring effect. The efficacy of this device has been reported in the control of torsional vibrations in a plan-asymmetric building structure.

Another type of circular TLCD configuration was developed for the vibration control of wind turbine blades by Basu et al. (2016) (see Figure 3.18). The device consists of a closed circular tube, and hence, in this book, this type of TLCD is referred to as the closed circular TLCD, in order to differentiate it from the configuration shown

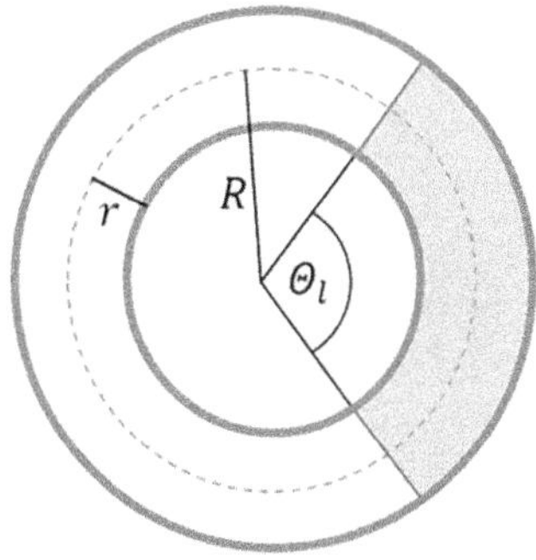

FIGURE 3.18   Closed circular tuned liquid column damper.

in Figure 3.16. The expressions for the frequency of oscillation of the liquid column, $\omega_{\text{CC-TLCD}}$, and the damping coefficient, $c_{\text{CC-TLCD}}$, are (Basu et al., 2016)

$$\omega_{\text{CC-TLCD}} = \sqrt{\frac{x_0}{L_{cc}}}\,\Omega \tag{3.68}$$

and

$$c_{\text{CC-TLCD}} = \frac{1}{2}\rho\xi r^2 R^3. \tag{3.69}$$

In Eqs. (3.68)–(3.69), x_0, Ω, L_{cc}, r and R denote the distance of the centre of the closed circular TLCD from the centre of rotation of the wind turbine blades, the rotational speed of the rotor, the equivalent length of the damper, the internal radius of cross section of the damper tube and the radius of the central axis of the circular tube, respectively. Further details on the mathematical formulation and the performance assessment of the damper are available in Basu et al. (2016). This type of damper has interesting nonlinear dynamic features and has demonstrated important contributions in the vibration control of wind turbine rotating blades.

3.3.5 S-Shaped TLCD

The S-shaped TLCD is an innovative modification of the conventional TLCD that is specially designed for application in very long-period structural systems (Zeng et al., 2015). The name of this specific damper is derived from the fact that its horizontal limb features one or more S-shaped segments. The advantage of the S-shaped TLCD chiefly lies in its compact shape. For a given frequency of the damper, the horizontal length requirement of the S-shaped TLCD is far less as compared to the conventional TLCD. To meet the desired damping from the device, the horizontal limb of the damper may be provided with one or multiple orifices. A schematic of the S-shaped TLCD is illustrated in Figure 3.19, on which the different key dimensions are marked.

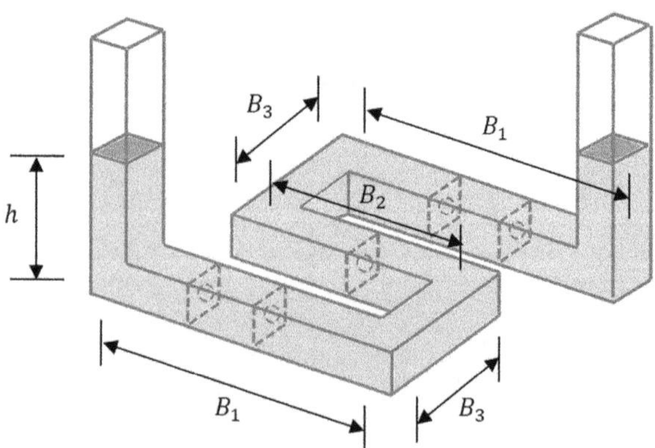

FIGURE 3.19 S-shaped tuned liquid column damper.

The structure-S-shaped TLCD system is shown in Figure 3.20. The structure is modelled as an SDOF system having properties as in the one considered in Section 3.2.2. The base excitation to the structure is represented by an acceleration time history, $\ddot{Z}_0(t)$. As in case of the TLCD configurations discussed in the earlier sections, the governing equations of the structure-damper system are again derived by making use of the Lagrange's equation. The expressions for the potential energy, V and the kinetic energy, T are

$$V = \frac{1}{2}K_S x^2 + \rho A g y^2 \tag{3.70}$$

and

$$T = \frac{1}{2}M_S(\dot{x}+\dot{Z}_0)^2 + \rho A B_1\left(\dot{x}+\dot{Z}_0+\dot{y}\right)^2$$

$$+\frac{1}{2}\rho A B_2\left(\dot{x}+\dot{Z}_0-\dot{y}\right)^2 + \rho A B_3\left[\dot{y}^2+\left(\dot{x}+\dot{Z}_0\right)^2\right] \tag{3.71}$$

$$+\frac{1}{2}\rho A(h-y)\left[\dot{y}^2+\left(\dot{x}+\dot{Z}_0\right)^2\right]+\frac{1}{2}\rho A(h+y)\left[\dot{y}^2+\left(\dot{x}+\dot{Z}_0\right)^2\right].$$

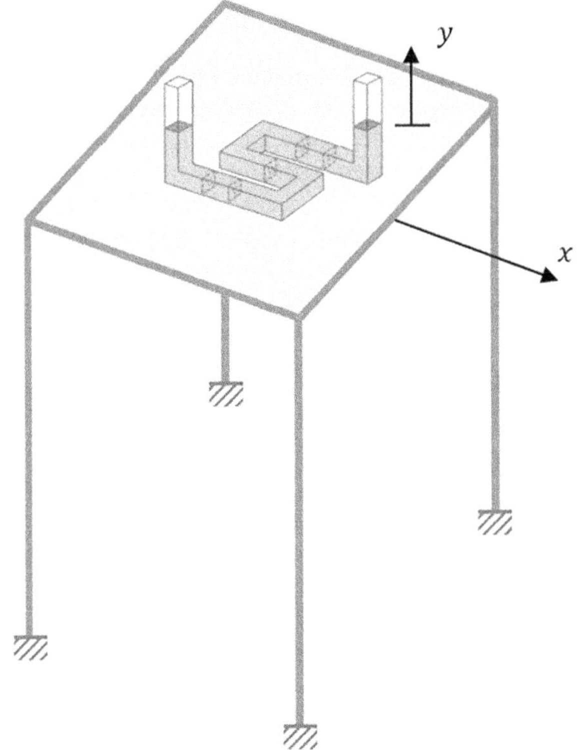

FIGURE 3.20 A single-degree-of-freedom structural system with S-shaped tuned liquid column damper.

The non-conservative forces in the structure and the damper are, respectively, given by

$$Q_x = -C_s \dot{x} \tag{3.72}$$

and

$$Q_y = -\frac{1}{2}\rho A \xi |\dot{y}| \dot{y} \tag{3.73}$$

where ξ is the head-loss coefficient of the orifice(s), similar to the conventional TLCD.

The governing equations of motion of the structure-S-shaped TLCD system are given by

$$\ddot{x} + \frac{2\zeta_s \omega_s}{(1+\mu)}\dot{x} + \frac{\omega_s^2}{(1+\mu)}x = -\ddot{Z}_0 - \frac{\mu}{(1+\mu)}\frac{2B_1}{L}\ddot{y} + \frac{\mu}{(1+\mu)}\frac{B_2}{L}\ddot{y} \tag{3.74}$$

and

$$\ddot{y} + \frac{1}{2}\frac{\xi}{L}|\dot{y}|\dot{y} + \omega_{\text{S-TLCD}}^2 y = -\left(\frac{2B_1}{L} - \frac{B_2}{L}\right)(\ddot{x} + \ddot{Z}_0) \tag{3.75}$$

where L denotes the length of the liquid column and $\omega_{\text{S-TLCD}} = \sqrt{2g/L}$ denotes the natural frequency of the S-shaped TLCD.

As mentioned, the S-shaped TLCD can be designed to possess a very long natural period. It may thus be considered for the vibration control of special structures, such as deep-sea floating platforms, which have unusually long natural periods. Further, the S-shaped TLCD may be considered for structures where the conventional TLCD cannot be accommodated due to the non-availability of a sufficiently long horizontal space. Performance-wise, the S-shaped TLCD is found to be comparable, if only slightly inferior to the conventional TLCD. This has been ascertained by Zeng et al. (2015) through experimental and numerical studies.

3.3.6 TUNED LIQUID COLUMN AND SLOSHING DAMPER

The tuned liquid column and sloshing damper (TLCSD) is essentially a combination of the conventional TLCD and the tuned sloshing damper (TSD). It is designed to provide bi-directional vibration control. As illustrated in Figure 3.21, a TLCSD comprises a long prismatic tank with a uniform U-shaped cross section. When subjected to excitation in the X-direction, the liquid in the damper undergoes oscillation, functioning akin to a conventional TLCD. Conversely, when excited in the Y-direction, the liquid within the damper's vertical limbs sloshes, transforming the device into a TSD.

Let the primary structure of a TLCSD be modelled as an SDOF system (see Figure 3.22), with the properties of the structure as described in Section 3.2.2. When the structure is excited at the base along the X-direction, the governing equations of motion of the structure-damper system are those of an SDOF structure with a conventional TLCD, as given by Eqs. (2.17) and (2.18) in Chapter 2. The natural frequency of liquid oscillation and the damping provided by the TLCSD, when excited along the X-direction, are therefore expressed by Eqs. (2.15) and (2.26) in Chapter 2, respectively.

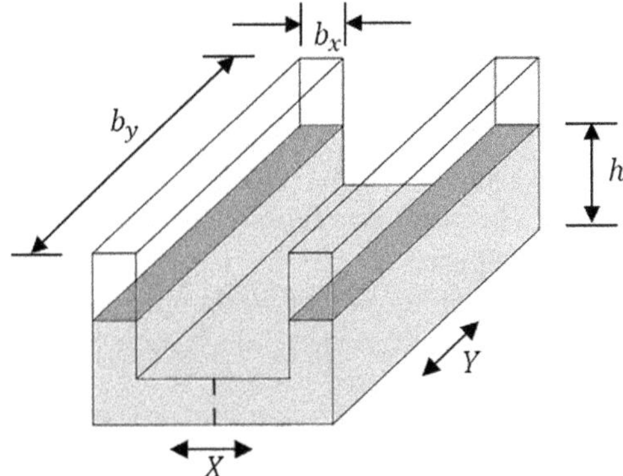

FIGURE 3.21 Tuned liquid column and sloshing damper.

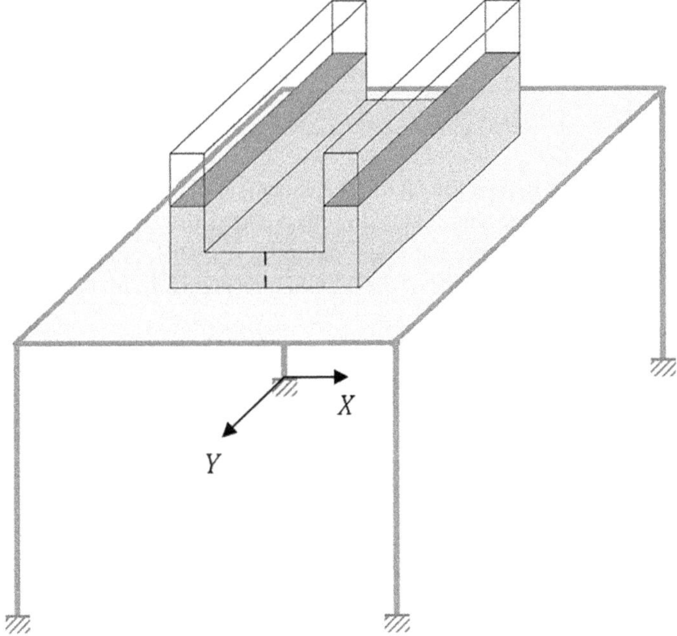

FIGURE 3.22 Tuned liquid column and sloshing damper with a single-degree-of-freedom structural system.

On the other hand, if the SDOF structure is excited along the Y-direction, the vertical limbs of the TLCSD would behave like two independent TSDs. In such a case, the equations of motion of the structure-damper system can be formulated by considering an equivalent mechanical model of the sloshing liquid in the vertical limbs of the damper container. The equivalent mechanical model is the simplest yet reasonably accurate model to represent the behaviour of laterally excited liquid tanks. In this model, there is a lumped mass that represents the impulsive liquid mass, which moves rigidly with the container. Additionally, there are several lumped masses, each associated with a convective or sloshing mode (Abramson et al., 1961; Graham and Rodriguez, 1952). However, typically the impact of higher sloshing modes is disregarded due to their relatively minor contributions (Das et al., 2023b; Konar, 2024; Konar and Ghosh, 2022). The equivalent mechanical model of the TLCSD when functioning as a TSD is presented in Figure 3.23. The impulsive liquid mass is not shown in Figure 3.23 to maintain the clarity of the figure.

The equations of motion of the structure-damper system, excited along the Y-direction under the action of base acceleration $\ddot{Z}_Y$, may be written as

$$\begin{bmatrix} M_s & 0 \\ 0 & 2m_{d1} \end{bmatrix} \begin{Bmatrix} \ddot{x}_Y \\ \ddot{y}_1 \end{Bmatrix} + \begin{bmatrix} C_s + 2c_{d1} & -2c_{d1} \\ -2c_{d1} & 2c_{d1} \end{bmatrix} \begin{Bmatrix} \dot{x}_Y \\ \dot{y}_1 \end{Bmatrix}$$

$$+ \begin{bmatrix} K_s + 2k_{d1} & -2k_{d1} \\ -2k_{d1} & 2k_{d1} \end{bmatrix} \begin{Bmatrix} x_Y \\ y_1 \end{Bmatrix}$$

$$= -\begin{Bmatrix} M_s \\ 2m_{d1} \end{Bmatrix} \ddot{Z}_Y(t). \tag{3.76}$$

In Eq. (3.76), x_Y and y_1 denote the lateral displacement of the structure and the lateral displacement of the fundamental sloshing mass along the Y-direction, respectively. Further, m_{d1}, c_{d1} and k_{d1} denote the mass, damping and stiffness of the first sloshing mode in one of the vertical limbs of the TLCSD, respectively, and are expressed as

$$m_{d1} = \rho b_x b_y^2 \frac{8}{\pi^3} \tanh\left(\frac{\pi h}{b_y}\right), \tag{3.77}$$

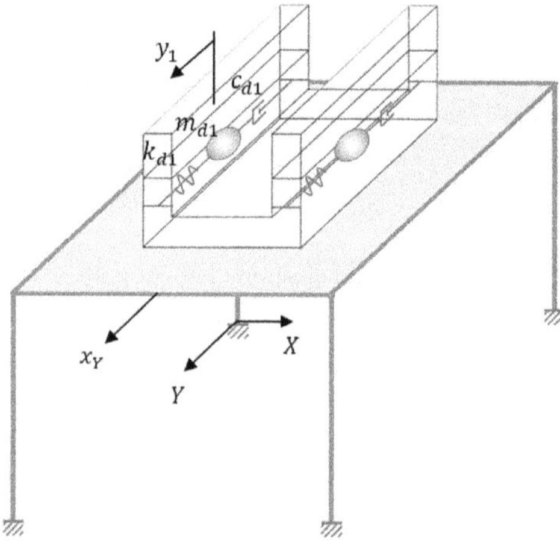

FIGURE 3.23 Equivalent mechanical model of tuned liquid column and sloshing damper for excitation of structure along Y-direction.

$$c_{d1} = \sqrt{2}\,\frac{m_{d1}}{(\eta+h)}\sqrt{\omega_{d1}V}\left[1+\frac{h}{b_x}\right] \qquad (3.78)$$

and

$$k_{d1} = m_{d1}\frac{\pi g}{b_y}\tanh\frac{\pi h}{b_y}. \qquad (3.79)$$

In Eqs. (3.77)–(3.79), h, b_x and b_y denote the height of liquid in the TLCSD and the width of the vertical limbs of the damper along the X and Y-directions, respectively. The maximum sloshing amplitude is represented by η and ω_{d1} the fundamental liquid sloshing frequency that would be tuned to the structural frequency along the Y-direction as defined by

$$\omega_{d1} = \left[\frac{\pi g}{b_y}\tanh\frac{\pi h}{b_y}\right]^{1/2}. \qquad (3.80)$$

The dynamics of the TLCSD discussed so far considered that the structure-damper system is excited along its principal directions. However, its behaviour under excitation from directions other than the principal axes warrants further study. Preliminary experimental studies conducted by Lee et al. (2011) have indicated that when the excitation is along a direction that is at an angle of 45° with the X-axis, about 50% liquid in the damper participates in the TLCD action, while about 50%–60% of the damper liquid participates in the TSD action. More detailed experimental studies with varying damper configuration and excitation are required for the development of a generalized relationship between the total mass of damper liquid and the mass of liquid that participates in the TLCD and the TSD action. However, experimental studies have confirmed that the frequencies of liquid oscillation and of sloshing remain practically unaffected by the direction of lateral excitation. Min et al. (2014b) have indicated that a practically feasible configuration of the TLCSD is possible for a real-world building structure. By means of a frequency domain study, significant damper performance under wind-induced vibration of the structure along the principal directions has been reported.

3.3.7 Tank-Pipe Damper

The tank-pipe damper (TPD) is another vibration suppression device that may be viewed as a combination of the TSD and the conventional TLCD. The TPD comprises two liquid tanks interconnected through one or more pipes near the base of the tanks. The pipes are fitted with orifice(s). A schematic of the TPD is presented in Figure 3.24.

Under lateral excitation along the X-direction, the liquid in the TPD exhibits predominantly oscillating behaviour, accompanied by additional liquid sloshing within the tanks. Thus, the damper can be tuned to suppress two distinct structural modes of vibration along the X-direction. Conversely, when the excitation is along the Y-direction, the liquid in the tanks will undergo only sloshing motion. By employing rectangular tanks in the TPD system, it is possible to achieve different sloshing

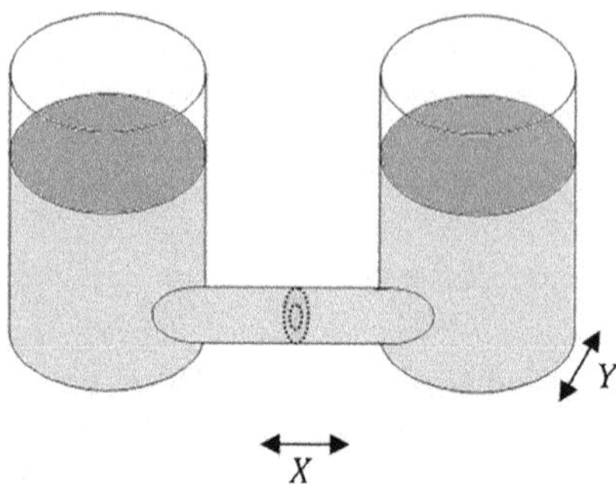

FIGURE 3.24   Tank-pipe damper.

frequencies along the X- and Y-directions. Thus, in total, the TPD may be designed to facilitate structural vibration control for a total of three vibration modes, two along the X-direction and one along the Y-direction.

When a TPD is employed to control a structural mode along the Y-direction, two tanks of the damper system would work as independent TSDs. In such a case, the equations of motion of the structure-damper system may be derived by considering the equivalent mechanical model of the damper and can be represented by Eq. (3.76). If the tanks of the TPD are of rectangular shape, the expressions for m_{d1}, c_{d1} and k_{d1}, as given in Eqs. (3.77)–(3.79), can be used. In case of circular tanks, the corresponding expressions are

$$m_{d1} = \frac{\pi \rho D_t^3}{17.6} \tanh\left(\frac{3.682h}{D_t}\right),$$ (3.81)

$$c_{d1} = 3.094\sqrt{\frac{m_{d1}k_{d1}V}{D_t^{3/2}\,g^{1/2}}}\left[1 + \frac{0.318}{\sinh(3.68h/D_t)}\left(1 + \frac{1 - 2h/D_t}{\cosh(3.68h/D_t)}\right)\right]$$ (3.82)

and

$$k_{d1} = m_{d1}\omega_{d1}^2.$$ (3.83)

In Eqs. (3.81)–(3.83), D_t is the tank diameter of the TPD. The fundamental sloshing frequency of liquid, ω_{d1} for circular tanks may be determined from

$$\omega_{d1} = \left[\frac{3.682g}{D_t}\tanh\left(\frac{3.682h}{D_t}\right)\right]^{1/2}.$$ (3.84)

When the TPD-structure system is excited along the X-direction (see Figure 3.25a), the main challenge in formulating the equations of motion of the structure-damper system lies in determining the respective proportions of liquid that participate in

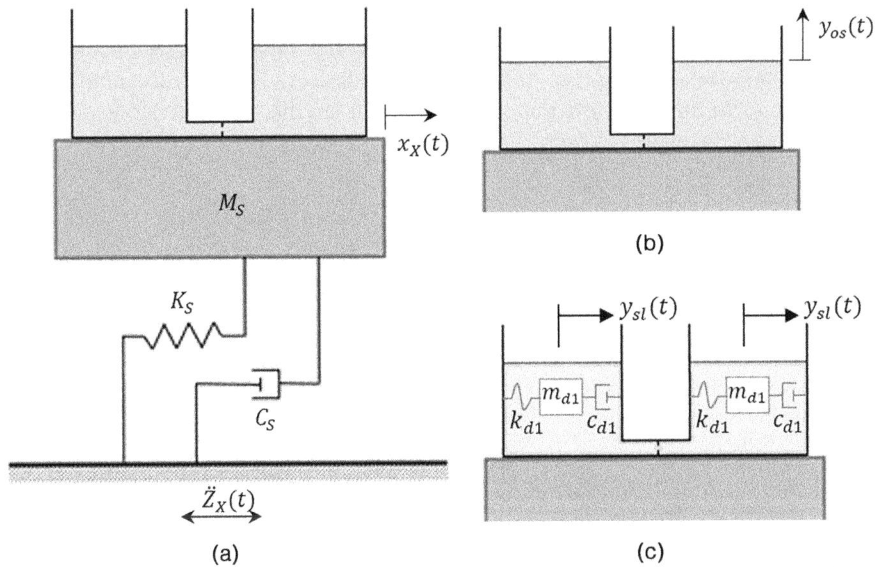

FIGURE 3.25 Tank-pipe damper attached to a single-degree-of-freedom structure excited along X-direction, (a) structural parameters; (b) oscillating mode of the damper; and (c) sloshing mode of the damper represented through equivalent mechanical model.

sloshing and oscillation. A simplistic method, which assumes the sloshing and oscillatory motion of liquid in the TPD to be uncoupled and linear, has been employed by Ghosh et al. (2012) (see Figure 3.25a–c). With this assumption, the potential and kinetic energy in the structure-damper system excited along the X-direction can be expressed as

$$V = \frac{1}{2}K_S x_X^2 + \rho A_t g y_{os}^2 + k_{d1} y_{sl}^2 \tag{3.85}$$

and

$$T = \frac{1}{2}M_S(\dot{x}_X + \dot{Z}_X)^2 + \frac{1}{2}\rho A_p b_p\left(\dot{x}_X + \dot{Z}_X + r_t \dot{y}_{os}\right)^2 + \frac{1}{2}\rho A_t\left(h - y_{os}\right)\dot{y}_{os}^2$$

$$+ \frac{1}{2}\rho A_t\left(h + y_{os}\right)\dot{y}_{os}^2 + \frac{1}{2}\left[\rho A_t\left(h - y_{os}\right) - m_{d1}\right]\left(\dot{x}_X + \dot{Z}_X\right)^2 \tag{3.86}$$

$$+ \frac{1}{2}\left[\rho A_t\left(h + y_{os}\right) - m_{d1}\right]\left(\dot{x}_X + \dot{Z}_X\right)^2 + m_{d1}\left(\dot{x}_X + \dot{Z}_X + \dot{y}_{sl}\right)^2.$$

In Eqs. (3.85)–(3.86), A_t, A_p, b_p and r_t denote the cross-sectional area of the tanks, the cross-sectional area of the pipe, the length of the pipe and the area ratio defined by A_t / A_p, respectively. The oscillatory displacement of liquid in the tanks along the vertical direction is denoted by y_{os}, while the lateral displacement of the lumped mass representing the first sloshing mode is y_{sl}. Further, the non-conservative forces in the structure, due to the orifice(s) ian the pipe and due to sloshing, are, respectively, expressed as

$$Q_x = -C_S \dot{x}_X, \tag{3.87}$$

$$Q_{y_{os}} = -\frac{1}{2} \rho r A_t \xi_{os} |\dot{y}_{os}| \dot{y}_{os} \tag{3.88}$$

and

$$Q_{y_{sl}} = -2c_{d1} \dot{y}_{sl}. \tag{3.89}$$

In Eq. (3.88), ξ_{os} is the coefficient of flow resistance or head-loss due to orifice(s) in the pipe of the TPD.

Using the Lagrange's equation, and following the methodology detailed in Section 2.2.2 of Chapter 2 for the conventional TLCD, the equations of motion of the structure-TPD system are derived as

$$\left(M_S + M_{tL}\right)\ddot{x}_X + C_S \dot{x}_X + K_S x_X = -\left(M_S + M_{tL}\right)\ddot{Z}_X - \rho A_p b_p r \ddot{y}_{os} - 2m_{d1} \ddot{y}_{sl}, \tag{3.90}$$

$$\left(b_p r_t + 2h\right)\ddot{y}_{os} + \frac{1}{2} r_t \xi_{os} |\dot{y}_{os}| \dot{y}_{os} + 2g y_{os} = -b_p \left(\ddot{x}_X + \ddot{Z}_X\right) \tag{3.91}$$

and

$$m_{d1} \ddot{y}_{sl} + k_{d1} y_{sl} + c_{d1} \dot{y}_{sl} = -m_{d1} \left(\ddot{x}_X + \ddot{Z}_X\right) \tag{3.92}$$

where M_{tL} is the total mass of liquid in the tanks. The equivalent linear damping coefficient corresponding to the orifice damping in Eq. (3.91) may be determined by following the procedure for the conventional TLCD given in Eqs. (2.21)–(2.26) of Chapter 2.

The degree of accuracy of tuned model of the structure-TPD system, as represented by Eqs. (3.90)–(3.92), is required to be ascertained through detailed experimental and numerical studies before a TPD can be considered for practical implementation. Some preliminary studies by Ghosh et al. (2012) have indicated that TPD could be a better option than the TLCD for the reduction of responses of structures that have a significantly higher mode participation. Further, TPD is found to have a relatively lower performance sensitivity and greater robustness to detuning.

3.4 TLCD CONFIGURATIONS TO IMPROVE TUNABILITY

3.4.1 SEALED TLCD

In this modified version of the conventional TLCD, the ends of the U-shaped container are securely sealed (see Figure 3.26), creating air chambers near the top of the vertical limbs of the damper. When the liquid column oscillates under lateral excitation of the damper, the air in the closed chambers above the liquid surface exerts a quasi-static compression/decompression pressure on the liquid surface in the vertical limbs of the TLCD. This induces a gas-spring effect on the oscillating liquid column. It may be noted that the overall length of the liquid column, L, the horizontal length of the liquid column, b, the cross-sectional area, A and the length ratio, α, of a sealed TLCD are the same as of a conventional TLCD.

To analyse the gas-spring effect, let P_0 and V_0, respectively, denote the pressure and volume of the gas in the chambers above the liquid surface in the vertical limbs of a sealed TLCD at rest condition. The stiffness of the air spring k, in each vertical limb may be expressed as (Mallik and Chatterjee, 2014)

$$k = A\frac{dP}{dy}. \tag{3.93}$$

In Eq. (3.93), P and y are the pressure in an air chamber at an instant of time and the vertical displacement of the surface of the liquid column, respectively. The assumption that the pressure–volume relationship of the gas in the air chamber follows a general polytropic relation leads to

$$P_0\, V_0^{\,n} = PV^n. \tag{3.94}$$

In Eq. (3.94), n denotes the polytropic index. The value of n is unity for an isothermal process, following Boyle's Law. For a quasi-adiabatic process, it ranges between 1 and γ, and for an isentropic process (i.e., adiabatic and reversible), it equals γ,

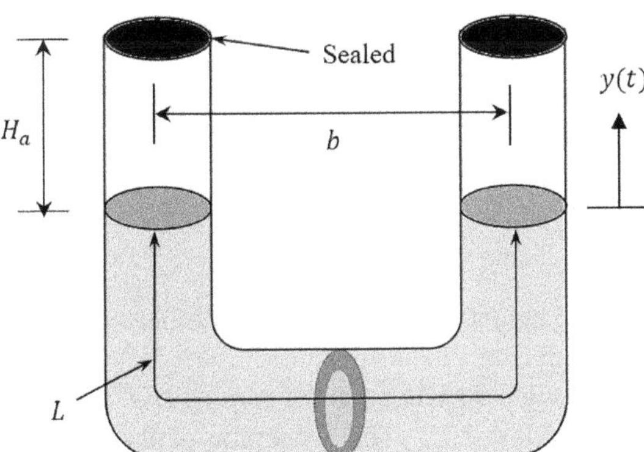

FIGURE 3.26 Sealed tuned liquid column damper.

where γ represents the specific heat ratio (approximately 1.4 for air). Various considerations of n have been made in previous studies on sealed TLCDs. Shum et al. (2008) assumed the process to be isothermal and n to be unity. Reiterer and Ziegler (2006) considered n as 1.2. However, while discussing the stiffness of the air spring in a pneumatic suspension, Mallik and Chatterjee (2014) have suggested that, unless the frequency is very low, the adiabatic gas law may be assumed to hold true for oscillatory movement. The experimental study by Bhattacharyya et al. (2017b) reported that in a laterally excited sealed TLCD, the liquid motion occurs rapidly, and hence, the process may be assumed as adiabatic with $n = 1.4$.

Substitution of Eq. (3.94) in Eq. (3.93) yields

$$k = -nAP_0\ V_0{}^n V^{-(n+1)}\frac{dV}{dy}. \tag{3.95}$$

Further,

$$V = V_0 - Ay \tag{3.96}$$

leading to

$$\frac{dV}{dy} = -A. \tag{3.97}$$

The substitution of Eq. (3.96) and (3.97) in Eq. (3.95) gives

$$k = nA^2\frac{P_0}{V_0}\left[1-\frac{Ay}{V_0}\right]^{-(n+1)}. \tag{3.98}$$

It is clear from Eq. (3.98) that the stiffness of the gas-spring is nonlinear. However, with an assumption that the vertical displacement of the liquid free surface y, is small in comparison to V_0, an expression for the linearized stiffness of the gas-spring may be obtained as

$$k = nA^2\frac{P_0}{V_0} \tag{3.99}$$

or, alternatively

$$k = nA\frac{P_0}{H_a}. \tag{3.100}$$

In Eq. (3.100), H_a is the height of the closed air chamber in each vertical limb of the damper at rest condition.

Let us consider a sealed TLCD attached to an SDOF structure having properties as described in Section 3.2.2. The schematic of the structure-damper system is shown in Figure 3.27. The base excitation to the structure is represented by the acceleration time history $\ddot{Z}_0(t)$.

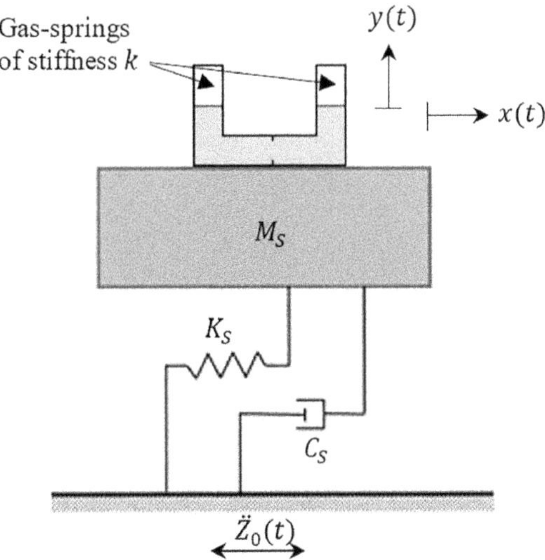

FIGURE 3.27 A single-degree-of-freedom structure with sealed tuned liquid column damper.

The potential energy, V and the kinetic energy, T of the structure-damper system are expressed as

$$V = \frac{1}{2}K_S x^2 + \rho A g y^2 + \frac{1}{2}k y^2 \tag{3.101}$$

and

$$T = \frac{1}{2}M_S(\dot{x}+\dot{Z}_0)^2 + \frac{1}{2}\rho A b\left(\dot{x}+\dot{Z}_0+\dot{y}\right)^2 + \frac{1}{2}\rho A(h-y)$$
$$\left[\dot{y}^2+\left(\dot{x}+\dot{Z}_0\right)^2\right] + \frac{1}{2}\rho A(h+y)\left[\dot{y}^2+\left(\dot{x}+\dot{Z}_0\right)^2\right]. \tag{3.102}$$

The dissipation of energy in a sealed TLCD takes place through orifice damping, as in case of the conventional TLCD. Thus, the non-conservative forces in the structure and the sealed TLCD are

$$Q_x = -C_S \dot{x} \tag{3.103}$$

and

$$Q_y = -\frac{1}{2}\rho A \xi |\dot{y}|\dot{y} \tag{3.104}$$

respectively.

In Eq. (3.104), ξ is the coefficient of head-loss due to orifice(s), similar to the conventional TLCD. The equivalent linear damping coefficient of a sealed TLCD may then be estimated by use of Eq. (2.26) in Chapter 2.

The governing equations of motion of the structure-sealed TLCD system are derived from the Lagrange's principle as

$$\ddot{x} + \frac{2\zeta_s \omega_s}{(1+\mu)}\dot{x} + \frac{\omega_s^2}{(1+\mu)}x = -\ddot{Z}_0 - \frac{\alpha\mu}{(1+\mu)}\ddot{y} \tag{3.105}$$

and

$$\ddot{y} + \frac{1}{2}\frac{\xi}{L}|\dot{y}|\dot{y} + \omega_{\text{sealed TLCD}}{}^2 y = -\alpha(\ddot{x} + \ddot{Z}_0). \tag{3.106}$$

In Eqs. (3.105) and (3.106), α is the length ratio, μ is the mass ratio given by the ratio of the mass of the damper liquid and the mass of the SDOF structural system, and $\omega_{\text{sealed TLCD}}$ is the linearized natural frequency of liquid oscillation in the sealed TLCD expressed as

$$\omega_{\text{sealed TLCD}} = \left[\frac{2g}{L}\left(1 + \frac{nP_o}{\rho g H_a}\right)\right]^{1/2}. \tag{3.107}$$

It may be observed from Eq. (3.107) that due to the gas-spring effect on the oscillating liquid column, a sealed TLCD has a higher natural frequency than a conventional TLCD. This makes the sealed TLCD particularly effective for short-period structures. It has been shown that a sealed TLCD can be tuned to a structure having a natural period as low as 0.2 s (Reiterer and Ziegler, 2006). On the other hand, a conventional TLCD is typically limited to addressing vibration concerns in structures with a fundamental period of 2.0 seconds or more (Corte et al., 2007).

The efficacy of the sealed TLCD in controlling vibrations induced by wind (Shum et al., 2008), earthquake (Bhattacharyya et al., 2017a) and pedestrian movement (Reiterer and Ziegler, 2006) is well documented in the literature. Experimental studies on the sealed TLCD have shown good agreement with numerical results (Achs, 2005; Bhattacharyya et al., 2017a). Here, it is pertinent to mention that this specific type of TLCD is occasionally denoted as a pressurized tuned liquid column damper (Shum et al., 2008) and also referred to as a tuned liquid column–gas damper (Mousavi et al., 2013).

Some configurational variations of the sealed TLCD have also been developed, and the air tuned damper (ATD) is a significant one among them. An ATD has a gas-spring arrangement at one end of the U-shaped damper container, and the other end of the damper container is open to the atmosphere. A schematic of the ATD is shown in Figure 3.28. The linearized natural frequency of liquid oscillation in the ATD is given by (Ghisbain et al., 2021)

$$\omega_{\text{ATD}} = \left[\frac{2g}{L}\left(1 + \frac{nP_{\text{atm}}A_A}{2\rho g(H_A A_A + H_B A_B)}\right)\right]^{1/2}. \tag{3.108}$$

In Eq. (3.108), P_{atm} denotes the atmospheric pressure. The height of the U-shaped damper container above the free liquid surface at rest condition and the length of the gas-spring are denoted by H_A and H_B, respectively. The cross-sectional area of the damper container and the gas-spring are represented by A_A and A_B, respectively.

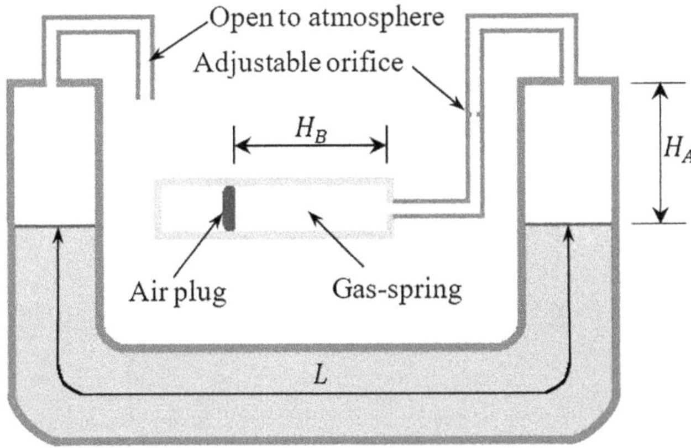

FIGURE 3.28 Air-tuned damper.

It may be observed from Eq. (3.108) that the frequency of the ATD can easily be adjusted by adjusting the position of the air plug, which in turn changes the length of the gas-spring. This is a major benefit of the ATD. An ATD has been installed at B2 tower, 461 Dean Street, Brooklyn, New York City, USA, for wind-induced vibration control of the building (see Section 8.10 of Chapter 8). Further details on the ATD are available at *https://hummingbirdkinetics.com/experience/* and Ghisbain et al. (2021).

Hokmabady et al. (2019) proposed a tuned liquid column ball gas damper (TLCBGD), which is essentially a combination of a sealed TLCD and a TLCBD. In the TLCBGD, both ends of the U-shaped damper container are sealed to create the gas-spring effect. The TLCBGD has been found to be effective in controlling wave-induced vibration of a fixed jacket platform (Hokmabady et al., 2019).

3.4.2 COMPLIANT LIQUID COLUMN DAMPER

In this configuration of the TLCD, the U-shaped damper container is attached to the primary structure through a flexible system comprising of a spring and a dashpot element, as illustrated in Figure 3.29. Essentially, a compliant liquid column damper (CLCD) combines the principles of the tuned mass damper (TMD) with those of the TLCD. This is the reason why some researchers, namely Xu et al. (1992a,b), have referred to the CLCD as the tuned liquid column mass damper (TLCMD). The CLCD has two DOFs, one that corresponds to the movement of the damper container and the other relates to the motion of the damper liquid. The frequency of movement of the whole damper container, and not the oscillating frequency of the damper liquid, is tuned to the frequency of the structural mode to be controlled.

Similar to the previous configurations of the TLCD discussed in this chapter, the equations of motion of the CLCD with the primary structure are derived using the energy principle. The potential energy, V and the kinetic energy T, of the system at any instant of time t, may be expressed as

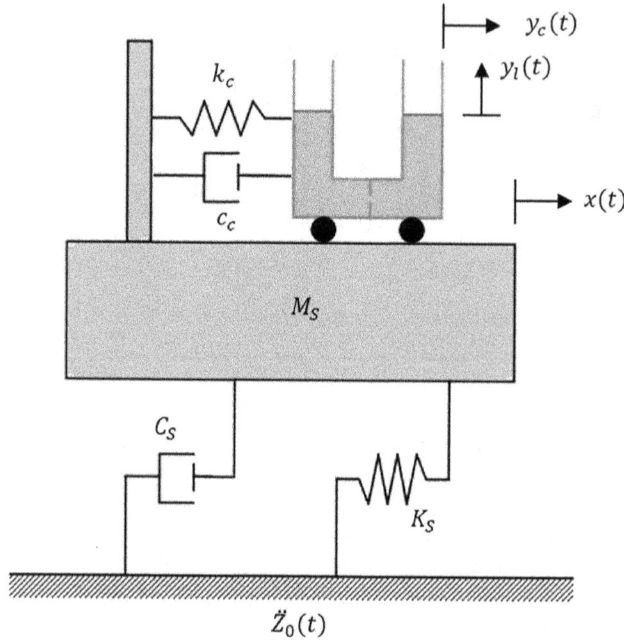

FIGURE 3.29　A compliant liquid column damper attached to a single-degree-of-freedom primary structure.

$$V = \frac{1}{2} K_S x^2 + \frac{1}{2} k_c y_c^2 + \rho A g y_l^2 \tag{3.109}$$

and

$$T = \frac{1}{2} M_S (\dot{x} + \dot{Z}_0)^2 + \frac{1}{2} m_c \left(\dot{x} + \dot{Z}_0 + \dot{y}_c\right)^2 + \frac{1}{2} \rho A b \left(\dot{x} + \dot{Z}_0 + \dot{y}_c + \dot{y}_l\right)^2$$

$$+ \frac{1}{2} \rho A \left(\frac{L-b}{2} - y\right)\left[\dot{y}_l^2 + \left(\dot{x} + \dot{Z}_0 + \dot{y}_c\right)^2\right] \tag{3.110}$$

$$+ \frac{1}{2} \rho A \left(\frac{L-b}{2} + y\right)\left[\dot{y}_l^2 + \left(\dot{x} + \dot{Z}_0 + \dot{y}_c\right)^2\right].$$

In Eqs. (3.109) and (3.110), m_c and k_c denote the mass of the damper container and the stiffness of the element connecting the U-shaped container with the structure, respectively. The lateral displacement of the damper container and the vertical displacement of the liquid column in the damper are, respectively, denoted by y_c and y_l. The SDOF structural system has the properties as described in Section 3.2.2.

In the structure-damper system, the non-conservative generalized forces corresponding to the degrees-of-freedom x, y_l and y_c are

$$Q_x = -C_S \dot{x}, \tag{3.111}$$

$$Q_{y_l} = -\frac{1}{2}\rho A \xi |\dot{y}_l| \dot{y}_l \tag{3.112}$$

and

$$Q_{y_c} = -c_c \dot{y}_c \tag{3.113}$$

respectively.

In Eq. (3.113), c_c is the damping coefficient of the element connecting the U-shaped container with the structure.

On using the Lagrange's equation, the equations of motion of the structure-damper system may be expressed as

$$\ddot{x} + 2\zeta_s\omega_s\dot{x} + \omega_s^2 x = -\ddot{Z}_0 + \mu_c\omega_{\text{CLCD}}^2 y_c + 2\mu_c\zeta_c\omega_{\text{CLCD}}\dot{y}_c, \tag{3.114}$$

$$\ddot{y}_c + 2\zeta_c\omega_{\text{CLCD}}\dot{y}_c + \omega_{\text{CLCD}}^2 y_c = -\frac{\alpha}{(1+\tau)}\ddot{y}_l - \ddot{x} - \ddot{Z}_0 \tag{3.115}$$

and

$$\ddot{y}_l + \frac{1}{2}\frac{\xi}{L}|\dot{y}_l|\dot{y}_l + \omega_l^2 y_l = -\alpha\left(\ddot{y}_c + \ddot{x} + \ddot{Z}_0\right). \tag{3.116}$$

In Eqs. (3.114)–(3.116), τ is the ratio of the mass of the damper container to the mass of the contained liquid and μ_c is the mass ratio of the CLCD defined by the ratio of the mass of the whole damper, which is the mass of the damper container as well as the damper liquid, to that of the structure. Further, ω_{CLCD}, ω_l and ζ_c denote the frequency of movement of the whole damper container, the frequency of liquid oscillation in the damper and the damping ratio of the element connecting the damper container to the structure, respectively, and are expressed as

$$\omega_{\text{CLCD}} = \left[\frac{k_c}{m_c + \rho AL}\right]^{1/2}, \tag{3.117}$$

$$\omega_l = \sqrt{\frac{2g}{L}} \tag{3.118}$$

and

$$\zeta_c = c_c / \left[2\omega_{\text{CLCD}}\left(\rho AL + m_c\right)\right] \tag{3.119}$$

It is clear from Eq. (3.117) that unlike a conventional TLCD, a CLCD can be tuned to the structural frequency through a proper choice of the stiffness of the element connecting the U-shaped container to the structure. This provides flexibility in the design of the damper. Further, by adjusting the value of k_c, CLCDs can be tuned to a wide range of structural frequencies, especially to those of short-period structures.

As can be observed from Eqs. (3.115) and (3.116), the energy dissipation in a CLCD occurs due to orifice damping of the oscillating liquid column and due to the damping element connecting the U-shaped container with the structure. In the design of the CLCD, the orifice damping is generally optimized numerically, while the optimum damping ratio of the damping element of the CLCD is obtained from the established expressions available in the literature for the optimum damping ratio of the TMD (Xu et al., 1992a). However, it has been shown through experimental studies that even without the damping element between the damper container and the structure, the CLCD exhibits substantial energy dissipation, resulting from orifice damping due to liquid oscillation in the damper container (Bhattacharyya et al., 2017a).

The effectiveness of the CLCD in suppressing structural vibration has been highlighted in several research works. Xu et al. (1992b) have shown that the CLCD can significantly reduce the along-wind vibration of MDOF structural systems and that its performance is even superior to that of conventional TLCD having the same mass ratio. In a subsequent work, Xu et al. (1992a) have considered the CLCD for across-wind vibration control of MDOF structural systems. Here, too, the CLCD has a superior performance to that of the conventional TLCD. The potential of the CLCD in the vibration control of short-period structures was first explored by Ghosh and Basu (2004). It was demonstrated that an optimally designed CLCD can significantly reduce the earthquake-induced vibration of a structure having a natural period equal to as low as 0.3 s. In contrast, it may be noted that conventional TLCDs are applicable to structures with a natural period of 2.0 s or more (Corte et al., 2007). Under pulse-type-near-fault earthquakes, which impose severe demands on structures, the CLCD is found to be effective in controlling the vibrations of both short- and long-period structures (Roy et al., 2023; Roy and Ghosh, 2015). Similar to the multiple conventional TLCD system, the multiple-CLCD system is also capable of improving the robustness of the damper performance when there is uncertainty in the values of the structural parameters (Ghosh et al., 2011).

Several configurational variations of the CLCD have been put forward. Gur et al. (2014) have replaced the linear spring element of the CLCD with a shape memory alloy spring and thereby improved the efficacy of the damper. Wei and Zhao (2020) have expanded the concept of the bi-directional TLCD to the bi-directional CLCD. A bi-directional CLCD is achieved by connecting a bi-directional TLCD to the primary structure through spring and damping elements in two perpendicular directions.

In the CLCD, through the flexible connection with the structure, the damper container moves in the direction along which the container is aligned. Departing from this original CLCD design, an alternative form was introduced for bi-directional vibration control. In this variant, the damper container is allowed to move perpendicular to the direction along which it is aligned. This modification was proposed by Heo et al. (2009) and further explored by Min et al. (2014a). With this setup, the damper functions as a conventional TLCD in one direction and as a TMD when subjected to excitation along the orthogonal direction.

3.4.3 PENDULUM-TYPE LIQUID COLUMN DAMPER

In another variation of the TLCD, the liquid-filled U-shaped damper container is connected to the primary structure similar to the case of a lumped mass of a pendulum-type TMD. Such a damping device is called pendulum-type liquid column damper (PLCD). The schematic of a PLCD attached to an SDOF structural system is illustrated in Figure 3.30. Here, p' and q' represent the local axes of the PLCD. In this configuration, the structure-damper system has three DOFs, namely the lateral displacement of the structure x, the oscillation of the liquid column in the U-shaped container y and the rotation of the whole damper system θ. The potential energy V, of the system at any instant of time t, is given by

$$V = \frac{1}{2}K_S x^2 + m_c g l_c (1-\cos\theta) + \rho A g b z_p (1-\cos\theta) + \rho A g (h-y) z_{v1}\left[1-\cos(\delta_1 - \theta)\right]$$

$$+\rho A g (h+y) z_{v2}\left[1-\cos(\delta_2 + \theta)\right]. \tag{3.120}$$

In Eq. (3.120), m_c and l_c denote the mass of the damper container and the distance of the centre of gravity (CG) of the empty damper container from the hinge, respectively. The notations z_p, z_{v1} and z_{v2} denote the distance of the hinge from the CG of the liquid in the horizontal limb, left vertical limb and right vertical limb of the damper, respectively. The height of the liquid in the vertical limb of PLCD at-rest condition is denoted by h. The angles made by the hanger with lines drawn from the hinge to the CG of the left vertical limb and the right vertical limb of the damper are δ_1 and δ_2, respectively. In Eq. (3.120), it is assumed that $z_{v1} \gg y$ and $z_{v2} \gg y$.

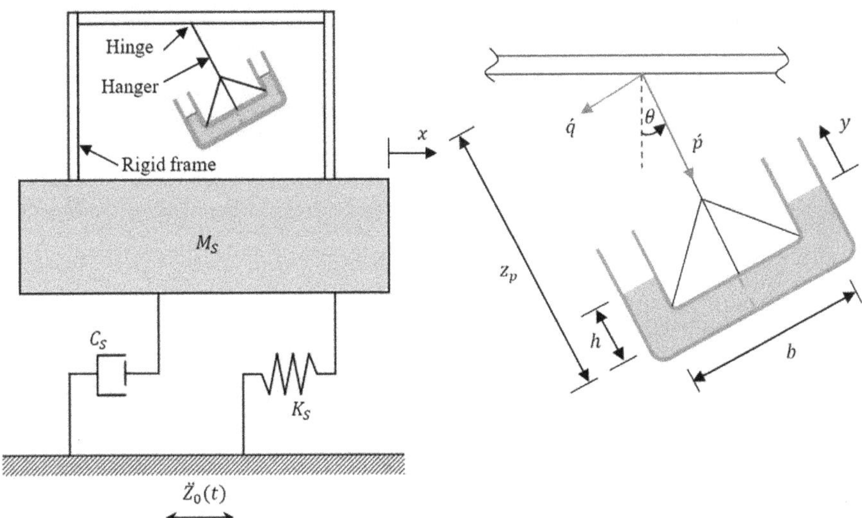

FIGURE 3.30 Pendulum-type tuned liquid column damper attached to a single-degree-of-freedom structure; right: enlarged view of pendulum-type tuned liquid column damper (PTLCD).

The kinetic energy T, of the structure damper system may be written as

$$T = \frac{1}{2} M_S \left(\dot{x} + \dot{Z}_0 \right)^2 + \frac{1}{2} m_c \left[\left(\dot{x} + \dot{Z}_0 + l_c \dot{\theta} \cos\theta \right)^2 + \left(l_c \dot{\theta} \sin\theta \right)^2 \right] + \frac{1}{2} J_c \dot{\theta}^2$$

$$+ \frac{1}{2} \rho A b \left[\dot{y} + z_p \dot{\theta} + \left(\dot{x} + \dot{Z}_0 \right) \cos\theta \right]^2 + \frac{1}{2} \rho A \int_{-b/2}^{b/2} \left[q' \dot{\theta} + \left(\dot{x} + \dot{Z}_0 \right) \sin\theta \right]^2 dq'$$

$$+ \frac{1}{2} \rho A (h - y) \left[\dot{y} + \frac{b}{2} \dot{\theta} + \left(\dot{x} + \dot{Z}_0 \right) \sin\theta \right]^2 + \frac{1}{2} \rho A \int_{z_c - h + y}^{z_c} \left[p' \dot{\theta} + \left(\dot{x} + \dot{Z}_0 \right) \cos\theta \right]^2 dp'$$

$$+ \frac{1}{2} \rho A (h + y) \left[\dot{y} + \frac{b}{2} \dot{\theta} - \left(\dot{x} + \dot{Z}_0 \right) \sin\theta \right]^2 + \frac{1}{2} \rho A \int_{z_c - h - y}^{z_c} \left[p' \dot{\theta} + \left(\dot{x} + \dot{Z}_0 \right) \cos\theta \right]^2 dp'.$$

$$(3.121)$$

where J_c denotes the moment of inertia of an empty damper container about its own CG.

The non-conservative generalized forces in the structure-damper system corresponding to the DOFs x, y and θ, respectively, are

$$Q_x = -C_S \dot{x} \tag{3.122}$$

$$Q_y = -\frac{1}{2} \rho A \xi |\dot{y}| \dot{y} \tag{3.123}$$

and

$$Q_\theta = -c_h \dot{\theta}. \tag{3.124}$$

In Eq. (3.124), c_h represents the coefficient of damping corresponding to the bearing friction at the hinge of the PLCD.

Using Eqs. (3.120)–(3.124) and with the help of Lagrange's equation, the equations of motion of the structure-damper system can be expressed in a generalized form as

$$\mathbf{M}\ddot{u} + \mathbf{C}\dot{u} + \mathbf{K}u = \mathbf{F}(t) \tag{3.125}$$

where $\mathbf{M}, \mathbf{C}, \mathbf{K}, u$ and $\mathbf{F}(t)$ denote the generalized mass, damping, stiffness, displacement and excitation matrices, respectively.

As the PLCD experiences simultaneous translational and rotational motions, and oscillation of the liquid column in the damper container as well, the system of equations represented by Eq. (3.125) involves several nonlinear terms. Thus, it would require a significant amount of numerical effort to evaluate the response of a structure with PLCD. So far, there is a single detailed study available on the PLCD (Sarkar and Gudmestad, 2013). Through numerical and experimental studies, the effectiveness of the PLCD in controlling the vibration of an SDOF structure under harmonic excitation has been established. An approximate expression for the natural frequency of the PLCD, ω_{PLCD}, may be written as

$$\omega_{\mathrm{PLCD}} = \left[\frac{g}{z_p + h} \right]^{1/2}.$$

(3.126)

3.4.4 Tuned Liquid Column Damper-Inerter

A tuned liquid column damper-inerter (TLCDI) may be viewed as an improvisation of the CLCD that taps into the benefit of an inerter to improve the performance of the damper. An inerter may be characterized as a one-port device with negligible mass and two independently movable terminals that can generate an internal force proportional to the relative acceleration between its terminals, resulting in the property of mass amplification. Further details on the inerter are available in Smith (2002, 2020).

The schematic of a TLCDI system attached to an SDOF structure is depicted in Figure 3.31. A comparison of Figure 3.31 with Figure 3.29 reveals that the TLCDI is identical to the CLCD with the only exception being the presence of the inerter. One terminal of the inerter is attached to the damper container, and the other terminal of the inerter is grounded, that is, attached to the ground. Here, it may be noted that an inerter has no stiffness or damping. Thus, it is neither associated with any non-conservative force nor does it contribute to the potential energy of the structure-damper system. However, due to its ability to augment inertia force despite having negligible mass, the inerter contributes to the kinetic energy of the structure-damper system.

The kinetic energy T, of the structure-TLCDI system at an instant of time t, may be written as

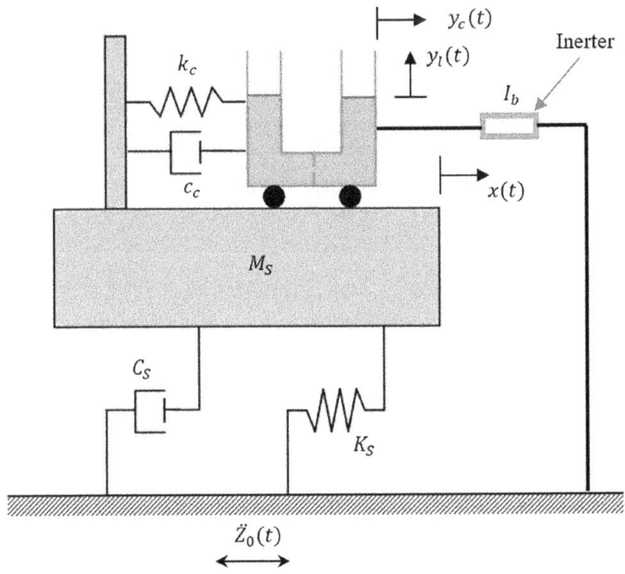

FIGURE 3.31   A tuned liquid column damper inerter attached to a single-degree-of-freedom primary structure.

$$T = \frac{1}{2}M_S(\dot{x}+\dot{Z}_0)^2 + \frac{1}{2}m_c\left(\dot{x}+\dot{Z}_0+\dot{y}_c\right)^2 + \frac{1}{2}I_b\left(\dot{x}+\dot{y}_c\right)^2 + \frac{1}{2}\rho Ab\left(\dot{x}+\dot{Z}_0+\dot{y}_c+\dot{y}_l\right)^2$$

$$+\frac{1}{2}\rho A\left(\frac{L-b}{2}-y\right)\left[\dot{y}_l^2+\left(\dot{x}+\dot{Z}_0+\dot{y}_c\right)^2\right]$$

$$+\frac{1}{2}\rho A\left(\frac{L-b}{2}+y\right)\left[\dot{y}_l^2+\left(\dot{x}+\dot{Z}_0+\dot{y}_c\right)^2\right]. \tag{3.127}$$

The expressions for potential energy and the non-conservative generalized force corresponding to the degrees of freedom of the structure-TLCDI system are the same as that for the structure-CLCD system presented in Section 3.4.2 (see Eqs. (3.109) and (3.111) to (3.113)).

By using the generalized Lagrange's equation, the equations of motion of the structure-TLCDI system may be derived as

$$\ddot{x}+2\zeta_s\omega_s\dot{x}+\omega_s^2 x = -\ddot{Z}_0 + \left(\mu_c+\mu_I\right)\omega_{\text{TLCDI}}^2 y_c + 2\left(\mu_c+\mu_I\right)\zeta_{\text{TLCDI}}\omega_{\text{TLCDI}}\dot{y}_c \tag{3.128}$$

$$\ddot{y}_c + 2\zeta_{\text{TLCDI}}\omega_{\text{TLCDI}}\dot{y}_c + \omega_{\text{TLCDI}}^2 y_c = -\frac{\mu_c}{\left(\mu_c+\mu_I\right)}\frac{\alpha}{\left(1+\tau\right)}\ddot{y}_l - \ddot{x} - \frac{\mu_c}{\left(\mu_c+\mu_I\right)}\ddot{Z}_0 \tag{3.129}$$

and

$$\ddot{y}_l + \frac{1}{2}\frac{\xi}{L}|\dot{y}_l|\dot{y}_l + \omega_l^2 y_l = -\alpha\left(\ddot{y}_c+\ddot{x}+\ddot{Z}_0\right). \tag{3.130}$$

In Eqs. (3.128)–(3.130), μ_I is the inertance ratio is, given by I_b / M_S. The frequency of movement of the whole damper container and the damping ratio of the element connecting the damper container to the structure are denoted by ω_{TLCDI} and ζ_{TLCDI}, respectively, and are expressed as

$$\omega_{\text{TLCDI}} = \left[\frac{k_c}{m_c+I_b+\rho AL}\right]^{1/2}, \tag{3.131}$$

$$\zeta_{\text{TLCDI}} = c_c / \left[2\omega_{\text{TLCDI}}\left(\rho AL+m_c+I_b\right)\right] \tag{3.132}$$

All the other parameters have the same meaning as defined in Eqs. (3.114)–(3.116). The nonlinear damping term in Eq. (3.130) may be linearized by following the procedure described in Section 2.2.2 of Chapter 2 for the conventional TLCD. Wang et al. (2020b) have reported that this linearization has a minor influence on the derived response quantities.

A detailed parametric and performance study on an SDOF structure-TLCDI system presented by Wang et al. (2020b) has highlighted the improved and robust displacement and acceleration response reduction capacity of the damper due to the inclusion of the inerter in the system. The efficacy of the TLCDI in controlling the response of MDOF structure has also been explored. As expected, the performance of the damper improves as the damper spans more floors (Pandey and Mishra, 2021). The possibility of vibration control of two linked tall buildings using a TLCDI has been demonstrated under wind (Wang et al., 2021b) and earthquake (Wang et al., 2020a) loading. The TLCDI has also been used to improve the effectiveness of base-isolation systems by locating the damper on the isolated floor (Masnata et al., 2023a,b). A variation of the TLCDI, in which a pendulum-type TLCD is used in conjunction with an inerter, has also been explored (Wang et al., 2021a). In summary, the TLCDI is an attractive choice as a structural vibration control device. Owing to the mass amplification property of the inerter, it can be designed as a highly efficient dynamic vibration absorber with a notably reduced mass. This is a significant advantage, as designers have to take care of space constraints and other limitations due to the addition of dead load on the building structure while designing dampers.

3.5 TLCD CONFIGURATIONS FOR VERTICAL VIBRATION CONTROL

3.5.1 Adaptive Tuned Liquid Column Damper

The adaptive tuned liquid column damper (Ad-TLCD), proposed by Wendner et al. (2007), is a variation of the sealed TLCD that is capable of controlling vertical vibration of structures. The unique feature of the Ad-TLCD is that it allows the reduction of vertical vibration by utilizing the unequal height of liquid in its vertical limbs. To ensure this, one vertical limb of the damper container is sealed at the top and a pressure higher than the atmospheric pressure is maintained over the liquid surface in that limb, while the other vertical limb of the damper container is open to the atmosphere (see Figure 3.32). The natural frequency of the Ad-TLCD is given by (Wendner et al., 2007)

$$\omega_{\text{Ad-TLCD}} = \left[\frac{2g}{L} \left\{ 1 + \frac{nP_{\text{atm}}}{2\rho g H_a} \left(1 + \frac{2\rho g H_0}{P_{\text{atm}}} \right) \right\} \right]^{1/2}. \tag{3.133}$$

In Eq. (3.133), H_a denotes the height of the air chamber at the sealed end of the U-shaped container, while the difference in elevation between the liquid surface in the vertical limbs of the damper is denoted by $2H_0$.

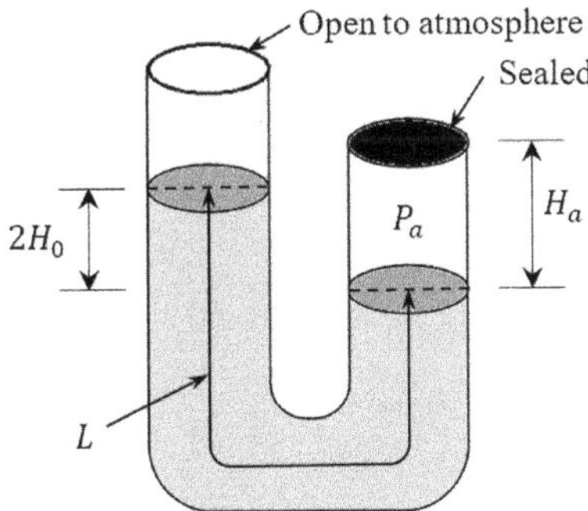

FIGURE 3.32 Adaptive tuned liquid column damper.

The equations of motion of a coupled structure-Ad-TLCD system may be obtained following the same procedure as described in Section 3.4.1 for a structure-sealed TLCD system.

An Ad-TLCD system having eight damper units has been successfully designed and installed in a steel railway bridge of the Austrian Federal Railways (Reiterer et al., 2008). The dampers are used to suppress vibration in the vertical direction induced by the passage of trains. The dampers have a total mass ratio of 2.1%. Field measurements have shown that the damper system has increased the effective damping ratio from 0.5% to 1.2% (Reiterer et al., 2008). More details regarding this installation of the Ad-TLCD are available in Section 6.4 of Chapter 6.

3.5.2 Vertical Sealed TLCD

The vertical sealed TLCD (VSTLCD) is also a variation of the conventional sealed TLCD. The concept of the VSTLCD was introduced by Liu et al. (2023). A VSTLCD has both ends of the U-shaped damper container sealed (see Figure 3.33). The air chambers in the vertical limbs of the VSTLCD have unequal pressure. As a result, the vertical limbs of the damper have unequal liquid height, which is required for effective vertical vibration control. The natural frequency of VSTLCD is given by (Liu et al., 2023)

$$\omega_{\text{VSTLCD}} = \left[\frac{2g}{L} \left\{ 1 + \frac{nP_1}{2\rho g H_1} + \frac{nP_2}{2\rho g H_2} \right\} \right]^{1/2}. \tag{3.134}$$

In Eq. (3.134), P_1 and P_2 denote the pressure of the left and right limbs of the VSTLCD, respectively. H_1 and H_2 are the height of the air chamber in the left and right limbs of the VSTLCD, respectively.

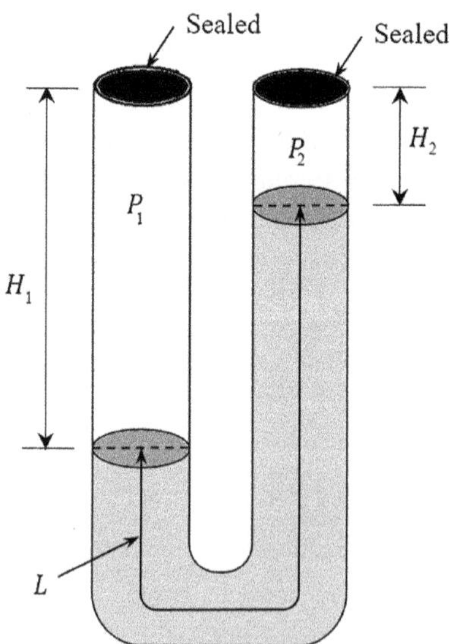

FIGURE 3.33   Vertical sealed tuned liquid column damper.

The process outlined in Section 3.4.1 for a structure-sealed TLCD system may be applied to obtain the equations of motion of a connected structure-VSTLCD system. The efficacy of the VSTLCD in controlling vertical vibrations in a building floor system caused by human walking has been demonstrated by Liu et al. (2023). The study also revealed that the performance of the damper improved with the increase in the difference in the liquid heights in the vertical limbs of the damper. Further details in this regard are available in Section 6.3 of Chapter 6.

3.5.3 Vertical TLCD (VTLCD)

Vertical TLCD (VTLCD) is a device specifically designed for the control of structural vibrations along the vertical direction. It has been developed by Ding et al. (2024). A VTLCD system may be described as a set of two compliant liquid column dampers (CLCD) fixed on two inclined planes on the structure (see Figure 3.34). The inclined planes have equal but opposite slope. With this arrangement, a component of the control force generated by the damper system would act along the vertical direction, leading to vertical vibration control.

The expression for the natural frequency of the VTLCD is the same as that of the CLCD (see Eq. 3.117). The equations of motion of the structure-VTLCD system may be obtained based on the Lagrange's equations by extending the procedure described in Section 3.4.2. The efficacy of the VTLCD under seismic excitation has been demonstrated by the work of Ding et al. (2024).

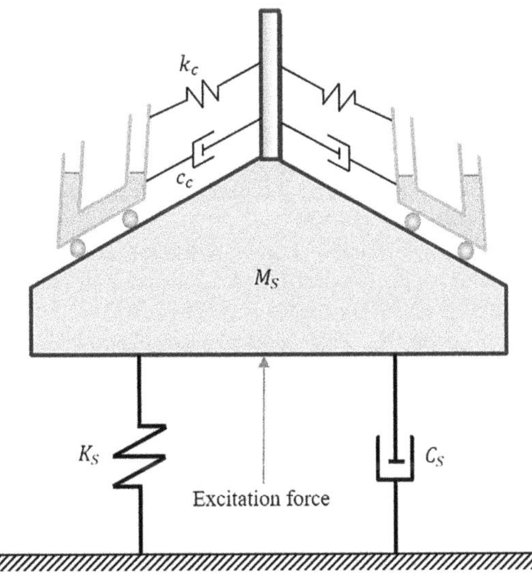

FIGURE 3.34 Vertical tuned liquid column damper.

REFERENCES

Abramson, H.N., Chu, W.-H., Ransleben Jr., G.E., 1961. Representation of fuel sloshing in cylindrical tanks by an equivalent mechanical model. *Am. Rocket Soc. J.* 31, 1697–1705. https://doi.org/10.2514/8.5896

Achs, G., 2005. *Anwendung von Flüssigkeitstilgern bei Schrägseilbrücken im Freivorbauzustand unter Windanregung.* Vienna University of Technology, Austria.

Al-saif, K.A., Aldakkan, K.A., Foda, M.A., 2011. Modified liquid column damper for vibration control of structures. *Int. J. Mech. Sci.* 53, 505–512. https://doi.org/10.1016/j.ijmecsci.2011.04.007

Basu, B., Zhang, Z., Nielsen, S.R.K., 2016. Damping of edgewise vibration in wind turbine blades by means of circular liquid dampers. *Wind Energy* 19, 213–226. https://doi.org/10.1002/we.1827

Bhattacharyya, S., Ghosh, A.D., Basu, B., 2017a. Nonlinear modeling and validation of air spring effects in a sealed tuned liquid column damper for structural control. *J. Sound Vib.* 410, 269–286. https://doi.org/10.1016/j.jsv.2017.07.046

Bhattacharyya, S., Ghosh, A.D., Basu, B., 2017b. Experimental investigations into CLCD with identification of tuning and damping effects. *J. Struct. Eng.* 143, 1–5. https://doi.org/10.1061/(ASCE)ST.1943-541X.0001788.

Cao, L., Gong, Y., Ubertini, F., Wu, H., Chen, A., Laflamme, S., 2020. Development and validation of a nonlinear dynamic model for tuned liquid multiple columns dampers. *J. Sound Vib.* 487, 115624. https://doi.org/10.1016/j.jsv.2020.115624

Chang, C.C., Hus, C.T., 1998. Control performance of liquid column vibration absorbers. *Eng. Struct.* 20, 580–586. https://doi.org/10.1016/S0141-0296(97)00062-X

Corte, W. De, Delesie, C., Bogaert, P. Van, 2007. Sealed tuned liquid column dampers: a cost effective solution for vibration damping of large arch hangers, in: *5th International Conference on Arch Bridges*, Madeira, Portugal. pp. 409–416.

Das, A., 2022. Characterization of liquid sloshing in u-shaped container with submerged cylinder to be used as dampers for structural vibration control. *Pr. Period. Struct. Des. Constr.* 27, 1–17. https://doi.org/10.1061/(ASCE)SC.1943-5576.0000621

Das, A., Konar, T., Maity, D., 2023a. Numerical investigation on sloshing phenomena in u-shaped containers subjected to near-fault strong ground motions. *J. Earthq.* https://doi.org/10.1142/s1793431123500331

Das, A., Konar, T., Maity, D., Bhattacharyya, S.K., 2023b. Revisiting equivalent tuned mass damper analogy for tuned liquid damper utilizing CFD-FEA framework. *Soil Dyn. Earthq. Eng.* 172, 108051. https://doi.org/10.1016/j.soildyn.2023.108051

Das, A., Maity, D., Bhattacharyya, S.K., 2020. Characterization of liquid sloshing in U-shaped containers as dampers in high-rise buildings. *Ocean Eng.* 210, 1–22. https://doi.org/10.1016/j.oceaneng.2020.107462

Ding, H., Altay, O., Wang, J.T., 2023a. Lateral vibration control of monopile supported offshore wind turbines with toroidal tuned liquid column dampers. *Eng. Struct.* 286, 116107. https://doi.org/10.1016/j.engstruct.2023.116107

Ding, H., Bi, K., Song, J., Fang, X., 2024. Vertical vibration control of structures with tuned liquid column dampers. *Int. J. Mech. Sci.* 279, 109502. https://doi.org/10.1016/j.ijmecsci.2024.109502

Ding, H., Das, A., Zhou, T.-Y., Wang, J.-T., Altay, O., 2022. Characterization of liquid oscillation and sloshing properties of toroidal TLCDs for multidirectional vibration control. *Int. J. Struct. Stab. Dyn.* 2250168, 1–25. https://doi.org/10.1142/s0219455422501681

Ding, H., Wang, J.T., Altay, O., Lu, L.Q., 2021a. Multilayer toroidal tuned liquid column dampers for seismic vibration control of structures. *Structures* 33, 406–422. https://doi.org/10.1016/j.istruc.2021.04.041

Ding, H., Wang, J.T., Lu, L.Q., Pan, J.W., 2021b. Experimental comparison of nonlinear damping performance of toroidal and conventional tuned liquid column dampers. *Nonlinear Dyn.* 104, 3365–3384. https://doi.org/10.1007/s11071-021-06552-7

Ding, H., Wang, J.T., Lu, L.Q., Zhu, F., 2020. A toroidal tuned liquid column damper for multidirectional ground motion-induced vibration control. *Struct. Control Heal. Monit.* 27, e2558. https://doi.org/10.1002/stc.2558

Ding, H., Wang, W., Liu, J.F., Wang, J.T., Le, Z.J., Zhang, J., Yu, G.M., 2023b. On the size effects of toroidal tuned liquid column dampers for mitigating wind- and wave-induced vibrations of monopile wind turbines. *Ocean Eng.* 273, 113988. https://doi.org/10.1016/j.oceaneng.2023.113988

Fu, C., 2011. Application of torsional tuned liquid column gas damper for plan-asymmetric buildings. *Struct. Control Heal. Monit.* 18, 492–509. https://doi.org/10.1002/stc.380

Ghisbain, P., Mendes, S., Pinto, M., Malsch, E., 2021. Innovative liquid damper for wind-induced vibration of buildings: performance after 4 years of operation, and next iteration. *Int. J. High-Rise Build.* 10, 117–121. https://doi.org/10.21022/IJHRB.2021.10.2.117

Ghosh, A., Gangopadhyay, A., Basu, B., 2011. Performance investigation of multiple compliant liquid column dampers for control of seismic vibrations, in: *Proceedings of the 8th International Conference on Structural Dynamics*, Lueven, Belgium, EURODYN 2011. pp. 1671–1677.

Ghosh, A.D., Basu, B., 2004. Seismic vibration control of short period structures using the liquid column damper. *Eng. Struct.* 26, 1905–1913. https://doi.org/10.1016/j.engstruct.2004.07.001

Ghosh, A.D., Saha, P.C., Basu, B., 2012. Study of a tank-pipe damper system for seismic vibration control of structures, in: *15 WCEE*, Lisbon, Portugal. pp. 1–10.

Graham, E.W., Rodriguez, A.M., 1952. The characteristics of fuel motion which affect airplane dynamics. *J. Appl. Mech.* 19, 381–388. https://doi.org/10.1017/CBO9781107415324.004

Gur, S., Mishra, S.K., Bhowmick, S., Chakraborty, S., 2014. Compliant liquid column damper modified by shape memory alloy device for seismic vibration control. *Smart Mater. Struct.* 23, 1–14. https://doi.org/10.1088/0964-1726/23/10/105009

Gur, S., Roy, K., Mishra, S.K., 2015. Tuned liquid column ball damper for seismic vibration control. *Struct. Control Heal. Monit.* 22, 1325–1342. https://doi.org/10.1002/stc.1740

Heo, J., Lee, Sung-kyung, Park, E., Lee, Sang-hyun, Min, K., Kim, H., Jo, J., Cho, B., 2009. Performance test of a tuned liquid mass damper for reducing bidirectional responses of building structures. *Struct. Des. Tall Build.* 18, 789–805.

Hitchcock, P.A., Glanville, M.J., Kwok, K.C.S., Watkins, R.D., Samali, B., 1999. Damping properties and wind-induced response of a steel frame tower fitted with liquid column vibration absorbers. *J. Wind Eng. Ind. Aerodyn.* 83, 183–196. https://doi.org/10.1016/S0167-6105(99)00071-9

Hitchcock, P.A., Kwok, K.C.S., Watkins, R.D., Samali, B., 1997a. Characteristics of liquid column vibration absorbers (LCVA)-I. *Eng. Struct.* 19, 126–134. https://doi.org/10.1016/S0141-0296(96)00042-9

Hitchcock, P.A., Kwok, K.C.S., Watkins, R.D., Samali, B., 1997b. Characteristics of liquid column vibration absorbers (LCVA)-II. *Eng. Struct.* 19, 135–144.

Hokmabady, H., Mojtahedi, A., Mohammadyzadeh, S., Ettefagh, M.M., 2019. Structural control of a fixed offshore structure using a new developed tuned liquid column ball gas damper (TLCBGD). *Ocean Eng.* 192, 106551. https://doi.org/10.1016/j.oceaneng.2019.106551

Hummingbird Kinetics, n.d. Hummingbird Kinetics Damping System Installation. https://hummingbirdkinetics.com/experience/ (accessed 18 October 23).

Huo, L., Li, H., 2005. *Torsionally Coupled Response Control of Offshore Platform Structures Using CTLCD.* In: *The Fifteenth International Offshore and Polar Engineering Conference*, Seoul, South Korea, 2005, pp. 296–303.

Huo, L., Li, H., 2004a. Torsionally coupled response control of offshore platform structures using circular tuned liquid column dampers. *China Ocean Eng.* 18, 173–183.

Huo, L., Li, H., 2004b. Torsionally coupled response control of structures using circular tuned liquid column dampers, in: *13 Th World Conference on Earthquake Engineering*. Vancouver, B.C., Canada, pp. 1–11.

Kareem, A., Kijewski, T., Tamura, Y., 1999. Mitigation of motions of tall buildings with specific examples of recent applications. *Wind Struct.* 2, 201–251. https://doi.org/10.12989/was.1999.2.3.201

Konar, T., 2024. Seismic vibration control of a building by overhead water tank designed as slender tuned sloshing damper. *Pract. Period. Struct. Des. Constr.* 29, 1–11. https://doi.org/10.1061/ppscfx.sceng-1393

Konar, T., 2008. *Passive Control of Seismically Excited Structures by the Liquid Column Vibration Absorber*. Bengal Engineering & Science University, Shibpur.

Konar, T., Ghosh, A.D., 2023. A review on various configurations of the passive tuned liquid damper. *J. Vib. Control* 29, 1945–1980. https://doi.org/10.1177/10775463221074077

Konar, T., Ghosh, A.D., 2022. Use of deep liquid-containing tanks as dynamic vibration absorbers for lateral vibration control of structures: a review. *Iran. J. Sci. Technol. Trans. Civ. Eng.* 46, 753–769. https://doi.org/10.1007/s40996-021-00679-8

Konar, T., Ghosh, A.D., 2013. Bimodal vibration control of seismically excited structures by the liquid column vibration absorber. *JVC/Journal Vib. Control* 19, 385–394. https://doi.org/10.1177/1077546311430718

Konar, T., Ghosh, A.D., 2010. Passive control of seismically excited structures by the liquid column vibration absorber. *Struct. Eng. Mech.* 36, 561–573. https://doi.org/10.12989/sem.2010.36.5.561

Konar, T., Ghosh, A.D., Basu, B., 2024. Real-world installations of tuned liquid column dampers for wind-induced vibration control of buildings: some important case studies. *Struct. Infrastruct. Eng.* in press. https://doi.org/10.1080/15732479.2024.2420174

Lee, S., Min, K., Lee, H., 2011. Parameter identification of new bidirectional tuned liquid column and sloshing dampers. *J. Sound Vib.* 330, 1312–1327. https://doi.org/10.1016/j.jsv.2010.10.016

Liang, S.G., 1996. Experiment study of torsionally structural vibration control using circular tuned liquid column dampers. *Spec. Struct.* 13, 33–35.

Liu, K., Shi, Q., Liu, Y., Liu, L., Zhou, F., 2023. Investigation of an improved tuned liquid column gas damper for the vertical vibration control. *Mech. Syst. Signal Process.* 196, 110340. https://doi.org/10.1016/j.ymssp.2023.110340

Mallik, A., Chatterjee, S., 2014. *Principles of Passive and Active Vibration Control*. Affiliated East-West Press Private Limited, New Delhi, India.

Masnata, C., Di Matteo, A., Adam, C., Pirrotta, A., 2023a. Efficient estimation of tuned liquid column damper inerter (TLCDI) parameters for seismic control of base-isolated structures. *Comput. Civ. Infrastruct. Eng.* 38, 1638–1656. https://doi.org/10.1111/mice.12929

Masnata, C., Matteo, A. Di, Adam, C., Pirrotta, A., 2023b. Nontraditional configuration of tuned liquid column damper inerter for base-isolated structures. *Mech. Res. Commun.* 129, 104101. https://doi.org/10.1016/j.mechrescom.2023.104101

Mehrkian, B., Altay, O., 2022. Omnidirectional liquid column vibration absorbers for multi-story buildings. *J. Build. Eng.* 62, 105306. https://doi.org/10.1016/j.jobe.2022.105306

Mehrkian, B., Altay, O., 2020. Mathematical modeling and optimization scheme for omnidirectional tuned liquid column dampers. *J. Sound Vib.* 484, 115523. https://doi.org/10.1016/j.jsv.2020.115523

Min, K.-W., Kim, J., Kim, Y.-W., 2014a. Design and test of tuned liquid mass dampers for attenuation of the wind responses of a full scale building. *Smart Mater. Struct.* 23, 1–10. https://doi.org/10.1088/0964-1726/23/4/045020

Min, K.-W., Kim, J., Lee, H.-R., 2014b. A design procedure of two-way liquid dampers for attenuation of wind-induced responses of tall buildings. *J. Wind Eng. Ind. Aerodyn.* 129, 22–30. https://doi.org/10.1016/j.jweia.2014.03.003

Mousavi, S.A., Bargi, K., Zahrai, S.M., 2013. Optimum parameters of tuned liquid column – gas damper for mitigation of seismic-induced vibrations of offshore jacket platforms. *Struct. Control Heal. Monit.* 20, 422–444. https://doi.org/10.1002/stc.505

Nakayama, Y., Boucher, R.F., 1999. *Introduction to Fluid Mechanics.* Butterworth-Heinemann, Oxford. https://doi.org/10.1093/biomet/60.1.125

Pandey, D.K., Mishra, S.K., 2021. Inerter assisted robustness of compliant liquid column damper. *Struct Control Heal. Monit.* 28, e2763. https://doi.org/10.1002/stc.2763

Pandey, D.K., Mishra, S.K., 2018. Moving orifice circular liquid column damper for controlling torsionally coupled vibration. *J. Fluids Struct.* 82, 357–374. https://doi.org/10.1016/j. jfluidstructs.2018.07.015

Reiterer, M., Altay, O., Wendner, R., Hoffmann, S., Strauss, A., 2008. Adaptive Flüssigkeitstilger für Vertikalschwingungen von Ingenieurstrukturen, Teil 2 – Feldversuche. *Stahlbau* 77, 205–212. https://doi.org/10.1002/stab.200810022

Reiterer, M., Ziegler, F., 2006. Control of pedestrian-induced vibrations of long-span bridges. *Struct. Control Heal. Monit.* 13, 1003–1027. https://doi.org/10.1002/stc.91

Roy, A.K., Ghosh, A., 2015. A study on the design parameters of the compliant LCD for structural vibration control under near fault earthquakes, in: Matsagar, V. (Ed.), *Advances in Structural Engineering: Dynamics, Volume Two (Select Proceedings of Structural Engineering Convention (SEC).* Springer, India, pp. 1243–1255. https://doi.org/10.1007/ 978-81-322-2193-7_97

Roy, A.K., Konar, T., Ghosh, A., 2023. Mitigation of structural vibrations due to pulse-type-near-fault earthquake by the compliant liquid column damper. *J. Earthq. Tsunami* 17, 2350004. https://doi.org/10.1142/S1793431123500045

Rozas, L., Boroschek, R.L., Tamburrino, A., Rojas, M., 2016. A bidirectional tuned liquid column damper for reducing the seismic response of buildings. *Struct. Control Heal. Monit.* 23, 621–640. https://doi.org/10.1002/stc.505

Sakai, F., Takaeda, S., Tamaki, T., 1991. Damping device for tower-like structure. US5070663.

Sarkar, A., Gudmestad, O.T., 2013. Pendulum type liquid column damper (PLCD) for controlling vibrations of a structure: theoretical and experimental study. *Eng. Struct.* 49, 221–233. https://doi.org/10.1016/j.engstruct.2012.10.023

Shah, M.U., Usman, M., Farooq, S.H., 2022. Applied sciences effect of tuned spring on vibration control performance of modified liquid column ball damper. *Appl. Sci.* 12, 318. https://doi.org/10.3390/ app12010318

Shum, K.M., Xu, Y.L., Guo, W.H., 2008. Wind-induced vibration control of long span cable-stayed bridges using multiple pressurized tuned liquid column dampers. *J. Wind Eng. Ind. Aerodyn.* 96, 166–192. https://doi.org/10.1016/j.jweia.2007.03.008

Smith, M.C., 2020. The inerter: a retrospective. *Annu. Rev. Control. Robot. Auton. Syst.* 3, 361–391. https://doi.org/10.1146/annurev-control-053018-023917

Smith, M.C., 2002. Synthesis of mechanical networks: the inerter. *IEEE Trans. Automat. Contr.* 47, 1648–1662. https://doi.org/10.1109/TAC.2002.803532

Tanveer, M., Usman, M., Khan, I.U., Farooq, S.H., Hanif, A., 2020. Material optimization of tuned liquid column ball damper (TLCBD) for the vibration control of multi-storey structure using various liquid and ball densities. *J. Build. Eng.* 32, 101742.

Teramura, A., Yoshida, O., 1996. Development of vibration control system using U-shaped water tank, in: *11th World Conferance on Earthquake Engineering.* Acapulco, Mexico, p. Paper No. 1343.

Wang, Q., Qiao, H., De Domenico, D., Zhu, Z., Tang, Y., 2021a. Seismic performance of optimal multi-tuned liquid column damper-inerter (MTLCDI) applied to adjacent high-rise buildings. *Soil Dyn. Earthq. Eng.* 143, 106653. https://doi.org/10.1016/j.soildyn.2021.106653

Wang, Q., Qiao, H., De Domenico, D., Zhu, Z., Tang, Y., 2020a. Seismic response control of adjacent high-rise buildings linked by the tuned liquid column damper-inerter (TLCDI). *Eng. Struct.* 223, 111169. https://doi.org/10.1016/j.engstruct.2020.111169

Wang, Q., Tian, H., Qiao, H., Tiwari, N.D., Wang, Quan, 2021b. Wind-induced vibration control and parametric optimization of connected high-rise buildings with tuned liquid-column-damper –inerter. *Eng. Struct.* 226, 111352. https://doi.org/10.1016/j.engstruct.2020.111352

Wang, Qinhua, Tiwari, N.D., Qiao, H., Wang, Quan, 2020b. Inerter-based tuned liquid column damper for seismic vibration control of a single-degree-of-freedom structure. *Int. J. Mech. Sci.* 184, Paper No. 105840. https://doi.org/10.1016/j.ijmecsci.2020.105840

Wang, Z., Cao, L., Ubertini, F., Laflamme, S., 2021. Numerical investigation and design of reinforced concrete shear wall equipped with tuned liquid multiple columns dampers. *Shock Vib.* 2021, 6610811. https://doi.org/10.1155/2021/6610811

Wei, X., Zhao, X., 2020. Vibration suppression of a floating hydrostatic wind turbine model using bidirectional tuned liquid column mass damper. *Wind Energy* 23, 1887–1904. https://doi.org/10.1002/we.2524

Wendner, R., Reiterer, M., Hoffmann, S., Strauss, A., Bergmeister, K., 2007. Adaptive Flüssigkeitstilger für Vertikalschwingungen von Ingenieurstrukturen: teil 1 – Laborversuche. *Stahlbau* 76, 916–923. https://doi.org/10.1002/stab.200710096

Wu, H., Cao, L., Chen, A., Laflamme, S., 2018. Development of an analytical model for a tuned liquid multi-column damper, in: *Earth and Space 2018: Engineering for Extreme Environments*. ASCE, Reston, Virginia, USA, pp. 833–843.

Xu, Y.L., Kwok, K.C.S., Samali, B., 1992a. The effect of tuned mass dampers and liquid dampers on cross-wind response of tall/slender structures. *J. Wind Eng. Ind. Aerodyn.* 40, 33–54. https://doi.org/10.1016/0167-6105(92)90519-G

Xu, Y.L., Samali, B., Kwok, K.C.S., 1992b. Control of along-wind response of structures by mass and liquid dampers. *J. Eng. Mech.* 118, 20–39.

Zeng, X., Yu, Y., Zhang, L., Liu, Q., Wu, H., 2015. A new energy-absorbing device for motion suppression in deep-sea floating platforms. *Energies* 8, 111–132. https://doi.org/10.3390/en8010111

MATLAB® CODE 3.1

```matlab
% Response of an SDOF structure with and without LCVA under harmonic
base excitation
clear all
clc
close all
% Definition of general constants
g=9.81; % gravitational acceleration, unit m/s^2
pi=3.141593;

% Input structural parameters
Ms=1000000; % mass of the structure, unit kg
ZETAs=0.01; % damping ratio of the structure
Ts=2; % natural period of the structure, unit s

% Input damper parameters
```

```matlab
tunr=1.0; % tuning ratio
mu=0.05; % mass ratio
alpha=0.5; % length ratio
zi=20; % head-loss coefficient of the orifice(s)
AR=1.25; % area ratio

OMEGAs=2*pi/Ts;
OMEGAd=tunr*OMEGAs;
L_e=2*g/OMEGAd/OMEGAd

disp('Effective length liquid column')
L_e=2*g/OMEGAd/OMEGAd
disp('Length of the horizontal limb')
b_L=alpha*L_e
disp('Liquid height in the vertical limbs')
h=(L_e-b_L*AR)/2

% Input harmonic base excitation in the form of a sine wave
dt=0.02; % time interval of base excitation, unit s
t=0:dt:120; % duration of excitation taken as 300 s
A=0.2; % amplitude of base acceleration, unit m/s^2
beta=0.99; % forced frequency ratio
omg=OMEGAs*beta;
abc=A*sin(omg*t);
[n,m]=size(abc);

Gabc=abc;

plot(t,(Gabc/9.81))
set(gca,{'FontName','FontSize'},{'Times New Roman',25})
set(gca,{'XMinorTick','YMinorTick'},{'on','on'})
xlabel('Time (s)')
ylabel('Base acceleration (g)')

% Determination of response of structure with LCVA
yy=zeros(4,1);
tt=0;
Gabcd=0;
Yin=[0 0 0 0];
time=[];
dispS=[];
velS=[];
dispD=[];
velD=[];
stateSD=[];
accelS=[];

for i=1:(m*n)

  ttf=tt+dt;

[T, Y]=ode45(@(tt,yy) New_rk_LCVA(tt,yy,Gabcd,OMEGAs,OMEGAd,ZETAs,L_e,...
zi,...
```

```
  mu,alpha),[tt ttf],Yin); % Solve ODE

  time=[time; T(:,1)];
  dispS=[dispS; Y(:,1)];
  velS=[velS; Y(:,2)];
  dispD=[dispD; Y(:,3)];
  velD=[velD; Y(:,4)];
  stateSD=[stateSD; Y(:,:)];

xdd=(1/(1+mu-alpha*alpha*mu))*(-(OMEGAs^2).*Y(:,1)...
 -(2*ZETAs*OMEGAs).*Y(:,2)...
 +(mu*alpha)*((OMEGAd^2).*Y(:,3)...
 +(0.5*zi/L_e).*abs(Y(:,4)).*Y(:,4)))...
 -Gabcd;

  accelS=[accelS; xdd];

  tt=ttf;
  Gabcd=Gabc(1,i);

  Yin=Y(end,:);
end

% Determination of response of uncontrolled structure
tts=0;
Gabcd=0;
Yin=[0 0];
times=[];
yys=zeros(2,1);
disp=[];
vel=[];
state=[];
accel=[];
for i=1:(m*n)
 ttsf=tts+dt;

 [T, Y]=ode45(@(tts,yys) rk_str(tts,yys,Gabcd,OMEGAs,ZETAs),[tts
ttsf],Yin); % Solve ODE

 times=[times; T(:,1)];
 disp=[disp; Y(:,1)];
 vel=[vel; Y(:,2)];
 state=[state; Y(:,:)];

 xdd=-(OMEGAs^2).*Y(:,1)-(2*ZETAs*OMEGAs).*Y(:,2)-Gabcd;
 accel=[accel; xdd];

 tts=ttsf;
 Gabcd=Gabc(1,i);
 Yin=Y(end,:);
end

% Determination of control effectiveness of LCVA
```

```matlab
rms=sqrt(mean(disp.^2));
rmsd=sqrt(mean(dispS.^2));
redrms_disp=(rms-rmsd)*100/rms

peak=max(abs(disp));
peakd=max(abs(dispS));
redpeak_disp=(peak-peakd)*100/peak

rms=sqrt(mean(vel.^2));
rmsd=sqrt(mean(velS.^2));
redrms_vel=(rms-rmsd)*100/rms

peak=max(abs(vel));
peakd=max(abs(velS));
redpeak_vel=(peak-peakd)*100/peak

rms=sqrt(mean(accel.^2));
rmsd=sqrt(mean(accelS.^2));
redrms_accel=(rms-rmsd)*100/rms

peak=max(abs(accel));
peakd=max(abs(accelS));
redpeak_accel=(peak-peakd)*100/peak

figure
plot(times,disp,'r-',time,dispS)
set(gca,{'FontName','FontSize'},{'Times New Roman',35})
set(gca,{'XMinorTick','YMinorTick'},{'on','on'})
legend(' without damper','with damper')
xlabel('Time (s)')
ylabel('Structural displacement (m)')
grid on

figure
subplot(3,1,1); plot(times,disp,'r-',time,dispS)
legend('without damper','with damper')
xlabel('Time (s)')
ylabel('Displacement (m)')
grid on
subplot(3,1,2); plot(times,vel,'r-',time,velS)
legend('without damper','with damper')
xlabel('Time (s)')
ylabel('Velocity (m/s)')
grid on
subplot(3,1,3); plot(times,accel,'r-',time,accelS)
legend('only structure','structure with damper')
xlabel('Time (s)')
ylabel('Acceleration (m/s^2)')
hold on
grid on

function[dydtt]= New_rk_LCVA(tt,yy,Gabcd,OMEGAs,OMEGAd,ZETAs,L_e,
zi,...
```

```
  mu,alpha)

dydtt=zeros(4,1); % a column vector

% structural displacement=yy(1), velocity=dydtt(1)=yy(2)
 dydtt(1)=yy(2);

%structural acceleration
dydtt(2)=(1/(1+mu-alpha*alpha*mu))*(-(OMEGAs^2)*yy(1)...
 -(2*ZETAs*OMEGAs)*yy(2)...
 +(mu*alpha)*((OMEGAd^2)*yy(3)...
 +0.5*zi/L_e*abs(yy(4))*yy(4)))...
 -Gabcd;

% liquid column displacement=yy(3), velocity=dydtt(3)=yy(4)
 dydtt(3)=yy(4);

% liquid column acceleration
dydtt(4)=(1/(1+mu-alpha*alpha*mu))*((-(OMEGAd^2)*yy(3)...
 -0.5*zi/L_e*abs(yy(4))*yy(4))*(1+mu)...
 +2*ZETAs*OMEGAs*alpha*yy(2)...
 +(OMEGAs^2)*alpha*yy(1));
end

  function[dydtt] = rk_str(tts,yys,Gabcd,OMEGAs,ZETAs)

dydtt=zeros(2,1);% a column vector

% Structural displacement=yys(1), velocity=dydtt(1)=yys(2)

dydtt(1)=yys(2);

% Structural acceleration
dydtt(2)=-(OMEGAs^2)*yys(1)-(2*ZETAs*OMEGAs)*yys(2)-Gabcd;

end
```

MATLAB® CODE 3.2

```
% Response of an SDOF structure with and without CLCD under harmonic
base excitation
clear all
clc
close all

% Definition of general constants
g=9.81; % gravitational acceleration, unit m/s^2
pi=3.141593;

% Input structural parameters
Ms=1000000; % mass of the structure, unit kg
ZETAs=0.01; % damping ratio of the structure
Ts=2; % natural period of the structure, unit s
```

```
% Input damper parameters
NU=1.0; % tuning ratio
MU=0.01; % mass ratio
alpha=0.5; % length ratio
zi=200; % head-loss coefficient of the orifice(s)
ZETA2=0.0; % damping ratio of spring element connecting liquid
container to structure
L=2; % length of liquid column, unit m
tau=1; % ratio of the mass of damper container to the mass of
contained liquid

OMEGAs=2*pi/Ts;
OMEGA2=OMEGAs*NU;
OMEGAs=2*pi/Ts;
OMEGA_L=sqrt(2*g/L);
B=L*alpha;
ziL=zi/L;

% Input harmonic base excitation in the form of a sine wave
dt=0.02; % time interval of base excitation, unit s
t=0:dt:300; % duration of excitation taken as 300 s
A=0.2; % amplitude of base acceleration, unit m/s^2
beta=0.99; % forced frequency ratio
omg=OMEGAs*beta;
abc=A*sin(omg*t);
[n,m]=size(abc);

Gabc=abc;

plot(t,(Gabc/9.81))
set(gca,{'FontName', 'FontSize'},{'Times New Roman',25})
set(gca,{'XMinorTick','YMinorTick'},{'on','on'})
xlabel('Time (s)')
ylabel('Base acceleration (g)')

% Determination of response of structure with CLCD
yy=zeros(6,1);

tt=0;
Gabcd=0;
F_bard=0;
Yin=[0 0 0 0 0 0];
time=[];
dispS=[];
velS=[];
dispD=[];
velD=[];
dispL=[];
velL=[];
stateSD=[];
accelS=[];
for i=1:(m*n)
```

```matlab
ttf=tt+dt;

[T, Y]=ode45(@(tt,yy) rk_clcd(tt,yy,Gabcd,OMEGAs,OMEGA2,ZETAs,MU,tau,
alpha,Ms,...
 OMEGA_L,L,F_bard,zi),[tt ttf],Yin); % Solve ODE

 time=[time; T(:,1)];
 dispS=[dispS; Y(:,1)];
 velS=[velS; Y(:,2)];
 dispD=[dispD; Y(:,3)];
 velD=[velD; Y(:,4)];
 dispL=[dispL; Y(:,5)];
 velL=[velL; Y(:,6)];
 stateSD=[stateSD; Y(:,:)];

 xdd=(-(OMEGAs^2)).*Y(:,1)-(2*ZETAs*OMEGAs).*Y(:,2)...
 +(MU*OMEGA2^2).*Y(:,3)+(MU*2*ZETA2*OMEGA2).*Y(:,4)-Gabcd;

 accelS=[accelS; xdd];

 tt=ttf;
 Gabcd=Gabc(1,i);

 Yin=Y(end,:);
end

% Determination of response of uncontrolled structure
tts=0;
Gabcd=0;
Yin=[0 0];
times=[];
yys=zeros(2,1);
disp=[];
vel=[];
state=[];
accel=[];
for i=1:(m*n)

 ttsf=tts+dt;

 [T, Y]=ode45(@(tts,yys) rk_str(tts,yys,Gabcd,OMEGAs,ZETAs),[tts
ttsf],Yin); % Solve ODE

 times=[times; T(:,1)];
 disp=[disp; Y(:,1)];
 vel=[vel; Y(:,2)];
 state=[state; Y(:,:)];

 xdd=-(OMEGAs^2).*Y(:,1)-(2*ZETAs*OMEGAs).*Y(:,2)-Gabcd;
 accel=[accel; xdd];

 tts=ttsf;
```

```
 Gabcd=Gabc(1,i);
 Yin=Y(end,:);
end

standard_deviation_liq_vel = std(velL)

 rms=sqrt(mean(disp.^2));
 rmsd=sqrt(mean(dispS.^2));
 redrms_disp=(rms-rmsd)*100/rms

 peak=max(abs(disp));
 peakd=max(abs(dispS));
 redpeak_disp=(peak-peakd)*100/peak

 rms=sqrt(mean(vel.^2));
 rmsd=sqrt(mean(velS.^2));
 redrms_vel=(rms-rmsd)*100/rms

 peak=max(abs(vel));
 peakd=max(abs(velS));
 redpeak_vel=(peak-peakd)*100/peak

 rms=sqrt(mean(accel.^2));
 rmsd=sqrt(mean(accelS.^2));
 redrms_accel=(rms-rmsd)*100/rms

 peak=max(abs(accel));
 peakd=max(abs(accelS));
 redpeak_accel=(peak-peakd)*100/peak

figure
plot(times,disp,'r-',time,dispS)
set(gca,{'FontName','FontSize'},{'Times New Roman',25})
set(gca,{'XMinorTick','YMinorTick'},{'on','on'})
legend('without damper','with damper')
xlabel('Time (s)')
ylabel('Structural displacement (m)')

figure
subplot(3,1,1); plot(times,disp,'r-',time,dispS)
legend('without damper','with damper')
xlabel('Time (s)')
ylabel('Displacement (m)')
grid on
subplot(3,1,2); plot(times,vel,'r-',time,velS)
legend('without damper','with damper')
xlabel('Time (s)')
ylabel('Velocity (m/s)')
grid on
subplot(3,1,3); plot(times,accel,'r-',time,accelS)
legend('only structure','structure with damper')
xlabel('Time (s)')
ylabel('Acceleration (m/s^2)')
hold on
```

```
grid on

function[dydtt] = rk_clcd(tt,yy,Gabcd,OMEGAs,OMEGA2,ZETAs,MU,tau,alpha,Ms,...
  OMEGA_L,L,F_bard,zi)
dydtt=zeros(6,1); % a column vector

  delta=(1+tau-alpha^2);
% structural displacement=yy(1), velocity=dydtt(1)=yy(2)

  dydtt(1)=yy(2);
%structural acceleration
dydtt(2)=-(OMEGAs^2)*yy(1)...
 -(2*ZETAs*OMEGAs)*yy(2)...
 +(OMEGA2^2*MU)*yy(3)...
 -(1/Ms)*F_bard...
 -Gabcd;

dydtt(3)=yy(4);

% whole damper acceleration
dydtt(4)=(OMEGAs^2)*yy(1)...
 +(2*ZETAs*OMEGAs)*yy(2)...
 -(OMEGA2^2*((1+tau+(delta*MU))/delta))*yy(3)...
 +(alpha*OMEGA_L^2/delta)*yy(5)...
 +alpha*zi*abs(yy(6))*yy(6)/(2*L*delta)...
 +(1+tau+(delta*MU))*F_bard/(Ms*delta*MU);

dydtt(5)=yy(6);

% liquid acceleration
dydtt(6)=(alpha*OMEGA2^2*(1+tau))*yy(3)/delta...
 -((1+tau)*OMEGA_L^2)*yy(5)/delta...
 -((1+tau)*(zi/(2*L)))*abs(yy(6))*yy(6)/delta...
 -alpha*(1+tau)*F_bard/(Ms*delta*MU);

end

function[dydtt] = rk_str(tts,yys,Gabcd,OMEGAs,ZETAs
dydtt=zeros(2,1); % a column vector

% structural displacement=yy(1), velocity=dydtt(1)=yy(2)

dydtt(1)=yys(2);

% structural acceleration
dydtt(2)=-(OMEGAs^2)*yys(1)-(2*ZETAs*OMEGAs)*yys(2)-Gabcd;
end
```

MATLAB® CODE 3.3

```
% Response of an SDOF structure with and without Sealed TLCD under
harmonic base excitation
```

```
clear all
clc
close all

% Definition of general constants
g=9.81; % gravitational acceleration, unit m/s^2
pi=3.141593;

% Input structural parameters
Ms=1000000; % mass of the structure, unit kg
ZETAs=0.01; % damping ratio of the structure
Ts=0.75; % natural period of the structure, unit s

% Input damper parameters
tunr=1.0; % tuning ratio
mu=0.05; % mass ratio
alpha=0.5; % length ratio
zi=20; % head-loss coefficient of the orifice(s)
nnn=1.4; % polytropic index
L=2; % length of liquid column in damper container
rho=1000; % mass density of damper liquid, unit kg/m^3
Ha=1; % height of closed air chamber in vertical limbs at rest
condition

OMEGAs=2*pi/Ts;
OMEGAd=tunr*OMEGAs;
disp('Pressure in air chambers of Sealed TLCD in Pa')
P0=(OMEGAd*OMEGAd*L/2/g-1)*rho*g*Ha/nnn

% Input harmonic base excitation in the form of a sine wave
dt=0.02; % time interval of base excitation, unit s
t=0:dt:120; % duration of excitation taken as 300 s
A=0.2; % amplitude of base acceleration, unit m/s^2
beta=0.99; % forced frequency ratio
omg=OMEGAs*beta;
abc=A*sin(omg*t);
[n,m]=size(abc);

Gabc=abc;

plot(t,(Gabc/9.81))
set(gca,{'FontName','FontSize'},{'Times New Roman',25})
set(gca,{'XMinorTick','YMinorTick'},{'on','on'})
xlabel('Time (s)')
ylabel('Base acceleration (g)')

% Determination of response tof structure with TLCD
yy=zeros(4,1);
tt=0;
Gabcd=0;
Yin=[0 0 0 0];
time=[];
dispS=[];
velS=[];
```

```matlab
dispD=[];
velD=[];
stateSD=[];
accelS=[];

for i=1:(m*n)

 ttf=tt+dt;
[T, Y]=ode45(@(tt,yy) New_rk_TLCD(tt,yy,Gabcd,OMEGAs,OMEGAd,ZETAs,L,
zi,...

 mu,alpha),[tt ttf],Yin); % Solve ODE

 time=[time; T(:,1)];
 dispS=[dispS; Y(:,1)];
 velS=[velS; Y(:,2)];
 dispD=[dispD; Y(:,3)];
 velD=[velD; Y(:,4)];
 stateSD=[stateSD; Y(:,:)];

xdd=(1/(1+mu-alpha*alpha*mu))*(-(OMEGAs^2).*Y(:,1)...
 -(2*ZETAs*OMEGAs).*Y(:,2)...
 +(mu*alpha)*((OMEGAd^2).*Y(:,3)...
 +(0.5*zi/L).*abs(Y(:,4)).*Y(:,4)))...
 -Gabcd;

 accelS=[accelS; xdd];

 tt=ttf;
 Gabcd=Gabc(1,i);

 Yin=Y(end,:);
end

% Determination of response of uncontrolled structure
tts=0;
Gabcd=0;
Yin=[0 0];
times=[];
yys=zeros(2,1);
disp=[];
vel=[];
state=[];
accel=[];
for i=1:(m*n)
 ttsf=tts+dt;

 [T, Y]=ode45(@(tts,yys) rk_str(tts,yys,Gabcd,OMEGAs,ZETAs),[tts
ttsf],Yin); % Solve ODE

 times=[times; T(:,1)];
 disp=[disp; Y(:,1)];
 vel=[vel; Y(:,2)];
```

```matlab
 state=[state; Y(:,:)];

 xdd=-(OMEGAs^2).*Y(:,1)-(2*ZETAs*OMEGAs).*Y(:,2)-Gabcd;
 accel=[accel; xdd];

 tts=ttsf;
 Gabcd=Gabc(1,i);
 Yin=Y(end,:);
end

% Determination of control effectiveness of TLCD

 rms=sqrt(mean(disp.^2));
 rmsd=sqrt(mean(dispS.^2));
 redrms_disp=(rms-rmsd)*100/rms

 peak=max(abs(disp));
 peakd=max(abs(dispS));
 redpeak_disp=(peak-peakd)*100/peak

 rms=sqrt(mean(vel.^2));
 rmsd=sqrt(mean(velS.^2));
 redrms_vel=(rms-rmsd)*100/rms

 peak=max(abs(vel));
 peakd=max(abs(velS));
 redpeak_vel=(peak-peakd)*100/peak

 rms=sqrt(mean(accel.^2));
 rmsd=sqrt(mean(accelS.^2));
 redrms_accel=(rms-rmsd)*100/rms

 peak=max(abs(accel));
 peakd=max(abs(accelS));
 redpeak_accel=(peak-peakd)*100/peak

 figure
plot(times,disp,'r-',time,dispS)
set(gca,{'FontName','FontSize'},{'Times New Roman',35})
set(gca,{'XMinorTick','YMinorTick'},{'on','on'})
legend(' without damper','with damper')
xlabel('Time (s)')
ylabel('Structural displacement (m)')
grid on

figure
subplot(3,1,1); plot(times,disp,'r-',time,dispS)
legend('without damper','with damper')
xlabel('Time (s)')
ylabel('Displacement (m)')
grid on
subplot(3,1,2); plot(times,vel,'r-',time,velS)
legend('without damper','with damper')
xlabel('Time (s)')
```

```matlab
ylabel('Velocity (m/s)')
grid on
subplot(3,1,3); plot(times,accel,'r-',time,accelS)
legend('only structure','structure with damper')
xlabel('Time (s)')
ylabel('Acceleration (m/s^2)')
hold on
grid on

function[dydtt]= New_rk_TLCD(tt,yy,Gabcd,OMEGAs,OMEGAd,ZETAs,L,zi,...
 mu,alpha)

dydtt=zeros(4,1); % a column vector
% structural displacement=yy(1), velocity=dydtt(1)=yy(2)

dydtt(1)=yy(2);

% structural acceleration
dydtt(2)=(1/(1+mu-alpha*alpha*mu))*(-(OMEGAs^2)*yy(1)...
 -(2*ZETAs*OMEGAs)*yy(2)...
 +(mu*alpha)*((OMEGAd^2)*yy(3)...
 +0.5*zi/L*abs(yy(4))*yy(4)))...
 -Gabcd;

% liquid column displacement=yy(3), velocity=dydtt(3)=yy(4)

dydtt(3)=yy(4);

% liquid column acceleration
dydtt(4)=(1/(1+mu-alpha*alpha*mu))*((-(OMEGAd^2)*yy(3)...
 -0.5*zi/L*abs(yy(4))*yy(4))*(1+mu)...
 +2*ZETAs*OMEGAs*alpha*yy(2)...
 +(OMEGAs^2)*alpha*yy(1));
end

function[dydtt] = rk_str(tts,yys,Gabcd,OMEGAs,ZETAs)
dydtt=zeros(2,1); % a column vector

% Structural displacement=yys(1), velocity=dydtt(1)=yys(2)

dydtt(1)=yys(2);

% Structural acceleration
dydtt(2)=-(OMEGAs^2)*yys(1)-(2*ZETAs*OMEGAs)*yys(2)-Gabcd;

end
```

4 Control of Wind-Excited Structures by TLCD

4.1 INTRODUCTION

Wind is one of the major causes of lateral load on structures. Wind flow dynamics, in general, is considered to be a complex phenomenon. It is composed of a multitude of eddies having varying sizes and rotational characteristics, carried along in a general stream of air flowing relative to the surface of the earth (Mendis et al., 2007). Wind derives its turbulent character from these eddies. The speed of wind varies randomly in time and can be expressed in a statistical sense by the sum of a mean component and a fluctuating component, the latter arising from the turbulence (see Figure 4.1). This summation of the mean and the fluctuating components provides the instantaneous wind speed, $U_I(t)$, as given by

$$U_I(t) = U + u(t) \tag{4.1}$$

where U and $u(t)$ denote the mean and fluctuating wind speeds, respectively. The randomness of strong winds is higher in the lower levels of atmosphere and largely arises from interaction with surface features. The average wind speed over a period of time of the order of 10 minutes or more tends to increase with height.

The total wind thrust acting on a structure may be expressed as

$$F_T(t) = 0.5\rho_a C_d A U_I^2 \tag{4.2}$$

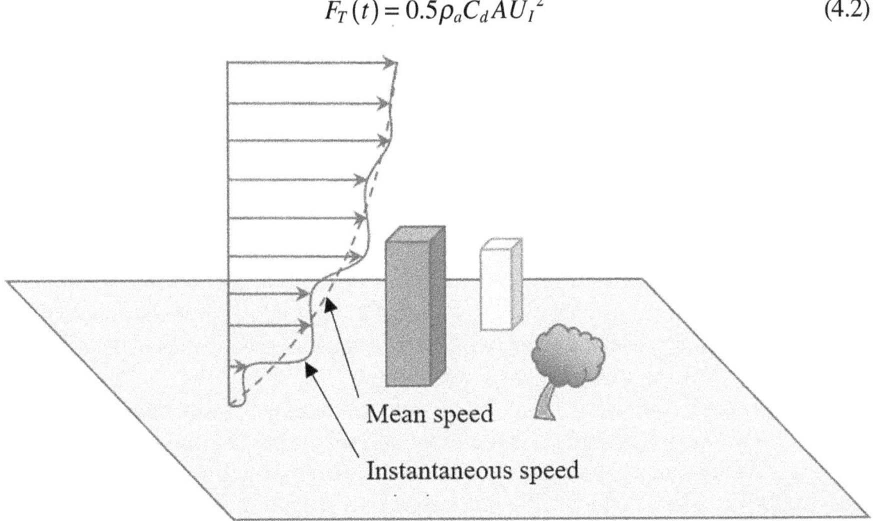

FIGURE 4.1 Wind speed profile.

DOI: 10.1201/9781003377894-4

where C_d, A and ρ_a respectively represent the drag coefficient, the projected area of the structure normal to the wind and the air density.

The motion of a structure due to wind consists of two components: first, a static or sustained deflection caused by the mean wind speed, which is not apparent to occupants but is included in the estimation of the building drift; second, oscillatory or resonant vibration, which is due to the dynamic and time-varying action of wind. In general, dynamic wind loading is characterized by narrow-banded, long-period frequencies. This causes wind-induced vibration to be more critical for flexible structures such as bridges, tall buildings, towers and chimneys. According to ASCE 7 (2002), structures with a fundamental time period greater than 1 s are designated as flexible structures. In recent times, to meet the demand for urbanization under space constraints on the ground and the urge to build tall landmark structures, as well as to enhance connectivity by constructing transmission towers, etc., many structures are becoming lighter and flexible, thereby making them more vulnerable to wind.

The vibration of a structure to wind excitation is significantly influenced by many factors such as site conditions and shape, height and dynamic characteristics of the structure. When a structure is subjected to wind loading, several different phenomena such as buffeting, vortex shedding, galloping and flutter occur, which are responsible for the dynamic response of structures to wind. Tall and slender structures are likely to be sensitive to dynamic response in the along-wind direction as a consequence of turbulence buffeting. However, along-wind vibration is relatively benign and easy to predict through code-based approaches. For more details, see Vickery and Basu (1983). Across-wind or transverse response normally arises from vortex shedding or galloping. If a dominant natural frequency of the structure is close to the vortex shedding frequency (Strouhal, 1878), across-wind excitation will produce substantial dynamic amplification that will govern the design of the structure (Vickery and Basu, 1983). As compared to the along-wind response of flexible structures, the across-wind response is more sensitive to wind speed. At lower wind speeds, the along-wind loads normally dominate but with an increase in wind speed, the across-wind loads take over. For instance, the wind tunnel test of the Jin Mao Tower in Shanghai, China, revealed that its maximum acceleration in the across-wind direction at its design wind speed is about 1.2 times of that in the along-wind direction (Jacobs, 2008). Other kinds of aerodynamic instability, such as flutter, occur under certain wind speeds in which aerodynamics and the vibrating structure jointly create dynamic amplification in the response. Wind can also lead to torsional vibration caused by the twisting torque about the vertical axis of the structure, which originates from the non-uniform distribution of wind load. Thus, under the influence of wind, a structure tends to vibrate in the along-wind, across-wind and torsional directions as illustrated in Figure 4.2. For highly flexible structures such as transmission lines, coupled along-wind and across-wind motions may develop, and these are referred to as wake flutter (Kwok, 2013).

Wind-induced vibrations in any of the abovementioned forms more commonly cause serviceability issues and, in the extreme event, may lead to major structural damage and even catastrophic failure. For many flexible structures such as tall buildings, the design against wind load is governed by serviceability conditions that chiefly intend to avoid damage to non-structural elements and to prevent occupant discomfort due to motion perception. To limit damage and cracking of non-structural elements,

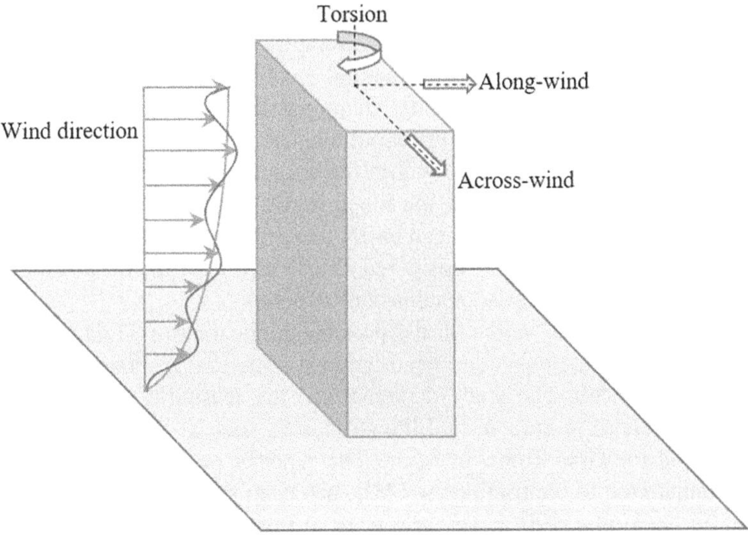

FIGURE 4.2   Components of the wind-induced response of a structure.

such as the facade, partition walls, ceilings, and equipment and machinery, control of the lateral deflection of structures (and storey drift for building structures) is imperative. Normally, the permissible lateral deflection of a civil engineering structure varies between $H_s/200$ and $H_s/1000$, depending upon the type of structure and its use, as discussed by Smith (2011). Here, H_s is the height of the structure. In case of building structures, this limit is generally restricted between $H_s/200$ and $H_s/600$ (Konar, 2023; Konar et al., 2024a; Mendis et al., 2007).

For tall buildings, occupant discomfort due to motion perception is an important issue. The lateral deflection of a structure due to the static component of wind load is not apparent to occupants, while the oscillatory or resonant vibration becomes perceptible to occupants and, if excessive, can cause possible discomfort including fear. Here it is pertinent to mention that the human response to vibration varies significantly from person to person. Several physiological and psychological parameters, such as occupants' expectations and experience, their activity, body posture and orientation and visual and acoustic cues affect the human perception of vibration. As expected, the structural response parameters, such as the frequency, displacement and accelerations for both the translational and torsional motions, also have a significant influence on the human perception of vibration. However, it has been shown in several studies, including those by Irwin (1978), and Smith and Coull (1991), that among the response parameters due to oscillatory or resonant vibration, acceleration is the predominant parameter in determining the nature of the human response to structural vibration. There exists a substantial amount of research on how human behaviour and responses are affected by various levels of building acceleration. Studies carried out by Goto (1975, 1983), Irwin (1978), Lamb and Kwok (2019) and Yamada and Goto (1975) are some examples in this regard. Based on the results reported in these studies, a simplified set of human perception levels indicating how human behaviour and responses are affected by various levels of building acceleration is shown in Table 4.1. This may serve as a preliminary guideline for determining the upper limit of the acceleration level to be achieved by the control of structural accelerations.

Installation of passive supplemental damping devices is one of the preferred alternatives to meet the serviceability requirements of flexible structures by reducing the wind-induced vibrations to within the acceptable limit. It has already been elaborated upon in Chapter 2 how the tuned liquid column damper (TLCD) is an inherently long-period system and can easily be tuned to flexible structures. Hence, researchers and designers have considered TLCD as one of the best-suited passive devices for wind-induced vibration control of structures.

One of the pioneering works on the possible application of TLCD as a supplemental damping device to reduce wind-induced structural response was taken up by Sakai et al. (1989). The study demonstrated the feasibility of TLCD through simplified models of prominent real-life structures, such as the Citicorp Centre in New York and the Gold Tower in Japan. The superior performance of the TLCD, even in comparison to the traditional TMD, has been reported by researchers. The research by Xu et al. (1990) is an early work in this direction. Within a few years of its invention in 1988, TLCDs were installed in the Higashi-Kobe Bridge, Japan, to control the wind-induced vibration of the free-standing pylons during the construction stage (Kitazawa et al., 1992). Subsequently, TLCD has been successfully installed in several real-life structures for suppressing vibration due to wind. The One Wall Centre, Vancouver, Canada; Comcast Center, Philadelphia, USA; Random House Tower, New York City, USA; and Songdo First World, Incheon, South Korea, are some of the prominent buildings to have TLCD as a means for wind-induced

---

TABLE 4.1

Human Perception Levels to Building Acceleration (Goto, 1983, 1975; Irwin, 1978; Yamada and Goto, 1975)

Perception Level	Peak Acceleration (m/s^2)	Effects
1-No Perception	<0.05	People cannot perceive motion
2-Perception threshold	0.05–0.1	Most people cannot perceive motion. Only sensitive people can perceive motion
3-Discomfort begins	0.1–0.25	Majority of people will perceive motion. May affect desk work, and long-term exposure may produce motion sickness
4-Increased discomfort	0.25–0.4	Desk work becomes difficult, but ambulation is still possible.
5-Difficulty in maintaining balance	0.4–0.5	People strongly perceive motion and walking naturally is difficult. Standing people may lose balance
6-Movement impossible	0.5–0.6	Most people cannot walk naturally
7-Intolerable	0.6–0.7	People cannot walk or tolerate motion
8-Extremely Intolerable	0.7–0.8	Level of intolerance to motion increases
9-Unsafe	>0.85	Objects begin to fall, and people may be injured

vibration suppression (Konar et al., 2024b; Konar and Ghosh, 2023). However, the design of TLCD for suppressing wind-induced vibration still remains an open field of research, and many new ideas are emerging in the field. In the following sections, the design, performance assessment and optimization of TLCD for wind-induced vibration control of structures, along with the present challenges, are discussed in detail.

4.2 FREQUENCY DOMAIN STUDIES

As already mentioned in Section 2.2.3 of Chapter 2, frequency domain analysis has the advantage of simplicity, ease of computation and fast execution. That is why several researchers, such as Xu et al. (1992a), Yalla and Kareem (2000) and Konar and Ghosh (2010), have adopted the frequency domain approach to study the structure-TLCD system in order to predict the generalized behaviour of the damper and to conduct parametric studies. The governing equations of motion of a single-degree-of-freedom (SDOF) structure in the frequency domain with attached TLCD and under the action of external force are given in Eqs. (2.31) and (2.32). To derive the frequency domain expression of the structure-damper system under wind loading, the square of $U_I(t)$ from Eq. (4.1) is expressed as

$$U_I^2(t) = U^2 + 2u(t)U \tag{4.3}$$

in which the term $u^2(t)$ is neglected, as generally the fluctuating wind speed is an order of magnitude smaller than the average wind speed (Dong et al., 2020; Kareem and Dalton, 1982). On substituting Eq. (4.3) in Eq. (4.2), the fluctuating wind loading on the structure may be expressed as

$$F_f(t) = \rho_a C_d A U u(t) \tag{4.4}$$

The Fourier transformation of Eq. (4.4) leads to

$$\mathcal{F}_f(\omega) = \rho_a C_d A U \mathcal{F}_u(\omega) \tag{4.5}$$

where $\mathcal{F}_f(\omega)$ and $\mathcal{F}_u(\omega)$ are the Fourier transforms of the time-dependent variables $F_f(t)$ and $u(t)$, respectively.

The substitution of Eq. (4.5) in Eqs. (2.31) and (2.32) yields

$$X(\omega) = \rho_a C_d A U H_X(\omega) \mathcal{F}_u(\omega) \tag{4.6}$$

and

$$Y(\omega) = \rho_a C_d A U H_Y(\omega) \mathcal{F}_u(\omega). \tag{4.7}$$

In Eqs. (4.6) and (4.7), $X(\omega)$ and $Y(\omega)$ denote the Fourier transforms of the lateral displacement of the SDOF structure and of the displacement of the free surface of the liquid column of the TLCD in the vertical direction, respectively. $H_X(\omega)$ and $H_Y(\omega)$ are the transfer functions relating the displacement of the structure and the displacement of the liquid in the vertical limb of the TLCD, to the external excitation force, with the expressions given in Eqs. (2.33) and (2.34), respectively. The root-mean-square (rms) displacement of the structure, σ_x and the rms acceleration of the structure, $\sigma_{\ddot{x}}$ may be obtained as (Newland, 1993)

$$\sigma_x^2 = \left(\rho_a C_d A U\right)^2 \int_{-\infty}^{\infty} |H_X(\omega)|^2 S_u(\omega)\, d\omega \tag{4.8}$$

and

$$\sigma_{\ddot{x}}^2 = \left(\rho_a C_d A U\right)^2 \omega^4 \int_{-\infty}^{\infty} |H_X(\omega)|^2 S_u(\omega)\, d\omega \tag{4.9}$$

respectively, where S_u denotes the power spectral density (PSD) of the fluctuating wind speed.

One of the important aspects of the frequency domain study of wind-induced vibration is the selection of the PSD of the wind force. The nature of the along-wind force spectrum differs from that in the across-wind direction (see Figure 4.3). While the along-wind force spectrum reflects the approaching wind turbulence properties, the across-wind force spectrum is largely determined by flow separation and vortex formation. The peak of the across-wind force spectrum corresponds to the effective Strouhal number of the structure.

There are several expressions that define the along-wind and across-wind spectra as proposed by different researchers, and these are provided in design codes. In this section, an overview of the more frequently used wind spectrum models in TLCD-related investigations is presented. It is noted that in some studies wind

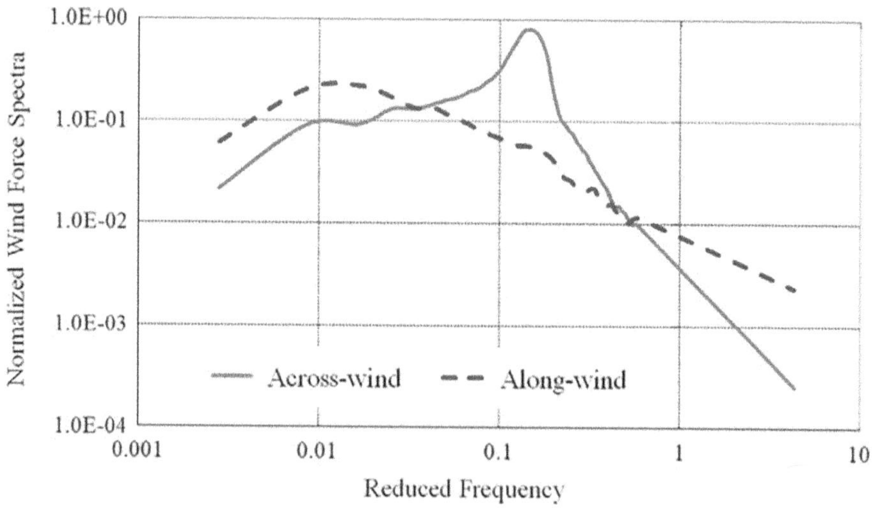

FIGURE 4.3 Components of the wind-induced response of a structure in along-wind and across-wind directions (Xie, 2014).

loading has been represented by a white noise excitation or even simply as a harmonic excitation (Chang and Hus, 1998; Shum and Xu, 2004; Wu et al., 2005, 2008), though they do not truly represent the characteristics of wind loading.

One of the earliest spectra of turbulence was proposed by von Kármán (1948). He provided a set of equations for PSD, developed based on the theory of isotropic turbulence, which has traditionally been used in wind engineering. ASCE has adopted the von Kármán spectrum for the along-wind direction in the following form (ASCE/SEI49-12, 2012)

$$S_u(z,f) = \frac{\sigma_u^2}{f} \frac{4\bar{x}}{\left(1+70.8\bar{x}^2\right)^{5/6}}. \tag{4.10}$$

In Eq. (4.10), S_u is the PSD of wind velocity in the along-wind direction, σ_u is the turbulence intensity in the along-wind direction, f is the frequency in Hz, and $\bar{x}$ is a dimensionless parameter expressed as

$$\bar{x} = \frac{fL_u^z}{U_z} \tag{4.11}$$

where U_z is the mean wind speed at height z, and L_u^z is the integral length scale of the longitudinal component of turbulence in the along-wind direction.

Davenport spectrum for horizontal gustiness in high winds has also been widely used in the along-wind analysis of structures. The expression is given by (Davenport, 1961)

$$S_u(f) = 4K_0 \frac{U_{10}^2}{f} \frac{x^2}{\left(1+x^2\right)^{4/3}}. \tag{4.12}$$

In Eq. (4.12), U_{10} is the mean wind speed at 10 m height, K_0 is the drag coefficient for the terrain surface, the value of which varies between 0.005 and 0.05, and x is a dimensionless parameter expressed as

$$x = L_s \, f/U_{10} \tag{4.13}$$

in which L_s denotes the length scale, the value of which is usually taken as 1200 m.

Another popular along-wind spectrum was proposed by Harris (1971), as described mathematically by the expression

$$S_u(f) = 4K_0 \frac{U_{10}^2}{f} \frac{x}{\left(2+x^2\right)^{5/6}}. \tag{4.14}$$

In the frequency range 0.1–1 Hz, Harris spectrum and Davenport spectrum almost overlap with each other (Kareem and Dalton, 1982). Subsequent studies highlighted that Davenport spectrum tends to somewhat overestimate the spectral density in the frequency range 0.1–0.5 Hz, which corresponds to the fundamental frequencies of many flexible civil engineering structures (Menon and Rao, 1997). This led to the development of several improved spectrum models. Simiu (1974) highlighted the dependency of spectral density on height, which was not taken into consideration by Davenport. The PSD of the along-wind spectra at frequency f and height z, as proposed by Simiu (1974) is

$$S_u(z,f) = \frac{K_0 U_{10}^2}{f} \frac{200\hat{x}}{\left(1+50\hat{x}\right)^{5/3}}$$

(4.15)

where

$$\hat{x} = \frac{fz}{U_z}.$$

(4.16)

Amongst some of the popularly used spectra to represent across-wind loading on a structure, the spectrum of vortex shedding for across-wind loading S_v, proposed by Vickery and Clark (1972) is one of the most well-established. It is expressed as

$$S_v(z,f) = \frac{C_L^2}{f} \frac{1}{B\sqrt{\pi}} \left[\frac{f}{f_{sh}}\right] \exp\left[-\left(\frac{1-\dfrac{f}{f_{sh}}}{B}\right)^2\right]$$

(4.17)

where C_L and B denote the rms of the total lift coefficient and spectral bandwidth parameter, respectively, and f_{sh} is the shedding frequency that is evaluated from

$$f_{sh} = \frac{S_t U_z}{D_z}.$$

(4.18)

In Eq. (4.18), S_t and D_z are the Strouhal number and the characteristic length of the structure at height z, respectively.

Another across-wind spectrum was put forward by Kaimal et al. (1972), which is reproduced in

$$S_v(z,f) = \frac{u_*^2}{f} \frac{17\hat{x}}{\left(1+9.5\hat{x}\right)^{5/3}}.$$

(4.19)

In Eq. (4.19), u_* denotes the friction velocity or shear velocity of the wind.

The across-wind spectrum developed by Ohkuma and Kanaya (1978), also well-used in structural analysis, is given by

$$S_v(z,f) = \frac{\sigma_v^2}{f} \frac{4B}{\pi} \frac{\left(x'/S_t\right)^2}{\left\{1-\left(x'/S_t\right)^2\right\}^2 + 4B^2\left(x'/S_t\right)^2}$$

(4.20)

where

$$x' = fD'/U_H.$$

(4.21)

In Eq. (4.21), D' and U_H represent the width of the structure perpendicular to the across-wind direction and the mean wind speed at the top of the structure, respectively.

One of the earliest detailed frequency domain studies on the structure-TLCD system was carried out by Xu et al. (1992b). They utilized the transfer matrix formulation to conduct a random vibration analysis of a multi-degree-of-freedom (MDOF) structural system with and without the TLCD (see Figure 4.4), under along-wind turbulence modelled as a stochastic process, which was non-homogeneous in space and stationary in time. In a separate work, they conducted a similar study for suppression of vibration of the MDOF structure against across-wind wake excitation (Xu et al., 1992a). For the along-wind loading in the frequency domain, the spectrum of longitudinal turbulence proposed by Davenport (1961) was considered, whereas for the across-wind loading, the Gaussian-type one-sided cross-wind force spectrum proposed by Vickery and Clark (1972) was utilized. In both studies, they reported that the performance of the TLCD in controlling wind-induced displacements as well as accelerations of the structural response is comparable to that of a traditional TMD. A comparison of the effectiveness of a TLCD and an equivalent TMD against along-wind and across-wind vibration of structures based on the work of Xu et al. (1992b) and Xu et al. (1992a) is presented in Table 4.2. Here, it may be observed that the efficiency of the TLCD as well as the TMD is higher in suppressing the across-wind response as compared to suppression of along-wind response, and this is more evident in case of the acceleration response reductions for the tall building. However, in the reduction of the displacement response of the tall building, both dampers are equally effective for along-wind and across-wind cases.

In another investigation into the effectiveness of the TLCD in controlling the wind-induced acceleration response of towers (Balendra et al., 1995), the two-sided Harris spectrum was considered to represent the along-wind loading on the structure.

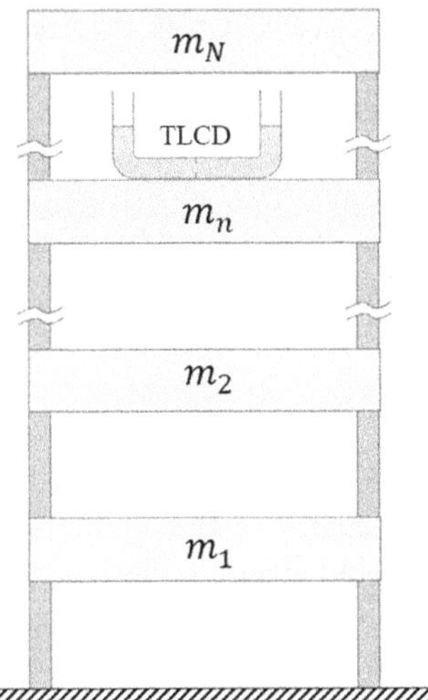

FIGURE 4.4 Multi-degree-of-freedom building structure with tuned liquid column damper at nth storey (Xu et al., 1992b).

TABLE 4.2

Comparison of Performance of TLCD and TMD in Controlling along-Wind and across-Wind Vibration of Two Example Structures

		TV Tower (Mass Ratio = 6%)		Tall Building (Mass Ratio = 3%)	
		Along-Wind (%)	Across-Wind (%)	Along-Wind (%)	Across-Wind (%)
Displacement response	by TLCD	39	48	37	38
reduction	by TMD	45	54	39	39
Acceleration response	by TLCD	35	44	21	53
reduction	by TMD	41	49	31	54

An important finding of the work was that the orifice opening ratio of the TLCD is required to be varied between 0.5 and 1.0, with smaller ratios applicable for shorter towers. Remarkably, it was demonstrated that even when the TLCD is not perfectly tuned, significant structural response reduction can be achieved by the control of the orifice opening ratio. A comparison of the normalized rms displacement of liquid in the vertical column of the TLCD derived from the original non-linear equation and the linearized equation of the damper revealed that the error induced due to the linearization of the damping term in the governing equation of the TLCD can be ignored in the practical design of the damper. In an extension of their earlier work, the same researchers studied wind-induced vibration control of different types of buildings using the TLCD. They considered a regular building, a building with a soft first storey, a tapered building and a building with a sudden mass-stiffness drop at the top one-fifth height of the building. These were designated as Models I to IV respectively. It was reported that for a given mass ratio, the acceleration response reductions achieved by the TLCD are comparatively less for Model III and IV due to a higher contribution of the second mode in the overall response of the structures (see Figure 4.5). In such cases, it was proposed to use two TLCDs, having a combined mass ratio equal to the mass ratio of the single TLCD used earlier, and tuning the dampers to the first two modes of the building.

Based on the results of wind tunnel experiments, a procedure for the analytical modelling of the vibration in a building with a TLCD, both in the along-wind and across-wind directions, was put forward. For the along-wind loading, Liu et al. (2003) used Davenport spectrum, while for the across-wind loading, the spectrum proposed by Ohkuma and Kanaya (1978) was considered appropriate. Results indicated that the efficiency of the TLCD in suppressing the along-wind response increases at higher wind velocity. Further, it was reported that the TLCD is highly effective in controlling the across-wind oscillation when the lock-in phenomenon occurs, that is, when the vortex shedding frequency is close to the natural frequency of the vibrating structure. Overall, it was observed that the TLCD is effective in suppressing both along-wind and across-wind vibrations, while the performance of the damper is better in the across-wind case. This corroborates the

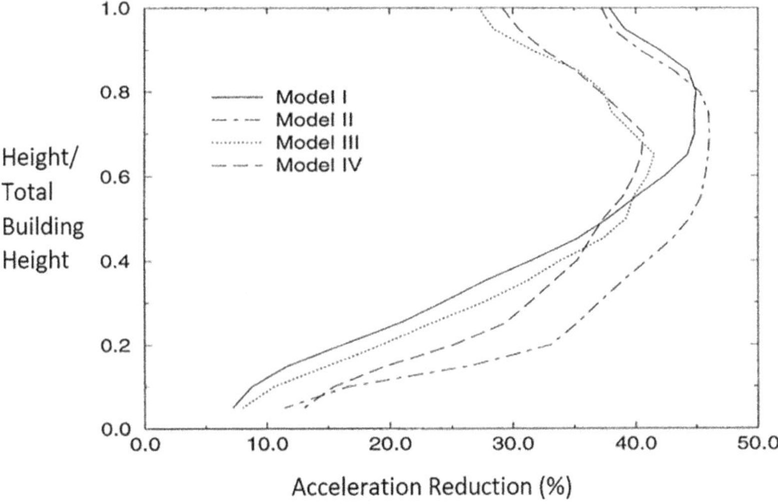

FIGURE 4.5 Variation of acceleration response reduction along the height of the buildings for different building models (Balendra et al., 1999).

findings of Xu et al. (1992b) and Xu et al., (1992a). Research findings have confirmed that even with a small mass ratio (say 1%), a TLCD can dissipate the structural response in the along-wind direction by up to two and a half times. This is valid for a length ratio of 0.7 (see Wu et al.(2005) for details).

Apart from buildings and tower-like structures, use of TLCDs have been studied for wind-induced vibration of long-span bridge decks. Research investigations carried out on bridges in the frequency domain are now discussed. Numerical analysis of both buffeting and flutter was carried out considering lateral, vertical as well as torsional vibration. Simiu spectrum was considered. The TLCD was found effective in reducing the buffeting response of the bridge deck, and here too the performance of the TLCD was reported to improve at higher wind speeds. Further, the presence of the TLCD increased the critical flutter wind velocity, thereby reducing the probability of the flutter instability of the bridge deck. It was observed that there is an optimum value of the orifice head-loss coefficient of the TLCD, for which critical flutter wind velocity increases the most. A detailed discussion on this is available in the work carried out by Suduo et al. (2002).

Multiple-TLCDs (MTLCDs) for mitigating the coupled lateral (along-wind) and torsional vibration of the deck of a long-span bridge have also been proposed (Shum and Xu, 2004). Two sets of TLCDs, with the first group tuned to the lateral frequency and the second group tuned to the torsional frequency of the bridge deck, were considered. In this study, von Kármán spectrum was used to characterize the wind load. Additionally, harmonic excitation and white noise excitation were considered. Under all loadings, the MTLCD could effectively reduce the coupled lateral and torsional vibration. The mathematical model included the provision of considering the motion-induced aeroelastic effect, which may be instigated in long-span bridges due to the interaction between fluctuating wind and bridge motion. It was observed that the consideration of the aeroelastic effects reduces the efficiency of the MTLCD system as compared to the case when aeroelastic effects are neglected. Significantly, when the aeroelastic effect is considered, the bridge response reduction by the TLCD remains almost unchanged with variation in the mean wind speed. This is attributed to changes in the aeroelastic damping of the bridge caused due to the changing mean windspeed.

Further, the TLCD in the form of the LCVA has been found successful in the control of wind-induced pitching motion or torsional motion of a bridge deck (Wu et al., 2008). In this work, the wind load causing the pitching motion was idealized as a harmonic excitation and optimal design of the damper was carried out.

It may be observed from the deliberations in this section that different wind spectra have been used by researchers. Thus, for the reference of future researchers in this field, a summary in this regard is presented in Table 4.3.

4.3 TIME DOMAIN STUDIES

Structure-TLCD systems have also been studied in time domain under wind loading. Generally, time domain studies provide more accurate solutions as compared to frequency domain studies, but at the expense of greater computational time (Brekke et al., 2005). However, wind excitation is random in nature, and multiple sets of time domain studies are essential to reach any generalized conclusion. Equations (2.12) and (2.14) govern the dynamics of an SDOF structure with attached TLCD in the time domain. These equations may be utilized for time domain study of a structure-TLCD system by replacing the generalized time-dependent force acting on the structure, $F(t)$, with the time-dependent wind force given in Eq. (4.2). The time-dependent wind force may be calculated using an actual record of wind speed time history or a synthetically generated wind speed time history. Unlike recorded seismic accelerograms of past events, open-source availability of actual records of wind speed time histories is quite limited. However, several simulation schemes are available that can synthetically generate wind-related data required for time domain numerical analysis of structures. One of the ways to synthetically generate wind-related data is to use the web-based NatHaz online wind simulator (NOWS), developed by the NatHaz Modeling Laboratory (Kwon and Kareem, 2006).

TABLE 4.3

Wind Spectra Considered by Researchers for Frequency Domain Studies on Structural Vibration Control Using TLCD

References	Wind Spectrum Considered	Type of Wind Loading
Xu et al. (1992b)	Davenport spectrum	Along-wind
Xu et al. (1992a)	Vickery and Clark spectrum	Across-wind
Balendra et al. (1995)	Harris spectrum	Along-wind
Liu et al. (2003)	Davenport spectrum	Along-wind
	Ohkuma and Kanaya spectrum	Across-wind
Shum and Xu (2004)	von Kármán spectrum	Along-wind
	White noise	

NOWS is available at *http://windsim.ce.nd.edu*, and it has a user-friendly interface where both metric and British units can be used for input and output. NOWS considers the exposure conditions as per ASCE 7-98. It allows users to choose among four available simulation schemes, namely discrete frequency function with Cholesky decomposition and FFT, Schur decomposition by AR model and polynomial approximation, ergodic spectral representation with Cholesky decomposition and FFT and conventional spectral representation method. An example wind velocity time history generated by NOWS at 10 m height is shown in Figure 4.6. For this particular case, exposure condition B and 3 s gust wind speed equal to 40 m/s have been considered, and the simulation scheme used is the discrete frequency function with Cholesky decomposition and FFT.

Probably, the earliest extensive time domain study on the performance assessment of the TLCD was taken up by Min et al. (2005), who considered a 76-storey benchmark building as the structure to be controlled. The damper was designed to mitigate only the across-wind directional responses, and the across-wind data provided by Yang et al. (2004) from wind tunnel tests were used for the design. Different sets of single objective parametric optimization of the TLCD were carried out to minimize the rms displacement and acceleration responses of the structure. The optimized TLCD achieved significant response reduction of the benchmark structure. The study was also extended to an MTLCD system, wherein it was found that the performance of the MTLCD is almost equivalent to that of a single TLCD when the structural parameters have a fixed value. However, when stiffness uncertainty exists in the structure, the MTLCD outperforms the single TLCD. In addition to actual wind load data, the behaviour of the structure-damper system considering harmonic wind loading and white noise excitation was studied. The optimum parameters of the damper, such as tuning ratio and head-loss coefficient, obtained under the measured wind loading were found to differ significantly from the optimum values determined for harmonic and white noise excitations. The difference is more when the mass ratio is higher.

Suthar and Jangid (2022) considered the same benchmark building and loading conditions as Min et al. (2005) in their study on across-wind response reduction by the TLCD. However, in contrast to the work of Min et al. (2005), the objective

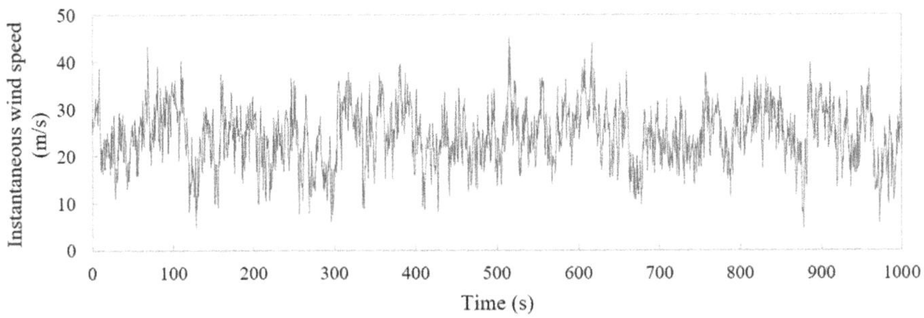

FIGURE 4.6   Time history of instantaneous wind speed generated by NatHaz online wind simulator (NOWS).

here was to minimize the absolute peak acceleration at the top floor of the building through a nonlinear constraint optimization technique, and the study was limited to a single TLCD system. A comparison of the top floor acceleration response time histories of the benchmark building, with and without the optimal TLCD, is reproduced in Figure 4.7. It is evident that the optimal TLCD effectively reduces the structural acceleration response. The effect of off-tuning due to uncertainties in the stiffness on the damper performance was also investigated. The optimal TLCD was found to be robust in acceleration control, while the displacement control capacity of the damper suffered significant deterioration under stiffness uncertainty.

In recent years, rigorous research efforts have been devoted to the control of wind-induced vibration of special structures, such as onshore and offshore wind turbines. A review on the topic is available in Awada et al. (2021). The application of the TLCD has been investigated for the vibration control of the wind turbine tower or the blades of the wind turbine (see Figure 4.8a and b). The pioneering work on the use of the TLCD in wind turbines for structural control was carried out by Colwell and Basu (2006, 2009). An offshore turbine tower was modelled as an MDOF system. The Kaimal spectrum was used to generate a time series for wind excitation. The MDOF structure was subjected to both "moderate" and "strong" wind. The response of the wind turbine under combined wind and wave loading was also examined. The JONSWAP wave spectrum was used to generate the wave loading. For a 1% mass ratio of the TLCD, the damper achieved a 55% reduction in structural displacement response and a 44% reduction in bending moments at the base of the wind turbine. A fatigue analysis of the structure was also carried out, and more than three and a half times enhancement in the fatigue life of the structure was reported due to the installation of the TLCD.

The study on wind turbine towers with TLCD by Mensah and Dueñas-Osorio (2014) demonstrated a 47% reduction in peak structural displacements by a TLCD with 1%

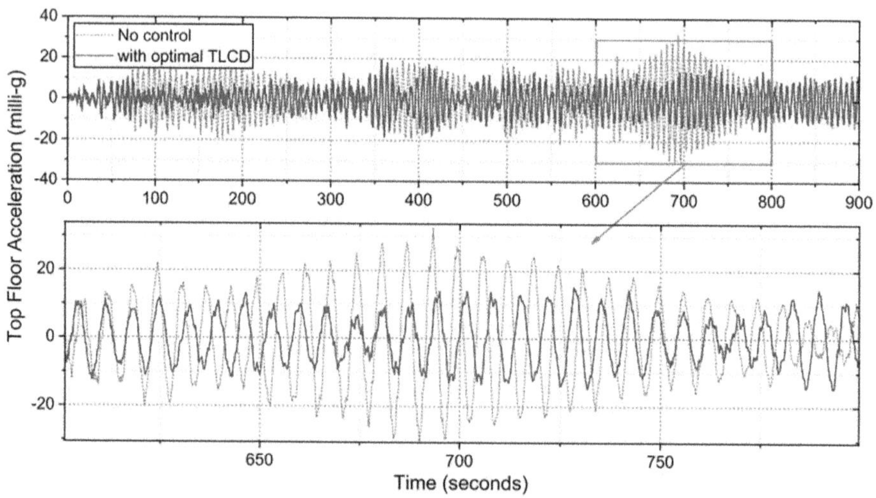

FIGURE 4.7 Comparison of top floor acceleration time histories of benchmark building with and without optimal tuned liquid column damper (Suthar and Jangid, 2022).

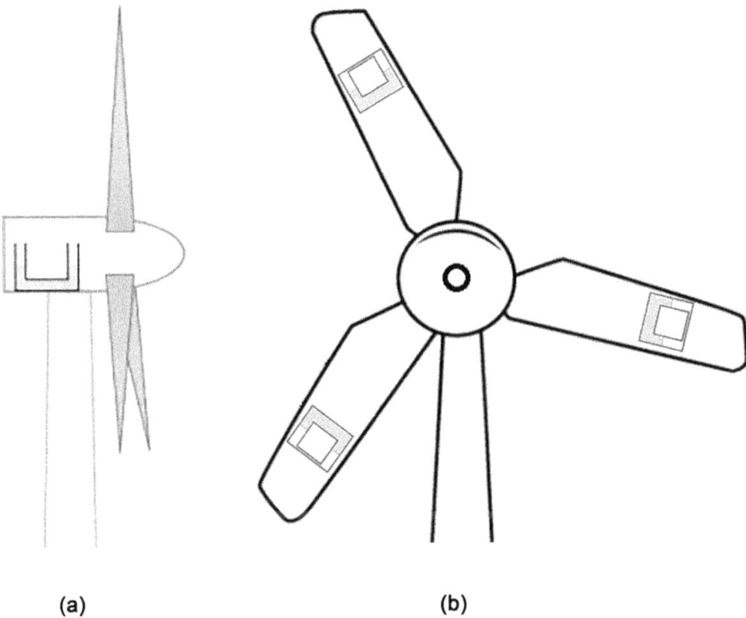

FIGURE 4.8 Application of tuned liquid column dampers in wind turbine for vibration control of (a) tower and (b) blades.

mass ratio, which is quite close to what was reported by Colwell and Basu (2009). Here, a modified version of the von Kármán spectrum was used to synthetically generate the wind load. A case with two TLCDs having a total mass ratio of 1.5% was also considered, which achieved only a marginal gain in peak structural displacement reduction. Thus, it was concluded that a single damper with increased mass is a more effective configuration than the two-TLCD system. It was indicated through fragility analysis that the reliability of the wind turbine towers gains up to 8% and 11% due to the installation of a single TLCD having a mass ratio of 1% and 1.5%, respectively.

Edgewise vibration mitigation of rotating wind turbine blades has been a focus of researchers in recent years due to low and even negative aerodynamic damping (Staino et al., 2012). The application of TLCD in this regard has been studied by researchers, such as Zhang et al. (2015) and Basu et al. (2016). In Zhang et al. (2015), as the TLCD container was mounted inside a rotating blade with a changing azimuthal angle, an extra slim tube connecting the two vertical limbs of the damper was considered (see Figure 4.9). This closed configuration of the TLCD prevented liquid from leaking out of the damper container and balanced the pressure above the liquid column during oscillation. The damper performance was evaluated under different turbulence intensities and rotational speeds of the rotor. The numerical study revealed that a TLCD having a mass ratio of 3% can reduce the maximum edgewise tip displacement of the rotating blades by 20%. As wind turbine blades are hollow and lightweight, the mass ratio of 3% translated into an actual mass of the TLCD of only 42.36 kg. The parametric study revealed that an improved control performance can be obtained by increasing the mass ratio and locating the damper closer to the blade tip.

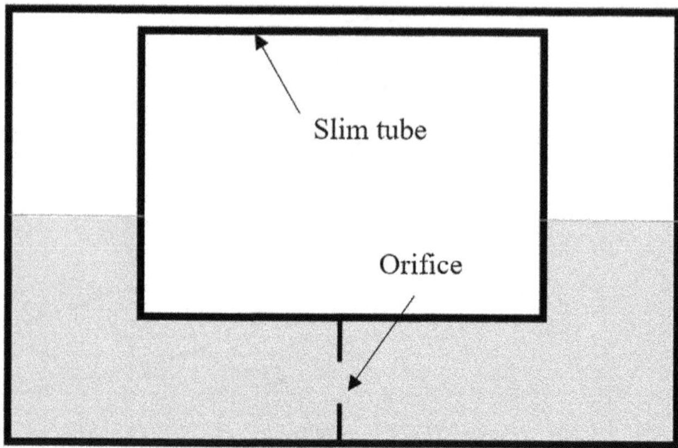

FIGURE 4.9 Tuned liquid column damper used by Zhang et al. (2015) for edgewise vibration mitigation of rotating wind turbine blades.

However, it is to be borne in mind that the damper mass ratio and the mounting position should be limited according to the installation capacity and the available space inside the blade. The study further indicated the independence of the optimal tuning ratio of the TLCD from the turbulence intensity. On the other hand, with the variation of turbulence intensity as well as the rotational speed of the rotor, the optimal orifice head-loss coefficient and subsequently the damper effectiveness varied significantly. This made it difficult to use a passive TLCD for edgewise vibration control of rotating wind turbine blades. Rather, a semi-active TLCD with varying head-loss coefficient would be more suitable. However, to push the state-of-the-art one step forward from a theoretical point of view, Chen et al. (2022) developed closed-form formulas for optimal calibration of the passive TLCD installed in a rotating blade for suppressing edgewise vibrations. A variant of the TLCD, the circular liquid damper, has also been developed by Basu et al. (2016) for edgewise vibration in wind turbine blades.

A combination of TMD and TLCD for suppressing the wind-induced vibration of a wind turbine tower with a jacket sub-structure has been examined (Hemmati et al., 2019). The structural model was subjected to stochastically generated wind loading based on Kaimal spectrum in operational, parked, startup and shutdown conditions. The TMD was found to be more efficient in operational conditions, whereas the TLCD performed better in the parked condition. In the shutdown condition, the TLCD aggravated the structural response as compared to the uncontrolled case. Notably, both at startup and shutdown conditions, the optimal TLCD-TMD combination performed better than when only the TLCD or TMD was used.

4.4 EXPERIMENTAL STUDIES

Due to the complex nature of wind loading and wind-induced structural response, wind tunnel tests are conducted to verify numerical computations and identify the scope of possible improvement in design. Wind tunnel tests provide a better understanding of the nature of airflow over and around a structure, as well as the effects it causes on that structure, especially the effects of aerodynamic forces. Further, due to the scarcity of land, many new structures are being designed for complex situations and surroundings. For such situations, wind tunnel tests are a necessity.

Broadly, two approaches are followed to conduct wind tunnel based experimental studies to evaluate the performance of TLCDs in suppressing wind-induced structural vibrations. In the first approach, in the wind tunnel test, an appropriately scaled TLCD model is installed in the scaled-down model of the structure, and the performance of the damper is estimated from the test results. In the second approach, a wind tunnel test is conducted on a scaled-down model of the structure, without any damper, to estimate the wind-induced forces and moments on the uncontrolled structure and to determine the maximum displacement and acceleration induced by wind. The TLCD is then analytically designed to reduce the structural response quantities to within the acceptable limit. This is followed by a shake table test on a scaled-down model of the designed TLCD or by a computational fluid dynamics (CFD) analysis on a full-scale model of the TLCD, to verify the key assumptions made in the analytical design. The first approach is suitable for studying the general behaviour of the TLCD or for a parametric study, while the second approach is more suitable for the design of the TLCD for real-life installations.

So far as the wind tunnel test of a structure-TLCD system is concerned, the earliest study was carried out by Liu et al. (2003). They followed the first approach. The test was conducted on a 1:200 scale model of a tall building (0.2 m × 0.1 m × 0.75 m). To study the effect of mass ratio on the damper performance, TLCDs of different sizes were made of acrylic plates and fixed on the scaled-down model of the structure. The tuning ratio and the length ratio of the TLCD were considered as 1 and 0.65, respectively. Both along-wind and across-wind vibrations of the structure were monitored during the test. The experimental along-wind spectra and across-wind spectra agreed well with Davenport and Ohkuma and Kanaya spectra, respectively. The experimentally obtained reductions in structural displacement and acceleration response validated the theoretical results. The trend of the wind tunnel test results was found to be consistent with the analytical results presented by Xu et al. (1992a,b).

Through wind tunnel experiments on a 1:200 scaled-down aeroelastic model of a tall building, You et al. (2012, 2013, 2014) investigated the effectiveness of a TLCD in mitigating the across-wind excitation of the building. TLCD models of mass ratio of 1.5% and 3%, respectively, were designed and attached to the topmost floor of the building model for testing. The experimental wind load spectra were very close to the von Karman spectra. They observed that the response reduction by the damper increases with an increase in mass ratio and length ratio while an increase in structural damping has a negative effect on the TLCD effectiveness.

Another experimental study involved in the design of the LCVA for the Songdo First World located in Incheon, South Korea, has been presented by Kim et al. (2008). The project consists of four 237 m tall identical buildings. Wind tunnel tests using high-frequency force balance technique were carried out to estimate the wind-induced accelerations at the top occupied floor of the buildings, which were found to be higher than the target values. In order to meet the target criteria for wind-induced motion along both principal directions, two perpendicular LCVAs were designed for each building. The frequencies of the partially completed buildings were measured on site when about 90% of the structural works was completed. The measured and the analytical frequencies of the partially completed buildings were compared, and the natural frequencies of the completed buildings were estimated based on the results of this comparison. The design of the LCVAs was modified based on the estimated frequencies of the buildings. Shake table tests were conducted on a 1:20 scaled-down model of the structure-LCVA system with the structure modelled as an SDOF system corresponding to the first mode of vibration. The shake table test results indicated that the LCVA increases the energy dissipation capacity of the building significantly.

Based on wind tunnel results, a pair of identical TLCDs was designed for a 180 m tall residential building located in the Middle East, to reduce the wind-induced acceleration of the highest occupied floors perpendicular to the weaker axis of the building to within the standard occupant comfort criteria (see Figure 4.10). The details of the design are available in Cammelli et al. (2016). Instead of a conventional orifice, the TLCDs were provided with multiple porous baffles (similar to lattice screens) in the horizontal limbs of the damper to generate greater damping. The full-scale measurements of frequencies and inherent damping of the structure took place at near-completion stage, and subsequent adjustment in the damper design was carried out. A 1:20 model of the TLCD was tested on a shake table. The motion of the shake table was programmed according to the first modal vibration of the actual building computed based on wind tunnel measurements. From the shake table test, the added damping by the TLCD for different cases with change in the number of baffles and their porosity was determined, and the optimum baffle configuration (three baffles with 75% porosity) was chosen. Further, to study the scale effects on the results of the shake table tests, a CFD analysis of the full-scale TLCDs was undertaken. The CFD analysis indicated about a 10% decrease in the damper effectiveness from what was determined through the shake table test. Despite that, the pair of TLCDs achieved about 30% improvement in the performance of the building perpendicular to the weaker axis against wind excitation. Later, Cammelli and Li (2016) also designed a TLCD for a 220 m tall building located within a large city on the East Coast of the USA using the above procedure.

FIGURE 4.10 Close-up of wind tunnel model of 180 m tall residential building used by Cammelli et al. (2016) for design of TLCD.

4.5 OPTIMALITY ISSUES

It can be recalled from Section 2.3 of Chapter 2 that the parameters that primarily influence the performance of a TLCD are the mass ratio μ, the length ratio α, the coefficient of head-loss of the orifice ξ and the tuning ratio $\gamma = \omega_{\text{TLCD}}/\omega_s$. As discussed earlier in Section 2.3.1 of Chapter 2, the mass ratio is normally determined based on practical constraints, such as the availability of space and the restriction on the additional dead load on the structure. Thus, there is practically no scope to provide optimum μ in a damper designed for a civil engineering structure (Konar and Ghosh, 2023). Further, the optimum value of the length ratio of a TLCD may be determined from Eq. (2.41), which ensures the condition of maintaining positive liquid height in the vertical limbs of the damper during vibration. This is essential for maintaining the validity of the governing equations of the structure-TLCD system. Thus, the coefficient of head-loss of the orifice ξ and the tuning ratio γ are the only two parameters that are more rigorously studied with the aim to provide an optimum performance of a TLCD.

The optimum values of the design parameters are dependent on the input excitation. A simple approach for finding optimal parameters considers the excitation as a harmonic force. An early detailed study by Gao et al. (1997) on the optimality issues of TLCDs assumed a harmonic external excitation on an SDOF system model of the structure. It was seen that the optimum value of the orifice head-loss coefficient ξ_{opt} increases with an increase in both the mass ratio of the damper and the structural damping ratio. The optimum value of the tuning ratio γ_{opt} is very close to unity when both the mass ratio of the damper and the structural damping ratio are low (e. g., $\mu = 1\%$ and $\zeta_s = 1\%$). However, as μ and/ or ζ_s increases, the value of γ_{opt} reduces from unity.

Idealization of wind load on a structure as a white noise excitation is a popular approach in the determination of the optimal damper parameters. Chang and Hus (1998) followed this approach, and they modelled the structure as an SDOF system. Their work indicated that ξ_{opt} decreases with the increase in the spectral intensity of the white noise loading and that the effectiveness of the TLCD increases with the increase in the loading intensity in the range considered. A subsequent study by Chang (1999) also confirmed the same trend.

For an SDOF undamped primary structure subjected to white noise excitation, the following closed-form solution for the optimum tuning ratio, γ_{opt}, that minimizes the rms displacement of the structure was given by Yalla and Kareem (2000) as

$$\gamma_{\text{opt}} = \frac{\sqrt{1 + \mu\left(1 - \dfrac{\alpha^2}{2}\right)}}{(1+\mu)}. \tag{4.22}$$

The expression for the optimum equivalent linearized damping ratio of the TLCD was also derived and from that the expression for the equivalent linearized damping coefficient $C_{p,\,\text{opt}}$ is given by

$$\frac{C_{p,\,\text{opt}}}{\omega_{\text{TLCD}}L} = \frac{\alpha}{2}\sqrt{\frac{2\mu\left(\alpha^2\dfrac{\mu}{4} - \mu - 1\right)}{\left(\alpha^2\mu^2 + \alpha^2\mu - 4\mu - 2\mu^2 - 2\right)}} \tag{4.23}$$

where ω_{TLCD} and L are the frequency of the TLCD and the overall length of the liquid column, respectively.

The optimum value of the orifice head-loss coefficient, ξ_{opt}, given by Yalla and Kareem (2000), may be evaluated from

$$\xi_{\mathrm{opt}} = \mu^{1.25} \sqrt{\frac{\left(1+\mu-\alpha^2\mu\right)}{S_0}\left(\frac{\mu+\alpha^2}{1+\mu}\right)^{1.5}} \, gL\omega_{\mathrm{TLCD}} \qquad (4.24)$$

where g is the gravitational acceleration, and S_0 is the intensity of white noise excitation.

With the same assumptions as Yalla and Kareem (2000), Wu et al. (2005) too developed expressions for γ_{opt} and ξ_{opt}. The expression for γ_{opt} presented by Wu et al. (2005) is identical to Eq. (4.22), while the expression for ξ_{opt} proposed by them is

$$\xi_{\mathrm{opt}}\sigma_{\dot{y}'} = 2^{0.5}\pi^{1.5}\frac{1}{\alpha}\sqrt{\frac{\mu\left(\mu+\alpha^2\right)}{1+\mu}} \qquad (4.25)$$

where $\sigma_{\dot{y}'}$ is the standard deviation of $\dot{y}'$, which is the differentiation of the non-dimensional liquid displacement $y' = y/b$, with respect to the non-dimensional time $t' = t/T_{\mathrm{TLCD}}$. T_{TLCD} is the natural period of the TLCD.

The variations of γ_{opt}, $(C_{p,\,\mathrm{opt}}/\omega_{\mathrm{TLCD}}L)$ and $(\xi_{\mathrm{opt}}\sigma_{\dot{y}'})$, calculated using Eqs. (4.22), (4.23) and (4.25) respectively, for a practically feasible range of α and μ are shown in Figures 4.11–4.13 respectively. Figure 4.11 demonstrates that γ_{opt} has a value close to unity when both α and μ are small. However, with the increase in α as well as μ, the value of γ_{opt} decreases from unity. On the other hand, it may be observed from Figures 4.12 and 4.13 that $C_{p,\,\mathrm{opt}}/\omega_{\mathrm{TLCD}}L$ and $\xi_{\mathrm{opt}}\sigma_{\dot{y}'}$ follow a similar trend as both increase with an increase in α as well as μ.

Here, it is pertinent to mention that Eqs. (4.22)–(4.25) may be used to obtain only an approximate value of the optimal TLCD parameters for real-life civil engineering structures where the structural damping ratio typically varies between 0.5% and 5%, depending on the type of structural construction, structural material and level of vibration in the structure. To obtain an accurate estimate of the optimal TLCD parameters, the primary system damping, ζ_s, must be included in the optimization study. It may be noted that both Yalla and Kareem (2000) and Wu et al. (2005) extended their studies to damped structures. In case of damped structures, as close-form solutions are not possible, the optimum parameters of the damper are determined numerically. When the structural damping ratio is within 5% and the excitation is white noise, ξ_{opt} remains practically unaltered from what has been obtained for the undamped case (Yalla and Kareem, 2000), while the variation in γ_{opt} is also very small from undamped case (see Figure 4.14a and b).

To study the effect of actual wind loading on the optimal parameters of the TLCD, Alkmim et al. (2017) carried out a detailed study in which the primary structure was subjected to stochastic load generated from different wind power spectral density

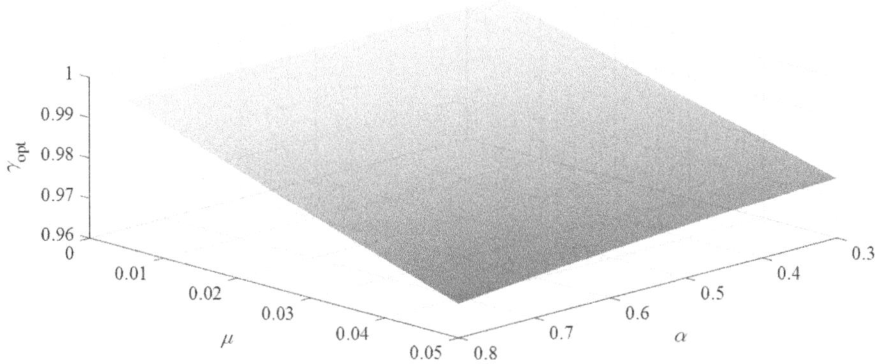

FIGURE 4.11 Variation of γ_{opt} with α and μ.

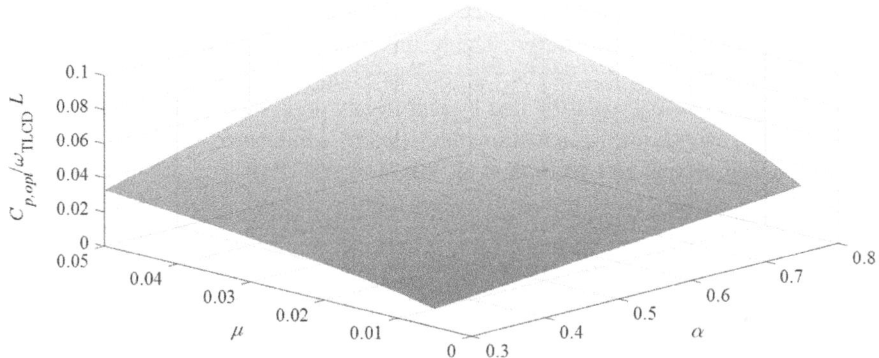

FIGURE 4.12 Variation of $\dfrac{C_{p,\text{opt}}}{\omega_{\text{TLCD}}L}$ with α and μ.

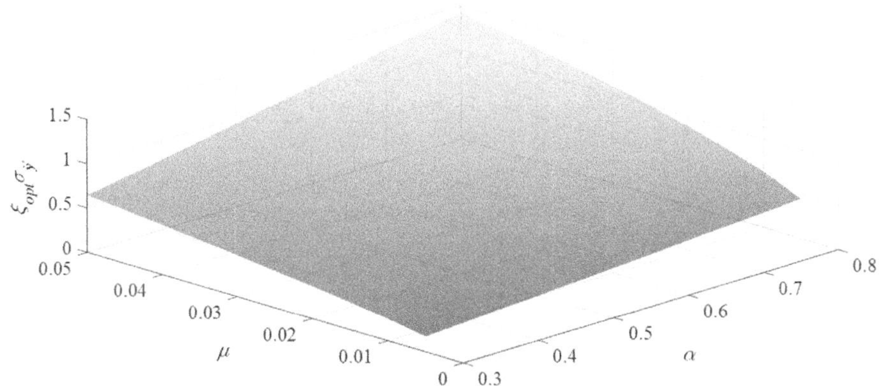

FIGURE 4.13 Variation of $\xi_{\text{opt}}\sigma_{\dot{y}'}$ with α and μ.

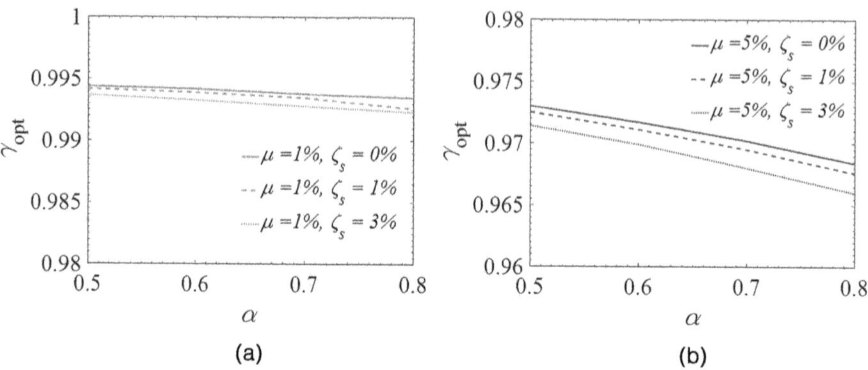

FIGURE 4.14 Variation of γ_{opt} with α for undamped and lightly damped primary structure for (a) $\mu = 1\%$ and (b) $\mu = 5\%$.

functions. A classical direct search method, namely the generalized pattern search, was employed for the optimization. The results indicate that different wind profiles can significantly affect γ_{opt}, while $C_{p,\,\mathrm{opt}}$ is negligibly influenced by the selection of the wind profile. Figure 4.15a and b respectively depict the variation of γ_{opt} and $C_{p,\,\mathrm{opt}}/\omega_{\mathrm{TLCD}}L$ with μ, for wind loading modelled by white noise, Kaimal, Davenport and Kanai-Tajime spectra (Alkmim et al., 2017).

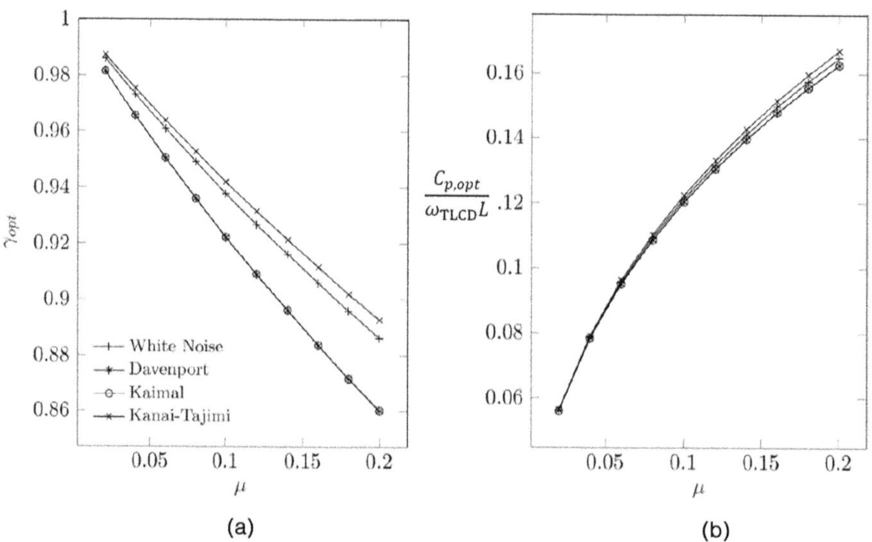

FIGURE 4.15 Optimum tuned liquid column damper parameters under different wind spectra for $\alpha = 0.8$, $\zeta_s = 1\%$ and different values of μ (a) γ_{opt} and (b) $C_{p,\,\mathrm{opt}}/\omega_{\mathrm{TLCD}}L$ (Alkmim et al., 2017).

REFERENCES

Alkmim, M.H., De Morais, M.V.G., Fabro, A.T., 2017. Optimum parameters of a tuned liquid column damper in a wind turbine subject to stochastic load. *IOP Conf. Ser. Mater. Sci. Eng.* 280, 012007. https://doi.org/10.1088/1757-899X/280/1/012007

ASCE/SEI49-12, 2012. *Wind Tunnel Testing for Buildings and Other.* American Society of Civil Engineers, Reston, VI.

ASCE 7, 2002. Minimum design loads for buildings and other structures, *ASCE* 7. https://doi.org/10.1520/G0154-12A

Awada, A., Younes, R., Ilinca, A., 2021. Review of vibration control methods for wind turbines. *Energies* 14, 1–35. https://doi.org/10.3390/en14113058

Balendra, T., Wang, C.M., Cheong, H.F., 1995. Effectiveness of tuned liquid column dampers for vibration control of towers. *Eng. Struct.* 17, 668–675. https://doi.org/10.1016/0141-0296(95)00036-7

Balendra, T., Wang, C.M., Rakesh, G., 1999. Vibration control of various types of buildings using TLCD. *J. Wind Eng. Ind. Aerodyn.* 83, 197–208. https://doi.org/10.1016/S0167-6105(99)00072-0

Basu, B., Zhang, Z., Nielsen, S.R.K., 2016. Damping of edgewise vibration in wind turbine blades by means of circular liquid dampers. *Wind Energy* 19, 213–226. https://doi.org/10.1002/we.1827

Brekke, J., Chakrabarti, S., Halkyard, J., 2005. Drilling and production risers, in: Chakrabarti, S. (Ed.), *Handbook of Offshore Engineering.* Elsevier Ltd, London, pp. 709–859. https://doi.org/10.1016/B978-0-08-044381-2.50016-3

Cammelli, S., Li, Y.F., 2016. Experimental and numerical investigation of a large tuned liquid column damper. *Proc. Inst. Civ. Eng. Struct. Build.* 169, 635–642. https://doi.org/10.1680/jstbu.14.00129

Cammelli, S., Li, Y.F., Mijorski, S., 2016. Mitigation of wind-induced accelerations using Tuned liquid column dampers : experimental and numerical studies. *J. Wind Eng. Ind. Aerodyn.* 155, 174–181. https://doi.org/10.1016/j.jweia.2016.06.002

Chang, C.C., 1999. Mass dampers and their optimal designs for building vibration control. *Eng. Struct.* 21, 454–463. https://doi.org/10.1016/S0141-0296(97)00213-7

Chang, C.C., Hus, C.T., 1998. Control performance of liquid column vibration absorbers. *Eng. Struct.* 20, 580–586. https://doi.org/10.1016/S0141-0296(97)00062-X

Chen, B., Zhang, Z., Hua, X., Basu, B., 2022. Optimal calibration of a tuned liquid column damper (TLCD) for rotating wind turbine blades. *J. Sound Vib.* 521, 116565. https://doi.org/10.1016/j.jsv.2021.116565

Colwell, S., Basu, B., 2009. Tuned liquid column dampers in offshore wind turbines for structural control. *Eng. Struct.* 31, 358–368. https://doi.org/10.1016/j.engstruct.2008.09.001

Colwell, S., Basu, B., 2006. Tuned liquid column dampers in offshore wind turbines, in: *Proceedings of the 5th World Wind Energy Conference*, New Delhi, India.

Davenport, A.G., 1961. The application of statistical concepts to the wind loading of structures. *Proc. Inst. Civ. Eng.* 19, 449–472. https://doi.org/10.1680/iicep.1961.11304

Dong, Y., Huang, X., Liang, W., Li, D., 2020. Mode-superposition response spectrum method for wind-induced vibration analysis of structures. *Vibroeng. Proc.* 33, 113–117. https://doi.org/10.21595/vp.2020.21690

Gao, H., Kwok, K.C.S., Samali, B., 1997. Optimization of tuned liquid column dampers. *Eng. Struct.* 19, 476–486. https://doi.org/10.1016/S0141-0296(96)00099-5

Goto, T., 1983. Studies on wind-induced motion of tall buildings based on occupants' reactions. *J. Wind Eng. Ind. Aerodyn.* 13, 241–252. https://doi.org/10.1016/0167-6105(83)90145-9

Goto, T., 1975. The criteria to motions in tall building: (part 1) factors affecting human perception and tolerance of motion. *Trans. Archit. Inst. Jpn.* 237, 109–119. https://doi.org/10.3130/aijsaxx.237.0_109

Harris, R.I., 1971. The nature of the wind, in: *The Modern Design of Wind Sensitive Structures*. Construction Industry Research and Information Association, London, UK, pp. 29–56.

Hemmati, A., Oterkus, E., Khorasanchi, M., 2019. Vibration suppression of offshore wind turbine foundations using tuned liquid column dampers and tuned mass dampers. *Ocean Eng.* 172, 286–295. https://doi.org/10.1016/j.oceaneng.2018.11.055

Irwin, A.W., 1978. Human response to dynamics motion of structures. *Struct. Eng.* 237–244.

Jacobs, W.P., 2008. Building periods: moving forward (and backward). *Structure*, 24–27.

Kaimal, J.C., Wyngaard, J.C., Izumi, Y., Coté, O.R., 1972. Spectral characteristics of surface-layer turbulence. *Q. J. R. Meteorol. Soc.* 98, 563–589. https://doi.org/10.1002/qj.49709841707

Kareem, A., Dalton, C., 1982. Dynamic effects of wind on tension leg platforms, in: *14th Annual Offshore Techonology Conferance*. Houston, USA, pp. 749–755.

Kármán, T. von, 1948. Progress in the statistical theory of turbulence. *Proc. Natl. Acad. Sci. U. S. A.* 34, 530–539. https://doi.org/10.1073/pnas.34.11.530

Kim, H., Kim, W., Lee, S.G., Jo, J.S., Kim, D., Lee, K.J., 2008. Mitigation of wind-induced vibration with liquid column vibration absorber. *Adv. Sci. Technol.* 56, 259–264. https://doi.org/10.4028/www.scientific.net/AST.56.259

Kitazawa, M., Nishimori, K., Noguchi, J., Shimoda, I., 1992. Earthquake resistant design of a long-period cable-stayed bridge, in: *10th World Conferance on Earthquake Engineering*, Madrid, Spain. pp. 4797–4802.

Konar, T., 2023. Design of intermediate - depth tuned liquid damper with horizontal baffles for seismic control and carbon footprint reduction of buildings. *J. Vib. Eng. Technol.* https://doi.org/10.1007/s42417-023-01005-4

Konar, T., Das, A., Ghosh, A.D., 2024a. Near-fault earthquake-induced vibration control of a supertall benchmark building using an inerter-assisted compliant liquid damper. *ASCE-ASME J. Risk Uncertain. Eng. Syst. Part A Civ. Eng* 10 (4), in press.

Konar, T., Ghosh, A.D., 2023. A review on various configurations of the passive tuned liquid damper. *J. Vib. Control* 29, 1945–1980. https://doi.org/10.1177/10775463221074077

Konar, T., Ghosh, A.D., 2010. Passive control of seismically excited structures by the liquid column vibration absorber. *Struct. Eng. Mech.* 36, 561–573. https://doi.org/10.12989/sem.2010.36.5.561

Konar, T., Ghosh, A.D., Basu, B., 2024b. Real-world installations of tuned liquid column dampers for wind- induced vibration control of buildings: some important case studies. *Struct. Infrastruct. Eng.* in press.

Kwok, K.C.S., 2013. Wind-induced vibrations of structures: with special reference to tall building aerodynamics, in: Tamura, Y., Kareem, A. (Eds.), *Advanced Structural Wind Engineering*. Springer, Tokyo, Japan.

Kwon, D., Kareem, A., 2006. *NatHaz On-Line Wind Simulator (NOWS) : Simulation of Gaussian Multivariate Wind Fields*. NatHaz Modeling Laboratory Report, Univ. of Notre Dame.

Lamb, S., Kwok, K.C.S., 2019. The effects of motion sickness and sopite syndrome on office workers in an 18-month field study of tall buildings. *J. Wind Eng. Ind. Aerodyn.* 186, 105–122. https://doi.org/10.1016/j.jweia.2019.01.004

Liu, M.Y., Chiang, W.L., Chu, C.R., Lin, S.S., 2003. Analytical and experimental research on wind-induced vibration in high-rise buildings with tuned liquid column dampers. *Wind Struct.* 6, 71–90. https://doi.org/10.12989/was.2003.6.1.071

Mendis, P., Ngo, T., Haritos, N., Hira, A., Samali, B., Cheung, J., 2007. Wind loading on tall buildings. *Electron. J. Struct. Eng.* 7, 41–54. https://doi.org/10.56748/ejse.641

Menon, D., Rao, P.S., 1997. Estimation of along-wind moments in RC chimneys. *Eng. Struct.* 19, 71–78. https://doi.org/10.1016/S0141-0296(96)00041-7

Mensah, A.F., Dueñas-Osorio, L., 2014. Improved reliability of wind turbine towers with tuned liquid column dampers (TLCDs). *Struct. Saf.* 47, 78–86. https://doi.org/10.1016/j.strusafe.2013.08.004

Min, K., Kim, Hyoung-seop, Lee, S., Kim, Hongjin, 2005. Performance evaluation of tuned liquid column dampers for response control of a 76-story benchmark building. *Eng. Struct.* 27, 1101–1112. https://doi.org/10.1016/j.engstruct.2005.02.008

Newland, D.E., 1993. *An Introduction to Random Vibrations, Spectral and Wavelet Analysis.* Longman, New York and London.

Ohkuma, T., Kanaya, A., 1978. On the correlation between the shape of rectangular cylinders and characteristics of fluctuating lifts on them, in: *5th Symposium on Wind Effects on Structures.* Tokyo, Japan, pp. 147–154.

Sakai, F., Takaeda, S., Tamaki, T., Takeda, S., Tamaki, T., 1989. Tuned liquid column dampers— new type device for suppression of building vibrations, in: *Proceedings of International Conference on High Rise Buildings.* Nanjing, China, pp. 926–931.

Shum, K.M., Xu, Y.L., 2004. Multiple tuned liquid column dampers for reducing coupled lateral and torsional vibration of structures. *Eng. Struct.* 26, 745–758. https://doi.org/10.1016/j.engstruct.2004.01.006

Simiu, E., 1974. Wind spectra and dynamic alongwind response. *J. Struct. Div.* 100, 1897–1910. https://doi.org/10.1061/jsdeag.0003880

Smith, B.S., Coull, A., 1991. *Tall Building Structures: Analysis and Design.* John Wiley & Sons, Inc., New York.

Smith, R.J., 2011. Deflection limits in tall buildings: are they useful ? in: Ames, D., Droessler, T. L., Hoit, M. (Eds.), *Structures Congress.* Las Vegas, Nevada, USA, pp. 515–527. https://doi.org/10.1061/41171(401)45

Staino, A., Basu, B., Nielsen, S.R.K., 2012. Actuator control of edgewise vibrations in wind turbine blades. *J. Sound Vib.* 331, 1233–1256. https://doi.org/10.1016/j.jsv.2011.11.003

Strouhal, V., 1878. Ueber eine besondere Art der Tonerregung. *Ann. der Phys. und Chemie* 3, 216–251.

Suduo, X., Ko, J.M., Xu, Y.L., 2002. Wind-induced vibration control of bridges using liquid column damper. *Earthq. Eng. Eng. Vib.* 1, 271–280. https://doi.org/10.1007/s11803-002-0072-3

Suthar, S.J., Jangid, R.S., 2022. Optimal design of tuned liquid column damper for wind-induced response control of benchmark tall building. *J. Vib. Eng. Technol.* 10, 2935–2945. https://doi.org/10.1007/s42417-022-00528-6

Vickery, B.J., Basu, R.I., 1983. Across-wind vibrations of structures of circular cross-section. Part I. Development of a mathematical model for two-dimensional conditions. *J. Wind Eng. Ind. Aerodyn.* 12, 49–73. https://doi.org/10.1016/0167-6105(83)90080-6

Vickery, B.J., Clark, and A.W., 1972. Lift or across-wind response of tapered stacks. *J. Struct. Div.* 98, Paper no. 8634.

Wu, J., Shih, M., Lin, Y., Shen, Y., 2005. Design guidelines for tuned liquid column damper for structures responding to wind. *Eng. Struct.* 27, 1893–1905. https://doi.org/10.1016/j.engstruct.2005.05.009

Wu, J.C., Wang, Y.P., Lee, C.L., Liao, P.H., Chen, Y.H., 2008. Wind-induced interaction of a non-uniform tuned liquid column damper and a structure in pitching motion. *Eng. Struct.* 30, 3555–3565. https://doi.org/10.1016/j.engstruct.2008.05.029

Xie, J., 2014. Aerodynamic optimization of super-tall buildings and its effectiveness assessment. *J. Wind Eng. Ind. Aerodyn.* 130, 88–98. https://doi.org/10.1016/j.jweia.2014.04.004

Xu, Y.L., Kwok, K.C.S., Samali, B., 1992a. The effect of tuned mass dampers and liquid dampers on cross-wind response of tall/slender structures. *J. Wind Eng. Ind. Aerodyn.* 40, 33–54. https://doi.org/10.1016/0167-6105(92)90519-G

Xu, Y.L., Samali, B., Kwok, K.C.S., 1992b. Control of along-wind response of structures by mass and liquid dampers. *J. Eng. Mech.* 118, 20–39.

Xu, Y.L., Samali, B., Kwok, K.C.S., 1990. Effect of passive damping devices on the response of wind-sensitive structures, in: *Second National Structural Engineering Conference.* Institution of Engineers. Adelaide, Australia, pp. 38–44.

Yalla, S.K., Kareem, A., 2000. Optimum absorber parameters for tuned liquid column dampers. *J. Struct. Eng.* 126, 906–915. https://doi.org/10.1061/(ASCE)0733-9445(2000)126:8(906)

Yamada, M., Goto, T., 1975. The criteria to motions in tall buildings, in: *Pan-Pacific Tall Buildings Conference*. Hawaii, pp. 233–244.

Yang, J.N., Agrawal, A.K., Samali, B., Wu, J.C., 2004. Benchmark problem for response control of wind-excited tall buildings. *J. Eng. Mech.* 130, 437–446. https://doi.org/10.1061/(ASCE)0733-9399(2004)130:4(437)

You, J., You, K., Kim, Y., 2012. Aeroelastic model test for a tall building using tuned liquid column damper, in: *The 2012 World Congress on Advances in Civil, Environmental, and Materials Research (ACEM' 12)*. Seoul, Korea.

You, J.Y., You, K.P., Kim, Y.M., 2014. Evaluation on aerodynamic across-wind response for a tall building using tuned liquid column damper. *Adv. Mater. Res.* 919–921, 1361–1370. https://doi.org/10.4028/www.scientific.net/AMR.919-921.1361

You, J.Y., You, K.P., Kim, Y.M., 2013. Aerodynamic across-wind response according to damping ratios of a tall building using tuned liquid column damper, in: *Proceedings of 8th Asia-Pacific Conference on Wind Engineering APCWE 2013*, pp. 289–298. https://doi.org/10.3850/978-981-07-8012-8_185

Zhang, Z., Basu, B., Nielsen, S.R.K., 2015. Tuned liquid column dampers for mitigation of edgewise vibrations in rotating wind turbine blades. *Struct. Control Heal. Monit.* 22, 500–517. https://doi.org/10.1002/stc.1689

MATLAB® CODE 4.1

```
% Vibration control of an SDOF structure with TLCD under wind-loading
clear all
clc
close all
% Definition of general constants
g=9.81; % gravitational acceleration, unit m/s^2
pi=3.141593;

% Input structural parameters
Ms=1000000; % mass of the structure, unit kg
ZETAs=0.01; % damping ratio of the structure
Ts=2; % natural period of the structure, unit s

% Input damper parameters
tunr=1.0; % tuning ratio
mu=0.05; % mass ratio
alpha=0.7; % length ratio
zi=20; % head-loss coefficient of the orifice(s)
OMEGAs=2*pi/Ts;
OMEGAd=tunr*OMEGAs;
L=2*g/OMEGAd/OMEGAd

% Input wind force
windforce=load('WL3_Final_50_B_ 921.dat'); % Input force, unit N
[n,m]=size(windforce); %number of rows & columns in input loading
abc=zeros(1,m*n);

z=1;
for y=1:n
```

```matlab
    for j=1:m
       abc(1,z)=(windforce(y,j));
        z=z+1;
    end
end
dt=0.5; % time interval of wind force data, unit s
t=0:dt:(dt*(m*n-1));
 %%%%%%%%%%%%%%%%%%%%%%%%%%%%%%%%%%%%%%%%
Gabc=abc;

plot(t,(Gabc/1000))
   xlim([0 1024])
set(gca,{'FontName','FontSize'},{,Times New Roman',35})
set(gca,{'XMinorTick','YMinorTick'},{'on','on'})
xlabel('Time (s)')
ylabel('Wind Force (kN)')
% Determination of response of structure with TLCD
yy=zeros(4,1);
tt=0;
Gabcd=0;
Yin=[0 0 0 0];
time=[];
dispS=[];
velS=[];
dispD=[];
velD=[];
stateSD=[];
accelS=[];
for i=1:(m*n)
  ttf=tt+dt;

[T, Y]=ode45(@(tt,yy) New_rk_TLCD(tt,yy,Gabcd,OMEGAs,OMEGAd,ZETAs,L
,zi,...
    mu,alpha,Ms),[tt ttf],Yin); % Solve ODE
  time=[time; T(:,1)];
  dispS=[dispS; Y(:,1)];
  velS=[velS; Y(:,2)];
  dispD=[dispD; Y(:,3)];
  velD=[velD; Y(:,4)];
  stateSD=[stateSD; Y(:,:)];
xdd=(1/(1+mu-alpha*alpha*mu))*(-(OMEGAs^2).*Y(:,1)...
    -(2*ZETAs*OMEGAs).*Y(:,2)...
    +(mu*alpha)*((OMEGAd^2).*Y(:,3)...
    +(0.5*zi/L).*abs(Y(:,4)).*Y(:,4))...
    +(Gabcd/Ms));
  accelS=[accelS; xdd];
  tt=ttf;
  Gabcd=Gabc(1,i);
  Yin=Y(end,:);
end
```

```
% Determination of response of uncontrolled structure
tts=0;
Gabcd=0;
Yin=[0 0];
times=[];
yys=zeros(2,1);
disp=[];
vel=[];
state=[];
accel=[];
for i=1:(m*n)
  ttsf=tts+dt;

  [T, Y]=ode45(@(tts,yys) rk_str(tts,yys,Gabcd,OMEGAs,ZETAs,Ms),[tts
ttsf],Yin); % Solve ODE

  times=[times; T(:,1)];
  disp=[disp; Y(:,1)];
  vel=[vel; Y(:,2)];
  state=[state; Y(:,:)];

  xdd=-(OMEGAs^2).*Y(:,1)-(2*ZETAs*OMEGAs).*Y(:,2)+(Gabcd/Ms);
  accel=[accel; xdd];

  tts=ttsf;
  Gabcd=Gabc(1,i);
  Yin=Y(end,:);
end

% Determination of control effectiveness of TLCD
        rms=sqrt(mean(disp.^2));
        rmsd=sqrt(mean(dispS.^2));
        redrms_disp=(rms-rmsd)*100/rms

        peak=max(abs(disp));
        peakd=max(abs(dispS));
        redpeak_disp=(peak-peakd)*100/peak

        rms=sqrt(mean(vel.^2));
        rmsd=sqrt(mean(velS.^2));
        redrms_vel=(rms-rmsd)*100/rms

        peak=max(abs(vel));
        peakd=max(abs(velS));
        redpeak_vel=(peak-peakd)*100/peak

        rms=sqrt(mean(accel.^2));
        rmsd=sqrt(mean(accelS.^2));
        redrms_accel=(rms-rmsd)*100/rms

        peak=max(abs(accel));
        peakd=max(abs(accelS));
        redpeak_accel=(peak-peakd)*100/peak
```

```matlab
        figure
plot(times,disp,'r-',time,dispS)
set(gca,{'FontName','FontSize'},{'Times New Roman',35})
set(gca,{'XMinorTick','YMinorTick'},{'on','on'})
legend(' without damper',' with damper')
xlabel('Time (s)')
ylabel('Structural displacement (m)')
grid on
figure

subplot(3,1,1); plot(times,disp,'r-',time,dispS)
legend('without damper','with damper')
xlabel('Time (s)')
ylabel('Displacement (m)')
grid on
subplot(3,1,2); plot(times,vel,'r-',time,velS)
legend('without damper','with damper')
xlabel('Time (s)')
ylabel('Velocity (m/s)')
grid on
subplot(3,1,3); plot(times,accel,'r-',time,accelS)
legend('only structure','structure with damper')
xlabel('Time (s)')
ylabel('Acceleration (m/s^2)')
hold on
grid on

function[dydtt]= New_rk_TLCD(tt,yy,Gabcd,OMEGAs,OMEGAd,ZETAs,L,zi,...
    mu,alpha,Ms)

dydtt=zeros(4,1);  % a column vector

% structural displacement=yy(1), velocity=dydtt(1)=yy(2)
dydtt(1)=yy(2);

%structural acceleration
dydtt(2)=(1/(1+mu-alpha*alpha*mu))*(-(OMEGAs^2)*yy(1)...
    -(2*ZETAs*OMEGAs)*yy(2)...
    +(mu*alpha)*((OMEGAd^2)*yy(3)...
    +0.5*zi/L*abs(yy(4))*yy(4))...
    +(Gabcd/Ms));

% displacement of liquid column=yy(3), velocity of liquid
column=dydtt(3)=yy(4)
dydtt(3)=yy(4);

% acceleration of liquid column
dydtt(4)=(1/(1+mu-alpha*alpha*mu))*((-(OMEGAd^2)*yy(3)...
    -0.5*zi/L*abs(yy(4))*yy(4))*(1+mu)...
    +2*ZETAs*OMEGAs*alpha*yy(2)...
    +(OMEGAs^2)*alpha*yy(1)...
    -(Gabcd/Ms)*alpha);
end
function[dydtt] = rk_str(tts,yys,Gabcd,OMEGAs,ZETAs,Ms)
dydtt=zeros(2,1);  % a column vector
```

```
% Structural displacement=yys(1), velocity=dydtt(1)=yys(2)
dydtt(1)=yys(2);

% Structural acceleration
dydtt(2)=-(OMEGAs^2)*yys(1)-(2*ZETAs*OMEGAs)*yys(2)+(Gabcd/Ms);
end
```

5 Seismic Vibration Control of Structures by TLCD

5.1 INTRODUCTION

Safety and serviceability of structures against earthquake-induced vibrations have been one of the greatest causes of concern to humanity since the days of early city-centric civilization. Due to its unpredictable nature, potential to cause widespread devastation and unavailability of reliable early warning systems, control of seismic vibrations is one of the most challenging problems for the structural engineer.

During an earthquake, a structure receives a large amount of external energy that is transformed into kinetic and strain energy of the structure. This energy is ultimately dissipated through heat and sound to the environment. When a structure lacks the intrinsic damping required to adequately dissipate the energy induced by the impacting earthquake, the structure is likely to fail from a serviceability and/or strength criterion. In such an instance, one of the options is to employ supplemental damping devices to enhance the energy dissipation capacity of the structure. Being a cost-effective and reliable alternative, researchers have taken up studies on the tuned liquid column damper (TLCD) as a supplemental damping device for aseismic design of structures. The earliest investigations in this regard were published in the year 1994 (Haroun et al., 1994; Sun, 1994). Subsequently, more research has been conducted on the TLCD and its configurational variations to suppress earthquake-induced vibration of structures.

Similar to wind loading, as discussed in Chapter 4, under seismic loading, TLCDs are chiefly considered to enhance the serviceability condition of the structure. Apart from displacement and acceleration of the structure, storey drift ratio (SDR) is also considered as an important serviceability parameter. The SDR is defined by the horizontal displacement at the top of a storey relative to the bottom of the storey divided by the storey height. A review of the relevant literature and design standards reveals that the serviceability limit state criterion is defined differently in different standards. For example, the Building Standard Law of Japan (2016) and the Indian seismic code (IS:1893(Part-1), 2016) recommend the value of the maximum allowable SDR as 0.5% and 0.4%, respectively. The seismic design code of New Zealand (NZS1170–5(S1), 2004) stipulates the acceptable serviceability limit state criterion for concrete columns in terms of maximum lateral displacement, specified as height/500. The UK National Annex to Eurocode 3 (NA+A1:2014 to BS EN 1993-1-1:2005+A1:2014, 2014) suggested storey height/300 as the maximum permissible limit for lateral displacement.

Under seismic excitation, it is expected that the structure would show considerable nonlinear behaviour and inelastic deformation beyond the serviceability limit. However, as studies on the use of TLCD for seismic vibration control chiefly aim to improve serviceability conditions, such studies normally consider that the structure

DOI: 10.1201/9781003377894-5

remains within the elastic limit. Similar to wind excitation, under earthquake excitations, TLCDs have been studied extensively both in the frequency domain and in the time domain. There are several experimental studies using shake table and real-time hybrid testing (RTHT) facilities as well. Despite all this, it is important to note that to date there is no real-world implementation of TLCDs exclusively for seismic vibration control of structures. One of the reasons for this is that since earthquake excitation is often broad-banded with considerable energy in the high-frequency band, control of the fundamental mode alone may not be sufficient. The other important reason is that any inertial passive damper requires some time to mobilize the control force and would hence be effective only after the first few cycles of vibration. Thus, if the peak of the earthquake excitation occurs early, the TLCD would not be able to achieve a substantial reduction in the peak structural response. This clearly indicates that the design of TLCD to mitigate seismic vibration is still an active field of research, and several researchers are working on this topic. In the succeeding sections of this chapter, a detail description of the theoretical and experimental studies on the seismic vibration control of different types of structures using TLCD is presented, in frequency and time domains.

5.2 FREQUENCY DOMAIN STUDIES

Frequency domain input–output relations of a system comprising of a base-excited single-degree-of-freedom (SDOF) structure with an attached TLCD are given in Eqs. (2.31) and (2.32), while the transfer functions relating the displacement of the structure to the base acceleration and the displacement of the liquid in the vertical limb of the TLCD to the base acceleration are given in Eqs. (2.35) and (2.36), respectively. Power spectral density function (PSD) is used to describe the frequency domain distribution of energy in the strong ground motion process. A simplistic form of representation of the ground motion input in the frequency domain is a white noise process, which has a constant PSD over all frequencies. Housner (1947) was the first to model seismic ground acceleration as a stationary white noise process, and since then, many researchers have considered this form of seismic input while examining the performance of TLCDs.

However, the amplitudes of PSD of real recorded accelerograms of ground motions vary with frequency (Sues et al., 1983), and for a more realistic modelling of the input ground motion characteristics, Kanai (1957) and Tajimi (1960) proposed a Gaussian-filtered white noise model popularly known as the Kanai–Tajimi (K-T) model. This model assumes the vibrating ground as an SDOF system. The K-T model idealizes the ground acceleration as a stationary random process having a spectral density of the form

$$S_{\text{KT}}(\omega) = S_0 \frac{1 + 4\xi_g^2 \left(\dfrac{\omega}{\omega_g} \right)^2}{\left[1 - \left(\dfrac{\omega}{\omega_g} \right)^2 \right]^2 + 4\xi_g^2 \left(\dfrac{\omega}{\omega_g} \right)^2}. \tag{5.1}$$

In Eq. (5.1), S_0 is the constant spectral density of the input white noise process. The natural frequency and the damping ratio of the surface soil deposit are denoted by ω_g and ξ_g, respectively. The parameters ω_g and ξ_g depend on the earthquake magnitude, the epicentral distance and the ground layer rigidity. Based on the Fourier analysis of a total of 247 ground motion records and using the least squares fit method, Moayyad (1982) suggested a set of values of these parameters, which are given in Table 5.1.

The K-T model has been used by many researchers to represent seismic ground motion in the frequency domain. However, the model exhibits the following notable shortcomings (Chen et al., 2022).

i. The model tends to overestimate the low-frequency components of the ground motion, potentially yielding unrealistic outcomes when applied for the seismic analysis of low-frequency structures.
ii. The model exhibits singularity at zero frequency and fails to meet the continuous quadratic integrability condition.
iii. The K-T model assumes that the ground acceleration of rock follows Gaussian white noise. This assumption does not adequately reflect the spectral characteristics of the motion of rock.

Clough and Penzien (1975) developed a method to adjust the low-frequency energy of the K-T model by introducing a second-order high-pass filter and proposed a modified model to idealize the ground acceleration as a stationary random process. The model proposed by Clough and Penzien (1975), popularly known as the C-P model, represents the PSD as

$$S_{CP}(\omega) = S_{KT}(\omega) \frac{\left(\dfrac{\omega}{\omega_f}\right)^4}{\left[1 - \left(\dfrac{\omega}{\omega_f}\right)^2\right]^2 + 4\xi_f^{\,2}\left(\dfrac{\omega}{\omega_f}\right)^2} \tag{5.2}$$

in which ω_f and ξ_f are the frequency and damping ratio of the second-order filter. The value of ω_f should be smaller than that of ω_g, and the recommended value of the ratio ω_f/ω_g is 0.1–0.2, while the value of ξ_f may be assumed to be equal to ξ_g (Chen et al., 2022).

TABLE 5.1

K–T Spectral Parameters Proposed by Moayyad (1982)

Ground Condition	ω_g (rad/s)	ξ_g
Soft	10.9	0.96
Intermediate	16.5	0.80
Hard	16.9	0.94

The C-P model is highly popular among researchers and has been extensively used (Adhikary et al., 2023; Jangid, 2022; Won et al., 1997, 1996). A comparison of the PSD function of white noise, K-T spectrum and C-P spectrum is shown in Figure 5.1. Some other models to represent seismic ground motions in the frequency domain have also been proposed. Examples of such models include the Hu-Zhou model (Hu and Zhou, 1962) and the Ou-Liu model (Ou and Liu, 1994). However, none of these models have gained the level of acceptance from the research community as has the C-P spectrum.

One of the earliest frequency domain studies on earthquake-induced vibration control of a structure by the TLCD was taken up by Samali (1990). In the study, the structure was modelled as an SDOF system, and the seismic excitation was modelled as a stationary random process. The study revealed that TLCD can achieve comparable response reductions as those offered by a tuned mass damper (TMD) with equivalent mass ratio. Later, Won et al. (1996, 1997) carried out detailed frequency domain studies on response control of seismically excited structures by TLCDs. Considering an SDOF structural system, they indicated the dependence of the optimum head-loss coefficient on the excitation amplitude for the cases when the input ground acceleration is represented through a stationary random process, using the C-P model, as well as white noise. It was also observed that even under non-optimum conditions, TLCDs are able to significantly reduce the response of structures having natural periods longer than about 3 s. They also examined the efficacy of the TLCD for a 30-storey multi-degree-of-freedom (MDOF) structural system having a damping ratio of 2%. A stochastic ground motion model, the parameters of which are so chosen that the mean response spectra of the input ground motions approximate the response spectra of selected real ground motion records, was considered. The results highlighted that a TLCD having a mass ratio of 2% can reduce

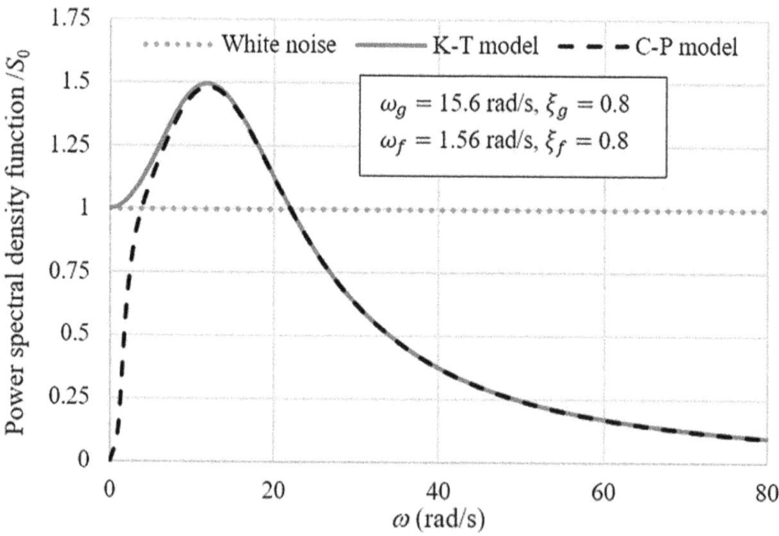

FIGURE 5.1 Comparison of different spectra used to represent earthquake ground motions.

the root-mean-square (rms) value of the inter-storey shear and the roof acceleration by about 40% and 30%, respectively.

A detailed frequency domain study on the use of the liquid column vibration absorber (LCVA), a configurational variation of the conventional TLCD (see Section 3.3.1 of Chapter 3), for seismic vibration control of structures, was taken up by Konar and Ghosh (2010). The structure was modelled as an SDOF system, and the base excitation was represented by a white noise. It was observed that the LCVA with a 2.5% mass ratio, 0.5 length ratio and 1.55 area ratio achieved a 30% reduction in the rms structural displacement, whereas an equivalent conventional TLCD could achieve a 24% reduction in the rms displacement of the structure, which demonstrates that the LCVA is more effective than the TLCD. In a subsequent work, Konar and Ghosh (2013) explored the possibility of controlling both modes of a 2-DOF system due to earthquake excitation by considering the sloshing action in the vertical limbs of the LCVA, in addition to liquid oscillation in the U-shaped damper container. The plot of uncontrolled and controlled displacement transfer function of the top mass of the example structure showed that with a properly designed LCVA, it is possible to achieve bimodal vibration control of a base-excited structure (see Figure 5.2). In Figure 5.2, the displacement transfer function of the top mass of the example structure with a conventional TLCD is also included. As anticipated, the TLCD is able to control only the fundamental mode of the structure. Due to its ability to control bimodal vibration, the LCVA achieved 1.2 times more reduction in the rms displacement of the top mass of the 2-DOF structure as compared to that achieved by the TLCD with the same liquid mass ratio.

Another configurational variation of the TLCD, namely the compliant liquid column damper (CLCD) (see Section 3.4.2 of Chapter 3), has been extensively studied for seismic vibration control. This is because, unlike the conventional TLCD, the CLCD can be tuned to short-period structures that are susceptible to seismic excitation. Here, it may be noted that the structures having a fundamental natural period of

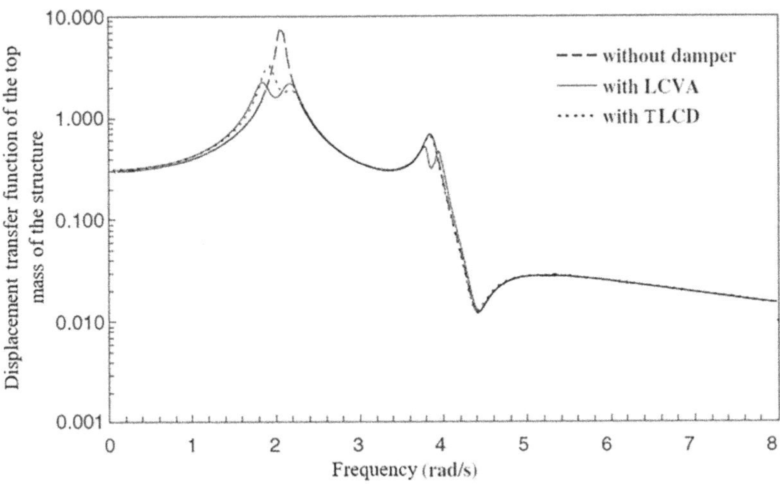

FIGURE 5.2 Displacement transfer function of the top mass of the structure under white noise input (Konar and Ghosh, 2013).

less than 1.0 s are normally referred to as short-period structures. On the other hand, structures having a fundamental natural period between 1.0 and 2.0 s and above 2.0 s are typically classified as mid and long-period structures, respectively (Cavdar et al., 2019). The research on the use of CLCD for seismic protection of short-period structures was pioneered by Ghosh and Basu (2004). A transfer function formulation of an SDOF structure with an attached CLCD was developed. The ground acceleration was characterized by a white noise. Two example structures having fundamental natural period of 0.3 and 0.7 s were considered in the study. A comparison of the displacement transfer function of the controlled and uncontrolled structures indicated that the designed CLCDs are highly effective in controlling the response of both the example structures (see Figure 5.3). It was also shown that the intensity of the white noise base excitation has practically no impact on the extent of control achieved by the CLCD. In a subsequent work, Ghosh et al. (2011) extended their study to a multiple-CLCD (MCLCD) system. Here also, the excitation was characterized as a white noise. The effectiveness of an MCLCD system in providing a robust vibration control solution against possible variation in the structural stiffness or natural frequency was demonstrated. It was also indicated that the performance of the MCLCD is practically uniform over a wide range of head-loss coefficient of the damper system.

5.3 TIME DOMAIN STUDIES

There exists a significant volume of time domain studies on the use of TLCD for seismic vibration mitigation of structures. The time domain analysis of an SDOF structure with attached TLCD can be carried out by solving Eqs. (2.17)

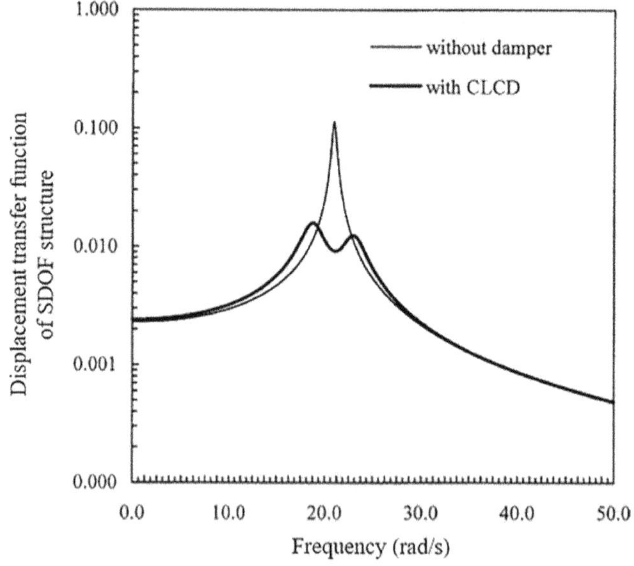

FIGURE 5.3 Displacement transfer function of a single-degree-of-freedom structure having a natural period = 0.3 s with compliant liquid column damper and without the damper under white noise input (Ghosh and Basu, 2004).

and (2.18). The input base acceleration, $\ddot{Z}_0(t)$ can be obtained from the accelerograms of previously recorded earthquakes, as available in the large set of digital records of past earthquakes in the web-based Pacific Earthquake Engineering Research Center (PEER, n.d.) and the Consortium of Organizations for Strong Motion Observation Systems (COSMOS, n.d.). Most of the studies available in literature on the performance assessment of TLCDs have used unscaled ground motions as input. However, for actual design, ground motions scaled to the seismic design spectrum may be utilized. Alternatively, there are several procedures to generate site-specific and spectral-compatible synthetic accelerograms for time domain studies. More details in this regard are available in the works of Wong and Trifunac (1979), Lam et al. (2000), Ferreira et al. (2020) and Adhikary et al. (2023), among others.

The first extensive time domain study on the TLCD-assisted seismic vibration control of structure was conducted by Sadek et al. (1998). The primary structure was modelled as an SDOF system, and 72 recorded ground motions, scaled to a peak ground acceleration of 0.25 g, were used in the analyses. In the first part of the study, a single TLCD system was considered, and an empirical formula was proposed to calculate the optimum head-loss coefficient, ξ_{opt}, of the TLCD given by

$$\xi_{\text{opt}} = 3.58\mu \frac{g}{\ddot{Z}_{0,\max}}. \tag{5.3}$$

In Eq. (5.3), μ and $\ddot{Z}_{0,\max}$ denote the mass ratio and the peak ground acceleration, respectively.

Sadek et al. (1998) also considered a multiple-TLCD (MTLCD) system for vibration control of a ten-storey example structure. The storey-wise displacement response reductions obtained under the 1994 Arleta earthquake and 1989 Capitola earthquake are shown in Figure 5.4a and b respectively. The results indicated that the MTLCD system is not always superior to a single-TLCD system. However, as expected, the former is shown to be more robust, if there are errors in estimating the structural parameters. Further, the performance of the TLCD was found to be comparable with that of a TMD having the same mass ratio.

Kavand and Zahrai (2006) conducted an extensive time domain study using 16 recorded ground motions to analyse the influence of seismic excitation characteristics on the performance of the TLCD. The ground motions were scaled to a peak acceleration of 0.35 g. A ten-storey example structure with a fundamental natural period of 2 s and a damping ratio of 2% was considered. The frequency content of the ground motion record was found to affect the performance of the TLCD significantly. When the predominant frequency range of the ground motion record was close to the natural frequency of the structure, the damper performance was found to be relatively better.

Ahadi et al. (2012) considered a ten-storey example building structure and obtained the optimum TLCD parameters using a genetic algorithm (GA)-based optimization technique under a synthetic base excitation close to white noise base acceleration. The efficacy of the optimally designed damper system was then assessed

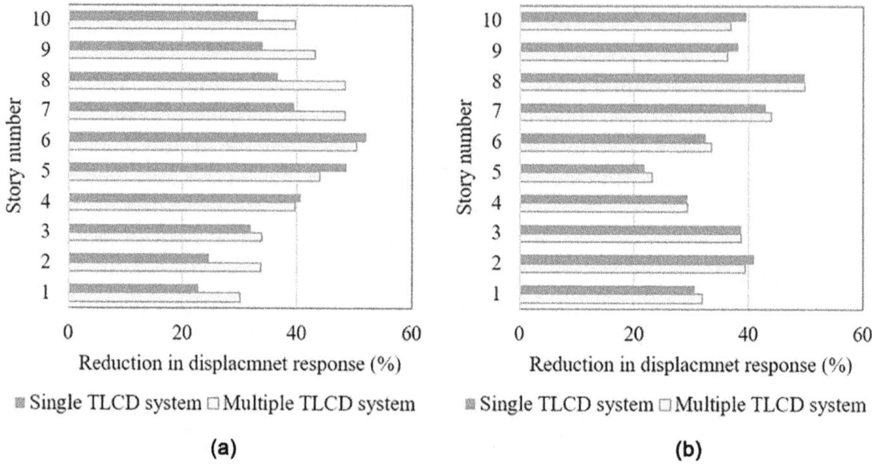

(a)   (b)

FIGURE 5.4   Storey-wise displacement response reduction of a ten-storey example building by single and multiple tuned liquid column damper systems under (a) the 1994 Arleta earthquake and (b) the 1989 Capitola earthquake (The figures are drawn based on the results provided by Sadek et al. (1998)).

under recorded ground motions. A sample set of storey-wise maximum displacements of the uncontrolled and controlled example structure subjected to the 1940 El Centro earthquake is shown in Figure 5.5. Here, the damper system consisted of five TLCD units having a total mass ratio of 2%. In a subsequent work by Mohebbi et al. (2015), the variation in the damper performance with the mass ratio was illustrated. Using GA-based optimization, two sets of optimum parameters of the MTLCD system were obtained, based on the minimization of the peak acceleration and the peak displacement responses, respectively. It was concluded from this study that MTLCDs were more effective in reducing the maximum displacement than the maximum acceleration of the structure. Further, an increase in the mass ratio led to an improvement in the performance of the MTLCD system (see Figure 5.6a and b).

Samali et al. (2002) examined the efficacy of a single LCVA in seismic vibration control of a five-storey benchmark building. Four recorded earthquake accelerograms were used for the numerical study. The performance of a multiple-LCVA (MLCVA) system was also investigated, and its robustness under damper mistuning due to the loss of structural stiffness during an earthquake, as compared to the performance of a single LCVA system, when mistuned, was highlighted. By comparing the displacement time histories of controlled and uncontrolled structures under recorded ground motions, Konar and Ghosh (2010) demonstrated that an optimally designed LCVA can outperform an equivalent conventional TLCD. The illustrative numerical study revealed that an LCVA having a mass ratio of 2.5% reduced the peak and rms displacement of the primary structure by 14% and 31%, respectively. In contrast, a mass-equivalent conventional TLCD achieved 10% and 24% reduction in the peak structural and rms displacements, respectively.

As already mentioned in Section 5.2, Ghosh and Basu (2004) initiated the study on CLCD for seismic vibration control of short-period structures. They also

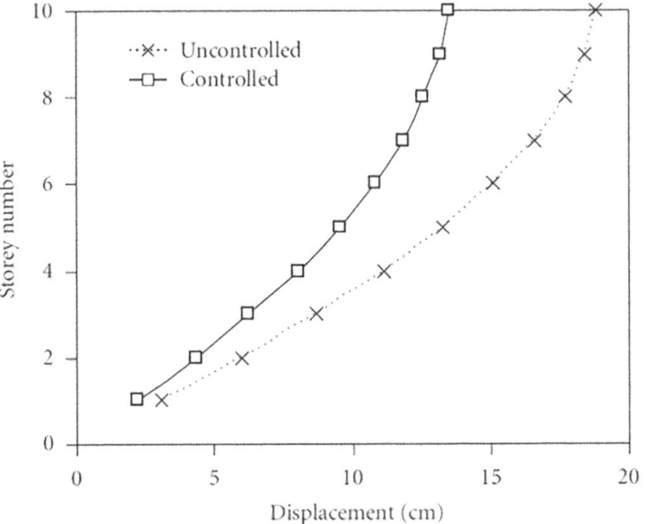

FIGURE 5.5 Storey-wise maximum displacement of an uncontrolled structure and a controlled structure with tuned liquid column damper under the effect of the 1940 El-Centro earthquake (Ahadi et al., 2012).

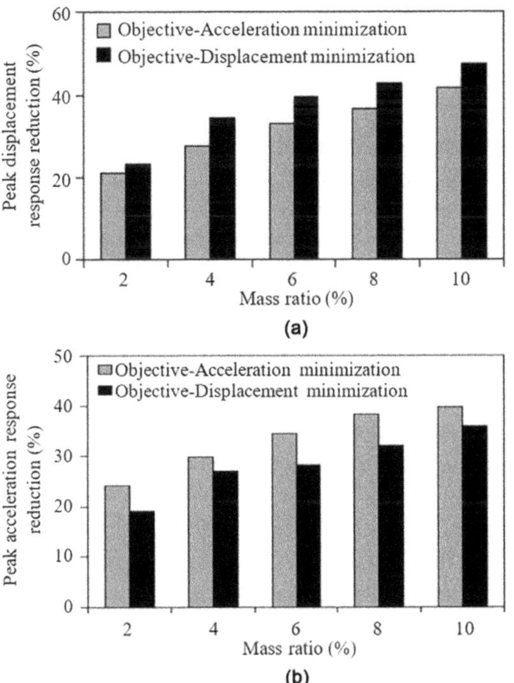

FIGURE 5.6 Reduction in structural response by using tuned liquid column damper for two different objective functions under the action of 1940 El Centro earthquake (a) maximum displacement and (b) maximum acceleration (Mohebbi et al., 2015).

conducted time history analysis of two example structures, having natural period of 0.3 and 0.7 s, respectively, subjected to the north-south component of the 1940 Imperial Valley earthquake at the El Centro site. The damping ratio of the structures was taken as 1%, and the mass ratio of the CLCD was taken as 2.5%. The CLCD produced 25% and 56% reduction in the peak and rms displacement of the structure, respectively, with the structure having natural period of 0.3 s. In case of the 0.7 s period structure, the corresponding reductions were 46% and 66%, indicating the possible high efficacy that can be achieved in mitigating the seismic response of SDOF structural systems. In a subsequent work, Ghosh and Basu (2008) examined the CLCD performance in the vibration control of a nonlinear structure having a shorter natural period in the linear range below yield, with subsequent period lengthening in the post-yield phase. The nonlinear structure was modelled by an SDOF system with bilinear hysteresis (Suzuki and Minai, 1988). The results indicated an improvement in the nature of the vibrational response of the nonlinear structure when the CLCD is employed. The damper eliminated the permanent set and reduced the number of crossings beyond yield by 50%. Roy and Ghosh (2015) demonstrated the control effectiveness of the CLCD under near-fault ground motion, which may be considered to be ground motions recorded within 20 km from the fault (Bray and Rodriguez-Marek, 2004). Compared to the ground motions observed in the far-off, or far-field ground motions, near-fault ground motions can exhibit distinct characteristics due to the heavy influence of the rupture mechanism, the slip direction relative to the site and the permanent ground displacement at the site resulting from tectonic movement. Sometimes, near-fault ground motions contain high-velocity pulses due to forward directivity (Baker, 2007). The pulse-type near-fault (P-N) ground motions impose large energy dissipation demands on the structures. Roy et al. (2023) extensively studied the performance of the CLCD in controlling the vibration of structures subjected to P-N ground motion and presented some interesting results. It was shown that the CLCD significantly reduced the rms displacement response of the structure, though the peak displacement response reductions were not always that significant. Further, the mass ratio, in general, had no correlatable effect on the optimum tuning ratio of the damper under P-N ground motions, and an increase in the length ratio did not always result in improved performance of the CLCD. Through an extensive numerical study, it was established that the linearization of the nonlinear damping term of CLCD did not significantly influence the numerically evaluated structural response reduction by the damper.

Another variation of the TLCD, which has the ends of the U-shaped damper container securely sealed (see Section 3.4.1 of Chapter 3), can also be tuned to short-period structures (Reiterer and Ziegler, 2006). With the use of numerical simulations on an SDOF structure having a natural period of 0.43 s and a damping ratio of 2%, Bhattacharyya et al. (2017a) demonstrated the efficacy of the sealed TLCD in reducing structural responses under recorded ground motions. The results indicated that a sealed TLCD having a mass ratio of 5% can achieve up to 34% and 44% reductions in the peak structural displacement and peak acceleration response, respectively (see Figure 5.7). The equivalent linear stiffness from the nonlinear air spring effect of the sealed TLCD may be used for the purpose of structure-damper

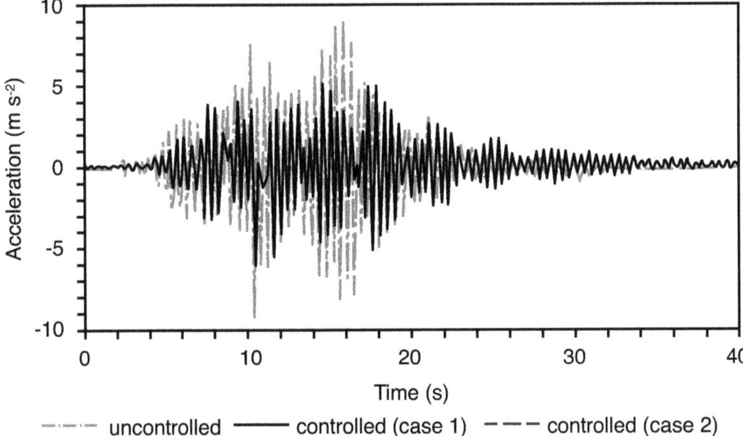

FIGURE 5.7 Time histories of acceleration of uncontrolled and controlled structure under the Landers earthquake (PGA 0.730 g) for 5% mass ratio with sealed tuned liquid column damper considering nonlinear (case 1) and linearized (case 2) stiffness of air spring (Bhattacharyya et al., 2017a).

response analysis. However, this may lead to a slight overestimation of the response reductions by the damper.

5.4 EXPERIMENTAL STUDIES

In view of the complex nature of the seismic excitation, some researchers have conducted experimental studies to simulate seismically excited structure-TLCD systems. Broadly, there are two methods for such experimental studies, namely the shake table test and the RTHT.

Shake table tests are normally conducted on a scaled-down model of a structure-TLCD system. The model of the structure used for the shake table system may be an SDOF or an MDOF system. A schematic of a shake table setup with a structure-TLCD model is shown in Figure 5.8. Park et al. (2018) conducted experiments on an SDOF structural model. The TLCD used for the study had mass ratio and length ratio equal to 3.6% and 0.47, respectively. Both free and forced vibration experiments were conducted. When the structure was excited at the base, the damper reduced the peak acceleration response of the structure by 43%. It was also observed that if the walls of the TLCD were provided with embossments, the effectiveness of the damper was increased by a factor of 1.2. Altunişik et al. (2018) conducted shake table tests to assess the impact of excitation direction on the performance of the TLCD. A table-shaped structural system was used for the experiment. The performance of the TLCD in terms of controlling the displacement and acceleration response was found to be practically unaffected if the angle between the direction of excitation and the longitudinal direction of the TLCD was within a maximum of 15°.

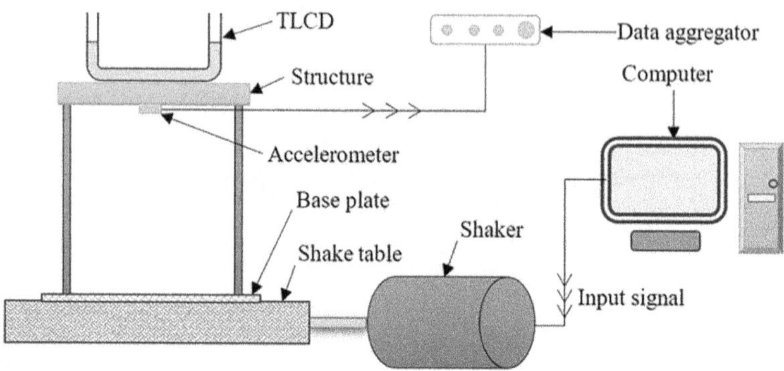

FIGURE 5.8 Schematic of a shake table set up with a structure–tuned liquid column damper model.

However, for angles greater than 15°, a significant drop in the efficacy of the damper was reported.

Bigdeli and Kim (2016) conducted a shake table test on a model of a TLCD-controlled MDOF structure, the MDOF system being a scaled model of a three-storey steel building that was excited by an artificially generated base acceleration. The results of the study indicated that the peak structural displacement response reduction achieved by employing the TLCD was slightly superior to that obtained by a tuned sloshing damper (TSD) and by a TMD having the same mass ratio. Shah and Usman (2022) also examined a model of a TLCD-controlled MDOF structure subjected to both harmonic and seismic base excitations. A four-storey building frame model having a fundamental natural period of 0.77 s and a damping ratio of 2.2% was experimentally studied. The TLCD had a 5% mass ratio and a 0.8 length ratio. The peak and rms accelerations of the TLCD-controlled and uncontrolled building

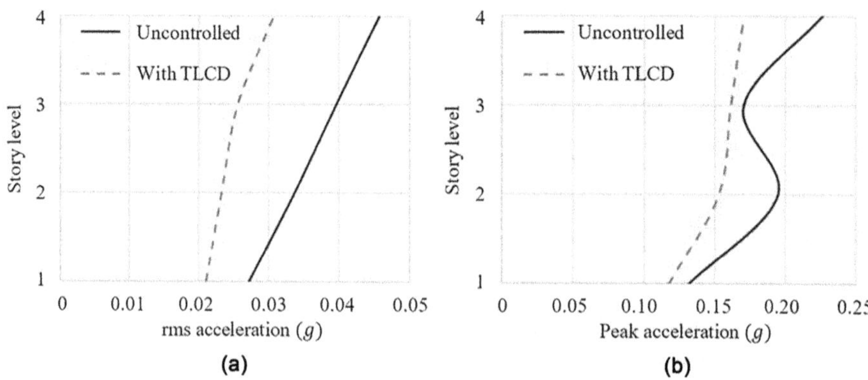

FIGURE 5.9 Acceleration response of tuned liquid column damper–controlled and uncontrolled building frame at different storey levels obtained from shake table test under the scaled 1940 El Centro seismic excitations (a) rms values and (b) peak values (Shah and Usman, 2022).

frame at different storey levels obtained from the experiment under the scaled 1940 El Centro seismic excitation are presented in Figure 5.9a and b.

In an innovative application of the TLCD, Adam et al. (2017) and Furtmüller et al. (2019) conducted shake table experiments on a base-isolated structure to assess the efficacy of the damper in the reduction of the displacement demand of the base isolator. The study reported that the strategy of a combined base isolation-TLCD system could reduce the isolator displacement demand as well as provide substantial reductions in the storey acceleration demand of the superstructure. Shake table tests have also been conducted on structural models with different configurational variations of the TLCD, such as the bi-directional TLCD (Rozas et al., 2016), CLCD (Bhattacharyya et al., 2017b; Masnata et al., 2024), toroidal TLCD (Ding et al., 2020) and tuned liquid column ball damper (Shah et al., 2023). All these studies aimed to control the dominant mode of the structural model by using the TLCD and reported successful achievement of the same.

RTHT is a hybrid experimental technique that combines computer simulation with simultaneous physical testing. To investigate the dynamics of a structural system using RTHT, the structural system is partitioned into numerical and physical substructures. While the numerical substructure is simulated on a computer, the physical substructure undergoes testing on a shaking table (Nakashima et al., 1992). The RTHT offers two distinctive benefits: first, the ability to test full-scale or large-scale models; and second, the capacity to investigate complex nonlinear structures. A schematic of an RTHT framework for structure-TLCD experiments is shown in Figure 5.10. In an attempt to achieve near real-time measurements of wave height, lateral motion and control forces applied by the TLCD in an RTHT, Min et al. (2016) developed a noncontact standalone vision sensing system based on binary

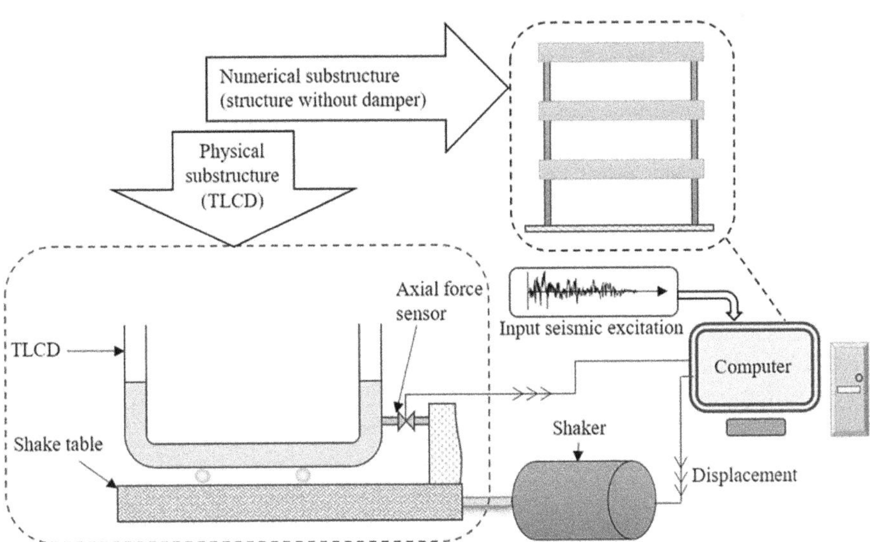

FIGURE 5.10 Schematic of a real-time hybrid testing framework for structure–tuned liquid column damper experiments.

pixel counting of the images streamed during the test, which would replace a series of conventional sensors required in the test.

The earliest research on the use of RTHT to study structure-TLCD systems under seismic excitation was conducted by Lee et al. (2007). Considering an SDOF structure having a natural period of 0.81 s, they conducted the RTHT under a couple of scaled recorded ground motions. They also performed a conventional shake table test on the physical model of the structure-TLCD system. A comparison of the acceleration response of the structure with TLCD obtained from the RTHT and shake table test showed that the performance of the TLCD can be accurately evaluated using the RTHT. Having established a satisfactory accuracy of RTHT, Zhu et al. (2015, 2016) employed the technique to investigate the influence of different damper design parameters on the performance of the TLCD. Both single- and multiple-TLCD systems were studied. The outcome of the tests was in line with established results, such as the damper performance generally improves with an increase in mass ratio and that the TLCD is more effective for low structural damping ratios (within 2%). The study also indicated that due to the nonlinear behaviour of the TLCD, the performance of the single TLCD was significantly dependent on the amplitude of the input excitation. Though initially in some instances, the effectiveness of the TLCD improved with the increase in the PGA of the earthquake input, the TLCD performance was found to degrade when the shaking intensity was relatively strong. Further, under the considered ground motions, better reductions in rms response were achieved by the single TLCD as compared to the peak response reductions (see Figure 5.11a–c). The results with multiple-TLCDs were found to be comparable and in some cases more robust and superior to that of the single-TLCD, again stressing the need for greater attention on the multiple-TLCD system for multi-modal seismic response control of structures. In a subsequent work, Zhu et al. (2017) conducted RTHT on a full-scale model of a nine-storey benchmark building. Different TLCD configurations and both single and multiple-TCLD systems were considered for the study. The results reinforced earlier observations on the input amplitude dependence of the TLCD performance and on the advantages of the multiple-TLCD systems over the single one for practical implementations. Further, the possibility of greater reduction in rms responses by the TLCD as compared to the reduction in peak responses was also exhibited.

A comparative study between the control effectiveness of the TLCD and the TSD using RTHT was conducted by Fei et al. (2019), considering a five-storey reinforced concrete building as the primary structure. The study demonstrated that though both the TLCD and the TSD could significantly reduce the structural response, the latter was found to be more effective. This is in contrast with the finding of Bigdeli and Kim (2016), reported earlier in this section. This contradiction may possibly be attributed to the differences in the experiment design, as the efficacy of both TSD and TLCD is strongly influenced by the parameters of the structure, damper and the frequency characteristics of the excitation. Thus, a more comprehensive study is essential to conclusively compare the performance of these two types of tuned liquid dampers.

The RTHT has also been used to assess the effectiveness of other configurational variations of TLCDs. These are briefly listed here. Heo et al. (2009) examined the bidirectional seismic control performance of a hybrid passive damper that combines a TMD and a TLCD (see Section 3.4.2 of Chapter 3). Ding et al. (2021b) investigated

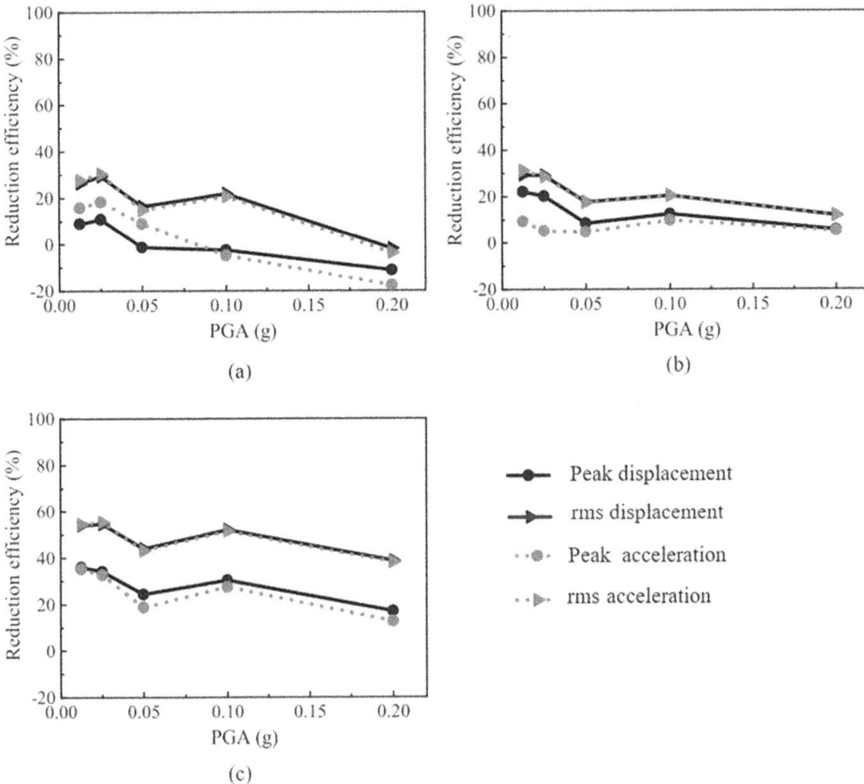

FIGURE 5.11 Comparison of reduction efficiency of a single tuned liquid column damper under ground motions scaled to different PGAs (a) Kobe earthquake, (b) El Centro earthquake, and (c) Taft earthquake (Zhu et al., 2016).

the nonlinear damping effects of conventional and toroidal TLCDs. The results demonstrate free surface contamination and wave breaking in the vertical limbs of the damper. Ding et al. (2021a) also used RTHT to examine the control efficiency of the multi-layer-toroidal TLCD (see Section 3.2.2 of Chapter 3). In these studies, RTHT has been utilized as a means to experimentally verify the theoretically developed working mechanism of some of the novel TLCD configurations.

5.5 OPTIMALITY ISSUES

The critical parameters of a TLCD that influence the damper performance are the mass ratio μ, the length ratio α, the coefficient of head-loss of the orifice ξ and the tuning ratio γ. As mentioned in Section 2.3.1 of Chapter 2 in the design of a TLCD, the mass ratio is often determined based on practical constraints, such as space availability and limitations on adding additional dead load to the structure. Within the practical range, the performance of a TLCD improves with an increase in the mass ratio (Sadek et al., 1998; Won et al., 1997). However, for real-world installations of

TABLE 5.2

Maximum Values of Mass Ratio Considered in Different Literature on the Use of TLCD for Seismic Vibration Control

Maximum Values of Mass Ratio Considered (%)	References
1.5	Konar and Ghosh (2013)
2	Debbarma et al. (2010a,b)
2.5	Konar and Ghosh (2010)
3	Chang (1999), Kavand and Zahrai (2006)
4	Sadek et al. (1998), Won et al. (1996)
5	Ghosh and Basu (2004, 2008), Bhattacharyya et al. (2017b), Roy et al. (2023)
8	Won et al. (1997)
10	Ahadi et al. (2012), Mohebbi et al. (2015)

TLCDs, μ is typically restricted to within 1.5% (Konar et al., 2024; Konar and Ghosh, 2023), though in some literature, higher values of μ have been considered in numerical simulations. The maximum values of mass ratio considered in some relevant literature on the use of TLCD for seismic vibration control are listed in Table 5.2.

Normally, the TLCD performance improves with an increase in length ratio. However, the optimal value of α can be determined using Eq. (2.41), ensuring the maintenance of a positive liquid height in the vertical limbs of the TLCD during vibration. The remaining two critical parameters, namely the coefficient of head-loss of the orifice ξ and the tuning ratio γ, have been rigorously studied to optimize the performance of a TLCD.

One of the earliest works on optimal design of a TLCD attached to a base-excited structure was carried out by Chang (1999). A generalized parametric study was conducted for both TMD and TLCD. The structure was modelled as an SDOF system, the base excitation was represented by a Gaussian white noise random process, and the structural damping was neglected. The expressions for optimum tuning ratio and optimum orifice head-loss coefficient of a TLCD obtained from the study are respectively given by

$$\gamma_{opt} = \frac{\sqrt{1+\mu-1.5\alpha^2\mu}}{1+\mu} \tag{5.4}$$

and

$$\xi_{opt} = \frac{2\omega_{TLCD}L}{|\dot{y}|}\sqrt{\frac{\mu\left(1+\mu-1.25\alpha^2\mu\right)}{\left(1+\mu\right)\left(1+\mu-1.5\alpha^2\mu\right)}}. \tag{5.5}$$

In Eq. (5.5), ω_{TLCD}, L and $\dot{y}$ respectively denote the natural frequency of the TLCD, the length of the liquid column in the damper and the velocity of the liquid column in the vertical limb of the damper.

In a subsequent work, Wu and Chang (2006) modified Eq. (5.5) to

$$\xi_{opt} = 2^{0.5}\pi^{1.5}\frac{1}{\sigma_{\ddot{y}}}\sqrt{\frac{\mu\left(1+\mu-1.25\alpha^2\mu\right)}{\left(1+\mu\right)\left(1+\mu-1.5\alpha^2\mu\right)}} \tag{5.6}$$

where $\sigma_{\ddot{y}}$ denotes the standard deviation of $\ddot{y}$.

Wu and Chang (2006) also presented a set of design tables for the optimum tuning ratio and orifice head-loss coefficient of a TLCD for a damped SDOF structural system. They considered the base excitation to the structure as a white noise and minimized the displacement response of the structure to obtain the optimum parameters. A set of design charts for the optimum parameters was provided. The presence of structural damping led to a lower value of the optimum parameters compared to what was obtained from Eqs. (5.4) and (5.6) for the undamped system. However, the differences are very small from a practical point of view, even when the structural damping ratio is as large as 10%. Considering a white noise base excitation, Di Matteo et al. (2015) developed a less computationally intensive approximate iterative method for obtaining optimal tuning ratio and head-loss coefficient. The results were in good agreement with the results presented by Wu and Chang (2006).

Ghosh and Basu (2007) proposed an alternate closed-form expression for the optimum tuning ratio of TLCD employed for controlling light to moderately damped structures under generalized seismic loading. The closed-form expression was derived based on the theory of fixed-point frequencies for a classical structure-damper system and is given by

γ_{opt}

$$= \sqrt{\frac{\mu\alpha^2\left\{2\left(1+\mu\right)-4\zeta_s^2\right\}\left[2\beta\left(1+\mu\right)+4\zeta_s^2\left\{\mu\alpha^2-2\left(1+\mu\right)\right\}\right]+\beta^2\left(1+\mu\right)^2-\left(1+\mu\right)^4}{\left(1+\mu\right)^3\left\{2\left(1+\mu\right)^2-2\beta\left(1+\mu\right)-8\mu\alpha^2\zeta_s^2\right\}}}. \tag{5.7}$$

In Eq. (5.7), ζ_s denotes the damping ratio of the structure and β is expressed as

$$\beta = 1+\mu\left(1-\alpha^2\right). \tag{5.8}$$

A comparison of the optimum tuning ratio obtained from Eq. (5.7) and from the design charts provided by Wu and Chang (2006) for $\alpha = 0.9$ is presented in Figure 5.12. It may be observed from Figure 5.12 that the optimum tuning ratios obtained from both methods are close to each other, with the values given by Eq. (5.7) being slightly on the higher side.

The optimum design of TLCD parameters under stochastic seismic-excitation considering uncertainty in the system parameters was studied by Debbarma et al. (2010a, 2010b) in the context of structural safety. However, it is important to note that structural safety is not influenced by the design of the passive TLCD and therefore will not be compromised by non-optimal TLCD parameters. Some other works on the optimization of TLCD parameters have been carried out for a

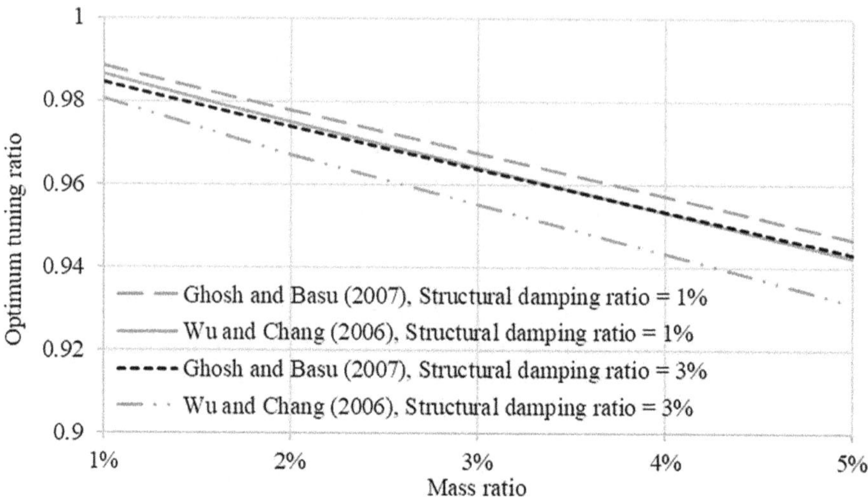

FIGURE 5.12 Comparison of the optimum tuning ratio obtained from the expression given by Ghosh and Basu (2007) and from the design charts presented by Wu and Chang (2006).

particular example structure, such as the one by Ahadi et al. (2012), who obtained optimum TLCD parameters using genetic algorithm (GA) for a ten-storey example building structure.

5.6 STUDIES CONSIDERING SOIL–STRUCTURE INTERACTION EFFECTS

The design of a control device necessitates precise knowledge of the dynamic properties of the structure into which it will be incorporated. Typically, these properties are assessed assuming a fixed-base structural system. However, in reality, dynamic interaction occurs between the foundation, soil and a structure (such as a building or a bridge) subjected to dynamic loads, especially earthquakes. The mutual influence of the soil response and the structural response on each other during the occurrence of these dynamic loads constitutes the soil–structure interaction (SSI). SSI may have a significant effect on the structural properties and may consequently impact the effectiveness of the damper (Takewaki, 2000). Though this is a significant issue, there are limited studies on this topic. Ghosh and Basu (2005) were the first to consider the effects of SSI while designing the TLCD for seismic vibration control of structures. In this work, a simplified model of the soil-structure-TLCD system has been considered. This model and the formulation of the equations of motion are presented in the following.

The structure is modelled as a linear SDOF system. A TLCD is rigidly attached to the structure. The foundation of the SDOF structural system is modelled as a rigid plate, anchored to the surface of a visco-elastic, homogeneous half-space through imaginary linear spring and viscous damping elements (see Figure 5.13). The mass, stiffness and damping coefficient of the structure are denoted by m, k

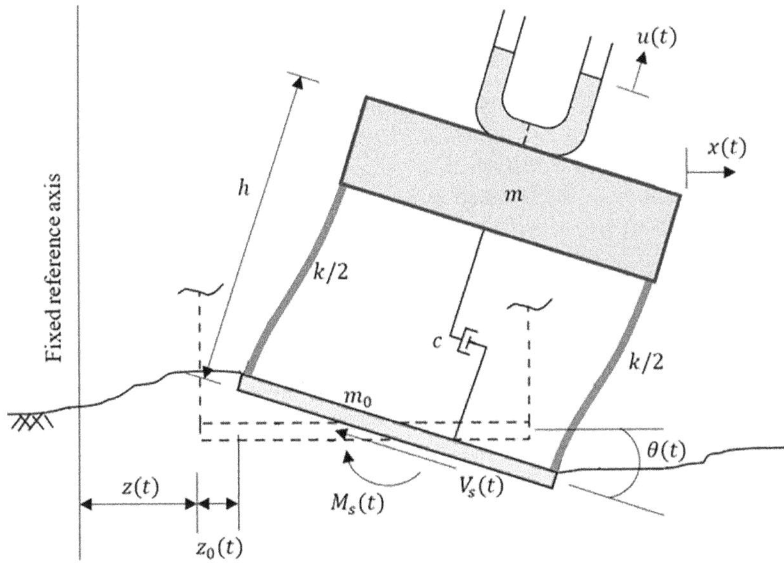

FIGURE 5.13 Soil-structure–tuned liquid column damper model.

and c, respectively. The height of the structure from the foundation is h. The mass of the foundation is denoted by m_0. The mass of the damper container is assumed to be lumped with the mass of the structure. Further, it is assumed that both the mass of the structure and the mass of liquid in the horizontal limb of the TLCD are concentrated at the roof level of the structure.

Neglecting the rocking component of the free-field seismic excitation and the impact of kinematic interaction, the input motion to the foundation is considered to be equal to the free-field ground translation, denoted by $z(t)$. The rotational and the translational displacements of the foundation, relative to the soil medium, are represented by $z_0(t)$ and $\theta(t)$ respectively. The lateral displacement of the structure relative to the foundation is denoted by $x(t)$. The vertical displacement of the liquid column in the TLCD is denoted by $u(t)$. Now, considering a small displacement assumption, the overall lateral displacement of the structure may be written as, $\left[z(t) + z_0(t) + x(t) + h\theta(t)\right]$. The base shear and the base moment acting on the foundation at the soil–structure interface are denoted by $V_s(t)$ and $M_s(t)$, respectively.

The consideration of the dynamic equilibrium of the SDOF superstructure yields

$$\ddot{x} + \frac{2\zeta_s \omega_s}{(1+\mu)}\dot{x} + \frac{\omega_s^2}{(1+\mu)}x = -\left(\ddot{z} + \ddot{z}_0 + h\ddot{\theta}\right) - \frac{\alpha\mu}{(1+\mu)}\ddot{u}. \tag{5.9}$$

In Eq. (5.9), ζ_s is the equivalent viscous damping ratio of the structure, and ω_s is the natural frequency of the structure.

The equation of motion of the liquid column in the TLCD may be written as

$$\ddot{u} + \frac{1}{2}\frac{\xi}{L}|\dot{u}|\dot{u} + \omega_{\text{TLCD}}{}^2 u = -\alpha\left(\ddot{x} + \ddot{z} + \ddot{z}_0 + h\ddot{\theta}\right) \tag{5.10}$$

with ω_{TLCD} representing the natural frequency of oscillation of the liquid column in the TLCD. Further, in Eq. (5.10), ξ is a dimensionless term that represents the head-loss coefficient of the orifice(s).

The consideration of the translational equilibrium of the entire structure–foundation system leads to

$$V_s + m_0\left(\ddot{z} + \ddot{z}_0\right) + m\left(\ddot{x} + \ddot{z} + \ddot{z}_0 + h\ddot{\theta}\right) + \rho AL\left(\ddot{x} + \ddot{z} + \ddot{z}_0 + h\ddot{\theta}\right) + \rho Ab\ddot{u} = 0 \tag{5.11}$$

in which L, b and A respectively denote the overall length, horizontal length and cross-sectional area of the liquid column of the TLCD. The density of the liquid residing in the TLCD is denoted by ρ. Further, the consideration of the rotational equilibrium of the whole structure–foundation system yields

$$M_s + I_T\ddot{\theta} + mh\left(\ddot{x} + \ddot{z} + \ddot{z}_0\right) + \rho ALh\left(\ddot{x} + \ddot{z} + \ddot{z}_0 + h\ddot{\theta}\right) + \rho Abh\ddot{u} = 0. \tag{5.12}$$

In Eq. (5.12), I_T is the mass moment of inertia of the structure–foundation system about a horizontal axis at the foundation level. The expression for I_T is given by

$$I_T = I_0 + I + mh^2 \tag{5.13}$$

where I_0 and I are the mass moment of inertia of the foundation and of the mass of the structure about the horizontal axis about their respective centres.

Equations (5.9)–(5.12) represent the governing equations of the soil-structure-TLCD system under seismic excitation. An extension of the above-mentioned procedure is available in the work of Farshidianfar and Soheili (2012), who developed the mathematical formulation of an MDOF shear building with TLCD, considering soil–structure interaction.

In the numerical analysis of the soil-structure-TLCD system, it is important to define the flexibility of the soil. Ghosh and Basu (2005) defined the soil flexibility in terms of the shear modulus G, mass density $\bar{\rho}$, Poisson's ratio and hysteretic damping ratio of the soil. On the other hand, Farshidianfar and Soheili (2012) modelled the soil flexibility through swaying stiffness, swaying damping, rocking stiffness and rocking damping of the soil. During actual analysis, these parameters may be determined from the soil parameters obtained from geotechnical investigations at the site. However, some reference values of these parameters for different soil conditions are available in the literature (Liu et al., 2008; Wolf, 1985).

It is well-known that when SSI is considered, due to soil flexibility, the fundamental natural period of the structure is lengthened as compared to the case when the structure is assumed to be fixed-base. Farshidianfar and Soheili (2012) reported that the fundamental natural period of a fixed-base structure may increase by more than 50% if the structure is supported on soft soil, leading to gross mistuning of the TLCD, if tuned to the host structure modelled as a fixed-based system. In Figure 5.14, the rms displacement response of the uncontrolled and the TLCD-controlled SDOF structural systems (the latter with and without consideration of SSI effects in obtaining the

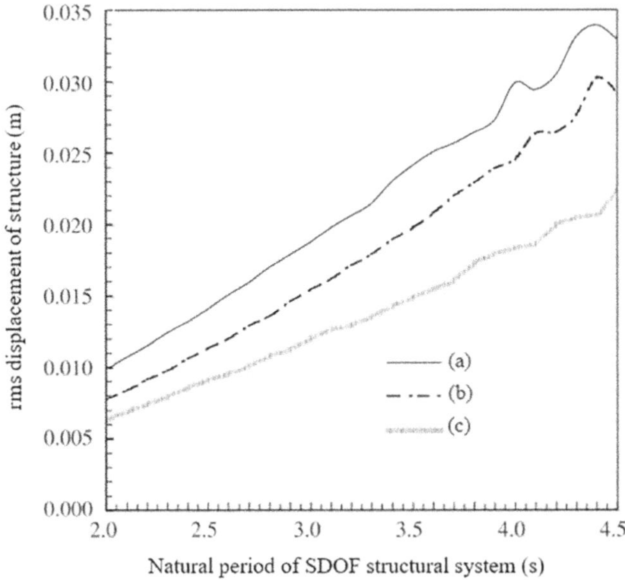

FIGURE 5.14   Root-mean-square (rms) displacement response of single-degree-of-freedom structural systems (a) without damper, (b) with tuned liquid column damper (TLCD) tuned to structural frequency determined considering a fixed support, and (c) with TLCD tuned to structural frequency determined considering soil–structure interaction (Ghosh and Basu, 2005).

tuning parameter) is presented (Ghosh and Basu, 2005). The structural systems were considered to be founded on soft soil having a shear wave velocity, $v_s\left[=\left(\sqrt{G/\bar{\rho}}\right)\right]$, equal to 100 m/s. Farshidianfar and Soheili (2012) also reported that apart from the tuning ratio, the optimum value of the other TLCD design parameters, especially the optimum head-loss coefficient of the orifice, may also be influenced by the SSI effects. These observations underline the importance of designing the TLCD properties by considering the SSI, especially when the structure is founded on soft soil.

Structural vibration control using CLCD, considering the effect of SSI, has been studied by Ghosh and Ghosh (2008). They showed that the rms displacement response of a short-period SDOF structural system (with damping ratio = 1%) resting on a soft soil (with v_s = 100 m/s) can be reduced by 41% using a CLCD having a mass ratio of 3% if the SSI effects are properly considered in the design of the damper. However, if that is not done, the damper would only produce a 4% reduction in the structural displacement. In a subsequent study, Ghosh and Ghosh (2009) considered an MDOF structural system and reported a similar result.

Mendes et al. (2019) investigated the SSI effects while designing sealed TLCD for seismic vibration control of a forty-storey building. Here also, when the structure was rested on soft soil, ignoring the SSI effect during damper design led to very poor performance of the damper.

REFERENCES

Adam, C., Matteo, A. Di, Furtmüller, T., Pirrotta, A., 2017. Earthquake excited base-isolated structures protected by tuned liquid column dampers: design approach and experimental verification. *Procedia Eng.* 199, 1574–1579. https://doi.org/10.1016/j.proeng.2017.09.060

Adhikary, A., Konar, T., Ghosh, A.D., 2023. Site-specific aseismic design of multi-storeyed buildings using optimum tuned mass damper inerter. *J. Earthq. Eng.* In press, 1–36. https://doi.org/10.1080/13632469.2023.2291127

Ahadi, P., Mohebbi, M., Shakeri, K., 2012. Using optimal multiple tuned liquid column dampers for mitigating the seismic response of structures. *ISRN Civ. Eng.* 2012, 1–6. https://doi.org/10.5402/2012/592181

Altunişik, A.C., Yetişken, A., Kahya, V., 2018. Experimental study on control performance of tuned liquid column dampers considering different excitation directions. *Mech. Syst. Signal Process.* 102, 59–71. https://doi.org/10.1016/j.ymssp.2017.09.021

Baker, J.W., 2007. Quantitative classification of near-fault ground motions using wavelet analysis. *Bull. Seismol. Soc. Am.* 97, 1486–1501. https://doi.org/10.1785/0120060255

Bhattacharyya, S., Ghosh, A.D., Basu, B., 2017a. Nonlinear modeling and validation of air spring effects in a sealed tuned liquid column damper for structural control. *J. Sound Vib.* 410, 269–286. https://doi.org/10.1016/j.jsv.2017.07.046

Bhattacharyya, S., Ghosh, A.D., Basu, B., 2017b. Experimental Investigations into CLCD with identification of tuning and damping effects. *J. Struct. Eng.* 143, 1–5. https://doi.org/10.1061/(ASCE)ST.1943-541X.0001788.

Bigdeli, Y., Kim, D., 2016. Damping effects of the passive control devices on structural vibration control: TMD, TLC and TLCD for varying total masses. *KSCE J. Civ. Eng.* 20, 301–308. https://doi.org/10.1007/s12205-015-0365-5

Bray, J.D., Rodriguez-Marek, A., 2004. Characterization of forward-directivity ground motions in the near-fault region. *Soil Dyn. Earthq. Eng.* 24, 815–828. https://doi.org/10.1016/j.soildyn.2004.05.001

Cavdar, E., Ozdemir, G., Bayhan, B., 2019. Significance of ground motion scaling parameters on amplitude of scale factors and seismic response of short- and long-period structures. *Earthq. Spectra* 35, 1663–1688. https://doi.org/10.1193/081718EQS204M

Chang, C.C., 1999. Mass dampers and their optimal designs for building vibration control. *Eng. Struct.* 21, 454–463. https://doi.org/10.1016/S0141-0296(97)00213-7

Chen, B., Sun, G., Li, H., 2022. Power spectral models of stationary earthquake-induced ground motion process considering site characteristics. *Emerg. Manag. Sci. Technol.* 2, 1–12. https://doi.org/10.48130/emst-2022-0011

Clough, R.W., Penzien, J., 1975. *Dynamics of Structures.* McGraw-Hill, New York, USA.

COSMOS, n.d. Strong-Motion Virtual Data Center: Consortium of Organizations for Strong Motion Observation Systems. https://www.strongmotioncenter.org/vdc/scripts/default.plx

Debbarma, R., Chakraborty, S., Ghosh, S., 2010a. Unconditional reliability-based design of tuned liquid column dampers under stochastic earthquake load considering system parameters uncertainties. *J. Earthq. Eng.* 14, 970–988. https://doi.org/10.1080/13632461003611103

Debbarma, R., Chakraborty, S., Ghosh, S., 2010b. Optimum design of tuned liquid column dampers under stochastic earthquake load considering uncertain bounded system parameters. *Int. J. Mech. Sci.* 52, 1385–1393. https://doi.org/10.1016/j.ijmecsci.2010.07.004

Di Matteo, A., Lo Iacono, F., Navarra, G., Pirrotta, A., 2015. Optimal tuning of tuned liquid column damper systems in random vibration by means of an approximate formulation. *Meccanica* 50, 795–808. https://doi.org/10.1007/s11012-014-0051-6

Ding, H., Wang, J.T., Altay, O., Lu, L.Q., 2021a. Multilayer toroidal tuned liquid column dampers for seismic vibration control of structures. *Structures* 33, 406–422. https://doi.org/10.1016/j.istruc.2021.04.041

Ding, H., Wang, J.T., Lu, L.Q., Pan, J.W., 2021b. Experimental comparison of nonlinear damping performance of toroidal and conventional tuned liquid column dampers. *Nonlinear Dyn.* 104, 3365–3384. https://doi.org/10.1007/s11071-021-06552-7

Ding, H., Wang, J.T., Lu, L.Q., Zhu, F., 2020. A toroidal tuned liquid column damper for multidirectional ground motion-induced vibration control. *Struct. Control Heal. Monit.* 27, e2558. https://doi.org/10.1002/stc.2558

Farshidianfar, A., Soheili, S., 2012. Optimized tuned liquid column dampers for earthquake oscillations of high-rise structures including soil effects. *Int. J. Optim. Civ. Eng.* 2, 221–234.

Fei, Z., Jinting, W., Feng, J., Liqiao, L., 2019. Control performance comparison between tuned liquid damper and tuned liquid column damper using real-time hybrid simulation. *Earthq. Eng. Eng. Vib.* 18, 695–701. https://doi.org/10.1007/s11803-019-0530-9

Ferreira, F., Moutinho, C., Cunha, Á., Caetano, E., 2020. An artificial accelerogram generator code written in Matlab. *Eng. Rep.* 2, 1–17. https://doi.org/10.1002/eng2.12129

Furtmüller, T., Di Matteo, A., Adam, C., Pirrotta, A., 2019. Base-isolated structure equipped with tuned liquid column damper: an experimental study. *Mech. Syst. Signal Process.* 116, 816–831. https://doi.org/10.1016/j.ymssp.2018.06.048

Ghosh, A., Basu, B., 2008. Seismic vibration control of nonlinear structures using the liquid column damper. *J. Struct. Eng.* 134, 146–153. https://doi.org/10.1061/(asce)0733-9445(2008)134:1(146)

Ghosh, A., Basu, B., 2007. Alternative approach to optimal tuning parameter of liquid column damper for seismic applications. *J. Struct. Eng.* 133, 1848–1852. https://doi.org/10.1061/(ASCE)0733-9445(2007)133:12(1848)

Ghosh, A., Basu, B., 2005. Effect of soil interaction on the performance of liquid column dampers for seismic applications. *Earthq. Eng. Struct. Dyn.* 34, 1375–1389. https://doi.org/10.1002/eqe.485

Ghosh, A., Gangopadhyay, A., Basu, B., 2011. Performance investigation of multiple compliant liquid column dampers for control of seismic vibrations, in: *Proceedings of the 8th International Conference on Structural Dynamics,* Lueven, Belgium, *EURODYN, 2011.* pp. 1671–1677.

Ghosh, A.D., Basu, B., 2004. Seismic vibration control of short period structures using the liquid column damper. *Eng. Struct.* 26, 1905–1913. https://doi.org/10.1016/j.engstruct.2004.07.001

Ghosh, R.K., Ghosh, A.D., 2009. Passive control of seismic response of soil-structure system by the compliant liquid column damper. *Int. J. Mater. Struct. Integr.* 3, 332–352. https://doi.org/10.1504/IJMSI.2009.029340

Ghosh, R.K., Ghosh, A.D., 2008. Soil interaction effects on the performance of compliant liquid column damper for seismic vibration control of short period structures. *Struct. Eng. Mech.* 28, 89–105. https://doi.org/10.12989/sem.2008.28.1.089

Haroun, M.A., Pires, J.A., Won, A.Y.J., 1994. Hybrid liquid column dampers for suppression of environmentally- induced vibrations in tall buildings, in: *Third Conference Tall Buildings in Seismic Regions.* Los Angeles, USA.

Heo, J., Lee, Sung-kyung, Park, E., Lee, Sang-hyun, Min, K., Kim, H., Jo, J., Cho, B., 2009. Performance test of a tuned liquid mass damper for reducing bidirectional responses of building structures. *Struct. Des. Tall Build.* 18, 789–805.

Housner, G.W., 1947. Characteristics of strong-motion earthquakes. *Bull. Seismol. Soc. Am.* 37, 19–31. https://doi.org/10.1785/BSSA0370010019

Hu, Y., Zhou, X., 1962. The response of the elastic system under the stationary and nonstationary ground motions, in: *Earthquake Engineering Research Report No. 1.* Institute of Civil Engineering, Chinese Academy of Sciences, Science Press, Beijing, China, pp. 33–55.

IS:1893(Part-1), 2016. *IS 1893 (Part 1)-Criteria for Earthquake Resistant Design of Structures : General Provisions and Buildings.* Bureau of Indian Standards, New Delhi.

Jangid, R.S., 2022. Performance and optimal design of base-isolated structures with clutching inerter.pdf. *Struct. Control Heal. Monit.* 29, e3000. https://doi.org/https://doi.org/10.1002/stc.3000

Kanai, K., 1957. Semi-empirical formula for the seismic characteristics of the ground. *Bull. Earthq. Res. Inst. Tokyo Univ.* 35, 308–325. https://doi.org/10.15083/0000033949

Kavand, A., Zahrai, S.M., 2006. Impact of seismic excitation characteristics on the efficiency of tuned liquid column dampers. *Earthq. Eng. Eng. Vib.* 5, 235–243. https://doi.org/10.1007/s11803-006-0639-5

Konar, T., Ghosh, A.D., 2023. A review on various configurations of the passive tuned liquid damper. *J. Vib. Control* 29, 1945–1980. https://doi.org/10.1177/10775463221074077

Konar, T., Ghosh, A.D., 2013. Bimodal vibration control of seismically excited structures by the liquid column vibration absorber. *J. Vib. Control* 19, 385–394. https://doi.org/10.1177/1077546311430718

Konar, T., Ghosh, A.D., 2010. Passive control of seismically excited structures by the liquid column vibration absorber. *Struct. Eng. Mech.* 36, 561–573. https://doi.org/10.12989/sem.2010.36.5.561

Konar, T., Ghosh, A.D., Basu, B., 2024. Real-world installations of tuned liquid column dampers for wind- induced vibration control of buildings: some important case studies. *Struct. Infrastruct. Eng.* in press. https://doi.org/10.1080/15732479.2024.2420174

Lam, N., Wilson, J., Hutchinson, G., 2000. Generation of synthetic earthquake accelerograms using seismological modelling: a review. *J. Earthq. Eng.* 4, 321–354. https://doi.org/10.1080/13632460009350374

Lee, S.K., Park, E.C., Min, K.W., Lee, S.H., Park, J.H., 2007. Experimental implementation of a building structure with a tuned liquid column damper based on the real-time hybrid testing method. *J. Mech. Sci. Technol.* 21, 885–890. https://doi.org/10.1007/BF03027063

Liu, M.Y., Chiang, W.L., Hwang, J.H., Chu, C.R., 2008. Wind-induced vibration of high-rise building with tuned mass damper including soil-structure interaction. *J. Wind Eng. Ind. Aerodyn.* 96, 1092–1102. https://doi.org/10.1016/j.jweia.2007.06.034

Masnata, C., Adam, C., Pirrotta, A., 2024. Optimal design of short-period structures equipped with sliding tuned liquid column damper and numerical and experimental control performance evaluation. *Acta Mech.* 235, 1603–1622. https://doi.org/10.1007/s00707-023-03832-8

Mendes, M.V., Ribeiro, P.M.V., Pedroso, L.J., 2019. Effects of soil-structure interaction in seismic analysis of buildings with multiple pressurized tuned liquid column dampers. *Lat. Am. J. Solids Struct.* 16, 1–21. https://doi.org/10.1590/1679-78255707

Min, K.W., Jang, S.J., Kim, J., 2016. A standalone vision sensing system for pseudo-dynamic testing of tuned liquid column dampers. *J. Sensors* 2016. https://doi.org/10.1155/2016/8152651

Moayyad, P., 1982. *A Study of Power Spectral Density of Earthquake Accelerograms.* Southern Methodist University, Dallas.

Mohebbi, M., Dabbagh, H.R., Shakeri, K., 2015. Optimal design of multiple tuned liquid column dampers for seismic vibration control of MDOF structures. *Period. Polytech. Civ. Eng.* 59, 543–558. https://doi.org/10.3311/PPci.7645

NA+A1:2014 to BS EN 1993-1-1:2005+A1:2014, 2014. NA+A1:2014 to BS EN 1993-1-1:2005+A1:2014 UK National Annex To Eurocode 3: Design of Steel Structures. EN

Nakashima, M., Kato, H., Takaoka, E., 1992. Development of real-time pseudo dynamic testing. *Earthq. Eng. Struct. Dyn.* 21, 79–92. https://doi.org/10.1002/eqe.4290210106

NZS1170–5(S1), 2004. NZS 1170–5 (S1) (2004): *Structural Design Actions-Part 5: Earthquake Actions - New Zealand Commentary.* Standards New Zealand, Wellington.

Ou, J., Liu, H., 1994. Random seismic response spectrum and its application based on the random seismic ground motion. *Earthq. Eng Eng Vib* 14, 14–23.

Park, B.J., Lee, Y.J., Park, M.J., Ju, Y.K., 2018. Vibration control of a structure by a tuned liquid column damper with embossments. *Eng. Struct.* 168, 290–299. https://doi.org/10.1016/j.engstruct.2018.04.074

PEER, n.d. *PEER Ground Motion Database*. Pacific Earthquake Engineering Research Center. https://ngawest2.berkeley.edu/

Reiterer, M., Ziegler, F., 2006. Control of pedestrian-induced vibrations of long-span bridges. *Struct. Control Heal. Monit.* 13, 1003–1027. https://doi.org/10.1002/stc.91

Roy, A.K., Ghosh, A. (Dey), 2015. A Study on the Design Parameters of the Compliant LCD for Structural Vibration Control Under Near Fault Earthquakes, in: Matsagar, V. (Ed.), *Advances in Structural Engineering: Dynamics: Select Proceedings of Structural Engineering Convention (SEC), 2014*, Vol. 2. Springer, India, pp. 1243–1255. https://doi.org/10.1007/978-81-322-2193-7_97

Roy, A.K., Konar, T., Ghosh, A., 2023. Mitigation of structural vibrations due to pulse-type-near-fault earthquake by the compliant liquid column damper. *J. Earthq. Tsunami* 17, 2350004. https://doi.org/10.1142/S1793431123500045

Rozas, L., Boroschek, R.L., Tamburrino, A., Rojas, M., 2016. A bidirectional tuned liquid column damper for reducing the seismic response of buildings. *Struct. Control Heal. Monit.* 23, 621–640. https://doi.org/10.1002/stc.91

Sadek, F., Mohraz, B., Lew, H.S., 1998. Single- and multiple-tuned liquid column dampers for seismic applications. *Earthq. Eng. Struct. Dyn.* 27, 439–463. https://doi.org/10.1002/(SICI)1096-9845(199805)27:5<439::AID-EQE730>3.0.CO;2-8

Samali, B., 1990. Dynamic response of structures equipped with Tuned Liquid Column Dampers, in: *Australian Vibration and Noise Conference 1990: Vibration and Noise-Measurement Prediction and Control*. Institution of Engineers, Australia, pp. 138–143.

Samali, B., Mayol, E., Mack, A., Kwok, K.C.S., Hitchcock, P.A., 2002. Vibration control of a five storey benchmark building excited by earthquake using liquid column vibration absorbers, in: *Proceedings of the Third Australasian Congress on Applied Mechanics*. World Scientific Publishing Company, Sydney, Australia, pp. 641–646. https://doi.org/10.1142/9789812777973_0104

Shah, M.U., Shah, S.W., Farooq, S.H., Usman, M., Ullah, F., 2023. Experimental investigation of tuned liquid column ball damper's position on vibration control of structure using different fluids. *Innov. Infrastruct. Solut.* 8, 111. https://doi.org/10.1007/s41062-023-01080-2

Shah, M.U., Usman, M., 2022. An experimental study of tuned liquid column damper controlled multi-degree of freedom structure subject to harmonic and seismic excitations. *PLoS One* 17, e0269910. https://doi.org/10.1371/journal.pone.0269910

Sues, R.H., Wen, Y.-K., Ang, A.H.-S., 1983. *Stochastic Seismic Performance Evaluation of Buildings*. University of Illinois at Urbana-Champaign, IL.

Sun, K., 1994. Earthquake responses of buildings with liquid column dampers, in: *Fifth US National Conference on Earthquake Engineering*. Chicago, IL, USA, pp. 411–420.

Suzuki, Y., Minai, R., 1988. Application of stochastic differential equations to seismic reliability analysis of hysteretic structures. *Probabilistic Eng. Mech.* 3, 43–52. https://doi.org/10.1016/0266-8920(88)90007-0

Tajimi, H., 1960. A statistical method for determining the maximum response of a building structure during an earthquake, in: *Proceedings of 2nd World Conference on Earthquake Engineering*. Tokyo and Kyoto, Japan, pp. 781–797.

Takewaki, I., 2000. Soil-structure random response reduction via TMD-VD simultaneous use. *Comput. Methods Appl. Mech. Eng.* 190, 677–690. https://doi.org/10.1016/S0045-7825(99)00434-X

The Building Standard Law of Japan, 2016. The Building Standard Law of Japan. Tokyo, Japan

Wolf, J.P., 1985. *Dynamic Soil-Structure Interaction.* Prentice-Hall, Inc., Englewood Cliffs, NJ, USA.

Won, A.Y.J., Pires, J.A., Haroun, M.A., 1997. Performance assessment of tuned liquid column dampers under random seismic loading. *Int. J. Non. Linear. Mech.* 32, 745–758. https://doi.org/10.1016/s0020-7462(96)00118-7

Won, A.Y.J., Pires, J.A., Haroun, M.A., 1996. Stochastic seismic performance evaluation of tuned liquid column dampers. *Earthq. Eng. Struct. Dyn.* 25, 1259–1274. https://doi.org/10.1002/(SICI)1096-9845(199611)25:11<1259::AID-EQE612>3.0.CO;2-W

Wong, H.L., Trifunac, M.D., 1979. Generation of artificial strong motion accelerograms. *Earthq. Eng. Struct. Dyn.* 7, 509–527. https://doi.org/10.1002/eqe.4290070602

Wu, J., Chang, C., 2006. Design table of optimal parameters for tuned liquid column damper responding to earthquake, in: *4th International Conference on Earthquake Engineering.* Taipei, Taiwan, p. Paper No. 165.

Zhu, F., Wang, J.T., Jin, F., Lu, L.Q., 2017. Real-time hybrid simulation of full-scale tuned liquid column dampers to control multi-order modal responses of structures. *Eng. Struct.* 138, 74–90. https://doi.org/10.1016/j.engstruct.2017.02.004

Zhu, F., Wang, J.T., Jin, F., Lu, L.Q., 2016. Seismic performance of tuned liquid column dampers for structural control using real-time hybrid simulation. *J. Earthq. Eng.* 20, 1370–1390. https://doi.org/10.1080/13632469.2016.1138170

Zhu, F., Wang, J.T., Prof, F.J., Altay, O., 2015. Real-time hybrid simulation of single and multiple tuned liquid column dampers for controlling seismic-induced response, in: *International Conference on Advances in Experimental Structural Engineering.* University of Illinois, Urbana-Champaign, IL, USA.

6 TLCD for Control of Structures Subjected to Other Types of Excitations

6.1 INTRODUCTION

Apart from wind and earthquakes, several other natural and man-made factors can induce vibration in civil engineering structures. These include ocean waves, pedestrian movement, vehicular traffic, etc. In many instances, the dynamic loading from these sources is the predominant load that guides the design of the structure. The use of supplemental damping devices in mitigating vibrations induced by these loads has been well documented in the literature. For example, the London Millennium Footbridge was closed on the day of its opening due to excessive lateral vibration due to pedestrian loading. Later, the bridge was fitted with a fluid viscous damper and tuned mass damper (TMD) to reduce the vibration within the acceptable limit (Dallard et al., 2001). There are several bridges where the vibration due to vehicular movement has required special attention. In an extraordinary event, the I-35 W Mississippi River bridge in Minnesota, USA, collapsed in the year 2007 due to traffic loading and related vibration during rush hour (Crawford, 2022). Further, as mentioned in Section 1.4 of Chapter 1, the development of the first tuned liquid column damper (TLCD)-like device, namely the U-shaped anti-roll tank (Frahm, 1911), had the objective of mitigating the rolling motion of ships due to ocean waves (Alujević et al., 2019; Konar and Ghosh, 2023). After a long span of time, several attempts have been devoted to the study of the potential of the TLCD in controlling structural vibrations induced by the aforementioned sources of dynamic loading. In this chapter, the focus is on the design, performance assessment and optimization of TLCD under excitations other than wind and earthquake, along with the present challenges and future prospects of this device.

6.2 CONTROL OF WAVE-INDUCED EXCITATION BY TLCD

Offshore structures such as floating platforms, jacket platforms, floating breakwater and wind turbines are continuously exposed to wave loading. The vibration induced by the wave loading alone or along with the effect of other environmental conditions, such as ocean currents, wind, and occasional earthquakes and tsunamis, may cause significant damage and even failure of offshore structures (Chakrabarti, 2005). Vibrations may also cause serviceability issues in offshore structures by affecting

DOI: 10.1201/9781003377894-6

functioning of the machinery and the working efficiency of the crew. Thus, for the survival and proper functioning of these structures, the effects of the wave and other environmental loadings on them are considered in their design. Sometimes, installation of supplemental damping devices is considered as a potential solution to reduce wave-induced vibrations in offshore structures (Kandasamy et al., 2016). Being a popular passive damping device, the use of the TLCD to dampen wave-induced vibration has been studied by many researchers, which is discussed in this section.

First, a brief discussion on the mathematical modelling of wave excitation is presented. Despite being random in nature, it is a common practice to represent an extreme wave by a regular wave of a specific height and period. The most popular theory on regular wave form in case of low sea states (with small wave amplitudes) is based on Airy's linear wave theory (Airy, 1841). This theory assumes sinusoidal waves (see Figure 6.1) and is applicable where the wave height, H is less than that of the wavelength, λ and the water depth, d. The free surface profile of water waves, $\eta(x,t)$ at a horizontal location, x and time, t is given by

$$\eta(x,t) = \frac{H}{2}\cos(kx - \omega t). \tag{6.1}$$

In Eq. (6.1), k is the wave number given by $2\pi/\lambda$, and ω is the wave frequency given by $2\pi/T$, where T is the period of the wave.

Consider a two-dimensional wave field with X and Y denoting the horizontal and vertical directions, respectively, and a flat sea bottom. Let the instantaneous velocities at (x,y) along the X and Y axes be denoted by $u(x,y,t)$ and $v(x,y,t)$, respectively, and the corresponding accelerations be denoted by $\dot{u}(x,y,t)$ and $\dot{v}(x,y,t)$, respectively. These are expressed as

$$u(x,y,t) = \frac{\pi H}{T}\frac{\cosh k(d+y)}{\sinh kd}\cos(kx - \omega t), \tag{6.2}$$

$$v(x,y,t) = \frac{\pi H}{T}\frac{\cosh k(d+y)}{\sinh kd}\sin(kx - \omega t), \tag{6.3}$$

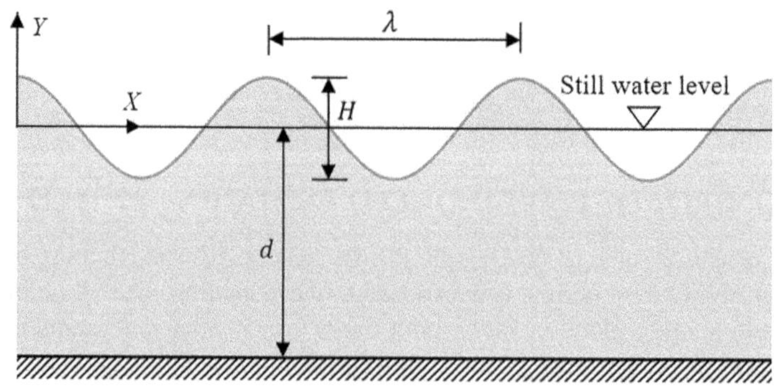

FIGURE 6.1 Airy's wave.

$$\dot{u}(x,y,t) = 2\pi \left(\frac{H}{T}\right)^2 \frac{\cosh k(d+y)}{\sinh kd} \sin(kx - \omega t) \qquad (6.4)$$

and

$$\dot{v}(x,y,t) = 2\pi \left(\frac{H}{T}\right)^2 \frac{\cosh k(d+y)}{\sinh kd} \cos(kx - \omega t). \qquad (6.5)$$

Since Airy's linear wave theory is applicable to small wave heights, it may be incapable of providing a satisfactory assessment of the water particle kinematics in case of storm waves. In such cases, Stokes wave theories (Stokes, 1847) are widely employed. The second-order Stokes wave theory delineates two components for the water wave kinematics: one at the wave frequency, and the other at twice the wave frequency. The contribution of the second-order component is typically smaller than that of the first-order. The superposition of these components shapes the wave profile, resulting in a steeper crest and shallower trough. The mathematical description of the free surface profile of water, $\eta(x,t)$ by this theory is expressed as

$$\eta(x,t) = \frac{H}{2}\cos(kx - \omega t) + \frac{\pi H^2}{8\lambda}\frac{\cosh kd}{\sinh^3 kd}[2 + \cosh 2kd]\cos 2(kx - \omega t). \qquad (6.6)$$

Further, the velocities, $u(x,y,t)$ and $v(x,y,t)$ and the accelerations, $\dot{u}(x,y,t)$ and $\dot{v}(x,y,t)$ in the horizontal and vertical directions are expressed as

$$u(x,y,t) = \frac{\pi H}{T}\frac{\cosh k(d+y)}{\sinh kd}\cos(kx - \omega t) + \frac{3}{4c}\left(\frac{\pi H}{T}\right)^2 \frac{\cosh 2k(d+y)}{\sinh^4 kd}\cos 2(kx - \omega t),$$

$$(6.7)$$

$$v(x,y,t) = \frac{\pi H}{T}\frac{\cosh k(d+y)}{\sinh kd}\sin(kx - \omega t) + \frac{3}{4c}\left(\frac{\pi H}{T}\right)^2 \frac{\sinh 2k(d+y)}{\sinh^4 kd}\sin 2(kx - \omega t),$$

$$(6.8)$$

$$\dot{u}(x,y,t) = 2\pi \left(\frac{H}{T}\right)^2 \frac{\cosh k(d+y)}{\sinh kd}\sin(kx - \omega t) + \frac{3\pi}{2L}\left(\frac{\pi H}{T}\right)^2$$

$$\frac{\cosh 2k(d+y)}{\sinh^4 kd}\sin 2(kx - \omega t) \qquad (6.9)$$

and

$$\dot{v}(x,y,t) = 2\pi \left(\frac{H}{T}\right)^2 \frac{\cosh k(d+y)}{\sinh kd}\cos(kx - \omega t)$$

$$+ \frac{3\pi}{4L}\left(\frac{\pi H}{T}\right)^2 \frac{\sinh 2k(d+y)}{\sinh^4 kd}\cos 2(kx - \omega t). \qquad (6.10)$$

In Eqs. (6.7) and (6.8), the wave speed c may be obtained from

$$c^2 = \frac{g}{k}\tanh kd. \tag{6.11}$$

Stokes third-order and fifth-order wave theories, cnoidal wave theory and solitary wave theory are the other notable wave theories (Chakrabarti, 2005; Fenton, 1990), all of which have their own advantages and disadvantages and are valid for specific sea conditions. The approximate range of validity of the different wave theories is provided by Le Méhauté (1976). It is worth pointing out that the studies so far had been limited to either on the linear waves or waves perturbed from the linear waves. Hence, large-amplitude nonlinear waves have not been studied or possible to be simulated for application to problems in science and engineering.

There has been some recent breakthrough in the research on large-amplitude nonlinear water waves due to the seminal work of Constantin and Strauss (2004) proving the existence of large-amplitude waves with vorticity on finite depth. Following this work, large-amplitude nonlinear waves have been simulated by Ko and Strauss (2008), Constantin et al. (2015) and Kalimeris (2018). Details of further theoretical and numerical developments on large-amplitude nonlinear waves are available in Basu (2020), Basu and Kolgelbauer (2021), Chen and Basu (2021a,b) with some recent applications in Nguyen and Manuel (2024), Paul et al. (2024).

In order to estimate the wave force on an offshore structure, such as the jacket platform, the Morison equation (Morison et al., 1950) given by

$$F(t) = C_m \rho \frac{\pi D^2}{4}\dot{u} + 0.5 C_d D \rho |u| u \tag{6.12}$$

is often used, which provides an expression for the hydrodynamic force per unit length, $F(t)$, exerted by surface waves on a cylindrical structural element. In Eq. (6.12), D denotes the diameter of the cylindrical element, ρ represents the density of sea water, C_d denotes the drag coefficient, and C_m is the coefficient of the hydrodynamic inertia. The first term on the right-hand side of Eq. (6.12) is essentially the sum of the Froude–Krylov force and the hydrodynamic mass force. The Froude–Krylov force represents the force due to the pressure field generated by the undisturbed waves and is given by $\rho\left(\pi D^2/4\right)\dot{u}$. The hydrodynamic mass force represents the force due to the inertia of the surrounding fluid that moves along with the structure. It accounts for the so-called "added mass" and is represented by $C_a\rho\left(\pi D^2/4\right)\dot{u}$, where C_a is the added mass coefficient. Thus, $C_m = 1 + C_a$. The second term on the right-hand side of Eq. (6.12) represents the viscous drag force acting on the structural element.

Over the years, an extensive amount of data on ocean waves has been measured and compiled with the objective to characterize sea states. Hogben et al. (1986) amassed a comprehensive dataset from 104 ocean areas covering major shipping routes. Despite the irregular nature of ocean waves, they exhibit frequency-dependent characteristics that adhere to a recognizable spectral description of the wave energy. The widely used spectrum formulas include the Bretschneider model (Bretschneider, 1959), the Pierson–Moskowitz model (Pierson and Moskowitz, 1964) and the Joint North Sea Wave Project (JONSWAP) model (Hasselmann et al., 1973).

Bretschneider (1959) described the wave spectrum in the form

$$S(\omega) = 0.1687 H_s \frac{\omega_s^{\,4}}{\omega^5} \exp\!\left(-0.675 \frac{\omega_s^{\,4}}{\omega^4}\right) \tag{6.13}$$

using two parameters, namely the significant wave height (H_s), given by the average height of the highest one third waves in a short-term record, and the modal wave frequency (ω_s), given by the average frequency corresponding to the significant waves in the short-term record.

The spectral description proposed by Pierson-Moskowitz (1964) for a fully developed sea state may be written as

$$S(\omega) = \alpha g^2 \omega^{-5} \exp\!\left(-1.25 \frac{\omega_0^{\,4}}{\omega^4}\right) \tag{6.14}$$

where, α denotes the Phillips constant, the value of which is usually taken as 8.1×10^{-3}, and ω_0 denotes the frequency at which the energy density spectrum peaks.

The JONSWAP spectrum, which may be viewed as a modified form of the Pierson–Moskowitz spectrum, is expressed as (Hasselmann et al., 1973)

$$S(\omega) = \bar{\alpha} g^2 \omega^{-5} \left[\exp\!\left(-1.25 \frac{\omega_m^{\,4}}{\omega^4}\right)\right] \gamma^{\,\exp\!\left[-(\omega-\omega_m)^2/\left(2\sigma^2\omega_m^2\right)\right]}. \tag{6.15}$$

In Eq. (6.15), $\bar{\alpha}, \sigma$ and γ are defined as the modified Phillips constant, spectral width parameter and the peak enhancement factor, respectively, while ω_m is the dominant frequency expressed as

$$\omega_m = 22\left(\frac{g^2}{U_{10}F}\right)^{1/3}. \tag{6.16}$$

In Eq. (6.16), U_{10} and F denote the mean wind speed at 10 m height from the sea surface and the fetch, defined by the uninterrupted distance over which the wind blows (measured in the direction of the wind) without a significant change of direction, respectively.

Apart from the three spectra described above, there are several other spectra that have been developed to characterize ocean waves. A more detailed discussion on the mathematical descriptions of wave spectra is available in Chakrabarti (2005). The choice of the wave spectrum would naturally depend on the region of applicability of the available spectra and should be done judiciously as there may be significant variations between them. This is evident in Figure 6.2 which presents a comparison of the Bretschneider spectrum and the JONSWAP spectrum for significant wave height equal to 0.5 m and modal period equal to 4 s. In the absence of a site-specific spectral form, selection guidelines for adoption of the wave spectrum are provided in Chakrabarti (2005).

A large volume of research on the TLCD-assisted control of wave-induced structural vibration is directed towards vibration control of offshore wind turbines. Over the past few decades, offshore wind turbines have been considered viable alternatives to fossil fuel–dependent electricity production. Offshore wind turbines may be fixed-type or floating-type (see Figure 6.3a–f). In the nacelle of a wind turbine, there

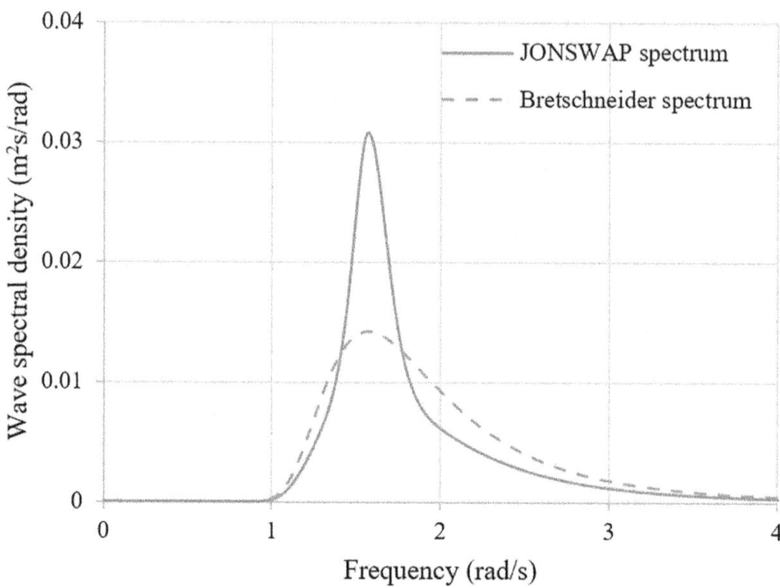

FIGURE 6.2 A comparison of an example Bretschneider spectrum and an example Joint North Sea Wave Project spectrum.

are several mechanical and electrical equipment that convert wind energy to electrical energy. Due to the high slenderness of offshore wind turbine structures, excessive vibration is highly detrimental to the functioning of the equipment in the nacelle.

The earliest study on the use of the TLCD to dampen joint wind-wave-induced vibration was undertaken by Colwell and Basu (2009), who examined an offshore wind turbine on a fixed monopile modelled as a multi-degree-of-freedom (MDOF) system under the combined action of wave and wind loadings. Two different approaches were followed for the modelling of the wind turbine. In the first model, the wind turbine tower was modelled, and the mass of turbine blades was considered to be lumped at the nacelle. In the second model, to capture the influence of rotating blades on the response of the wind turbine tower, a sub-structure approach was followed (Murtagh et al., 2005). Initially, the response of the rotating blades to wind loading was calculated. The interaction between the blades and the tower was then modelled by considering the motion at the top of the tower and the transfer of shear forces from the blades to the nacelle level of the tower. Wave excitation was modelled using the JONSWAP spectrum, while wind excitation was represented by the Kaimal spectrum (see Section 4.2 of Chapter 4) A TLCD having a mass ratio of 1% and a length ratio of 0.2 was considered. Here, such an unusually low length ratio was considered to minimize the horizontal space occupied by the damper, as horizontal space in the nacelle is at a premium. The results indicated that when an offshore wind turbine is equipped with a designed TLCD, significant reductions, of about 60%, in peak response were achieved as compared to systems without TLCDs (see Figure 6.4a and b). Additionally, it was observed that the installation of a TLCD enhanced the fatigue life of the wind turbine by more than 3.5 times. Hemmati et al. (2019) also considered monopile-supported wind turbines and developed a numerical

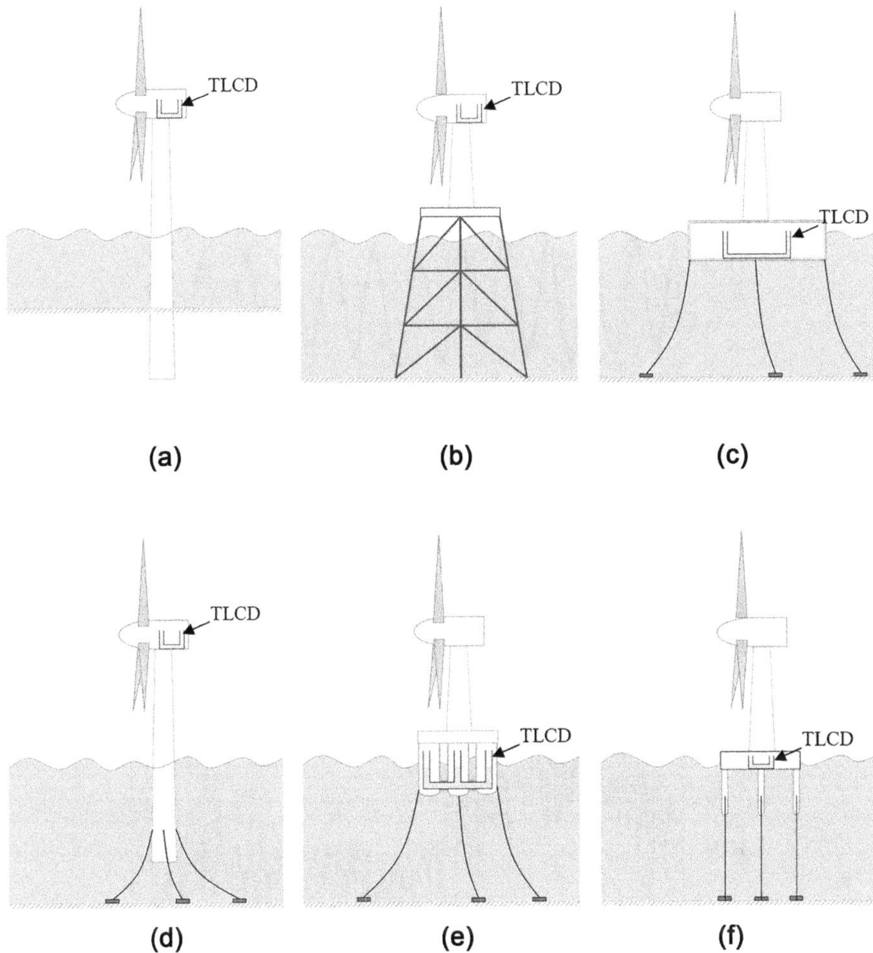

FIGURE 6.3 Tuned liquid column damper in different types of offshore wind turbines (a) fixed-type on monopile, (b) fixed-type on jacket, (c) barge-type floating, (d) spar-type floating, (e) semi-submersible-type floating and (f) tension leg platform-type floating.

model of the wind turbine–TLCD system that can take care of the nonlinearities of TLCDs as well as soil–pile interaction. The study considered the TLCD for controlling wave, wind and earthquake-induced vibration of a fixed-type offshore wind turbine. Wave and wind excitations were again modelled using the JONSWAP and Kaimal spectra, respectively. For the seismic loading, 30 scaled recorded ground motions were considered. The results indicated that under the combined action of wave-wind-seismic loading, a TLCD with a 2.5% mass ratio reduced the peak displacement at the top of the wind turbine by 13% in the operational condition and by 50% in the parked condition. Further, fragility curves under multi-hazard conditions were established following the

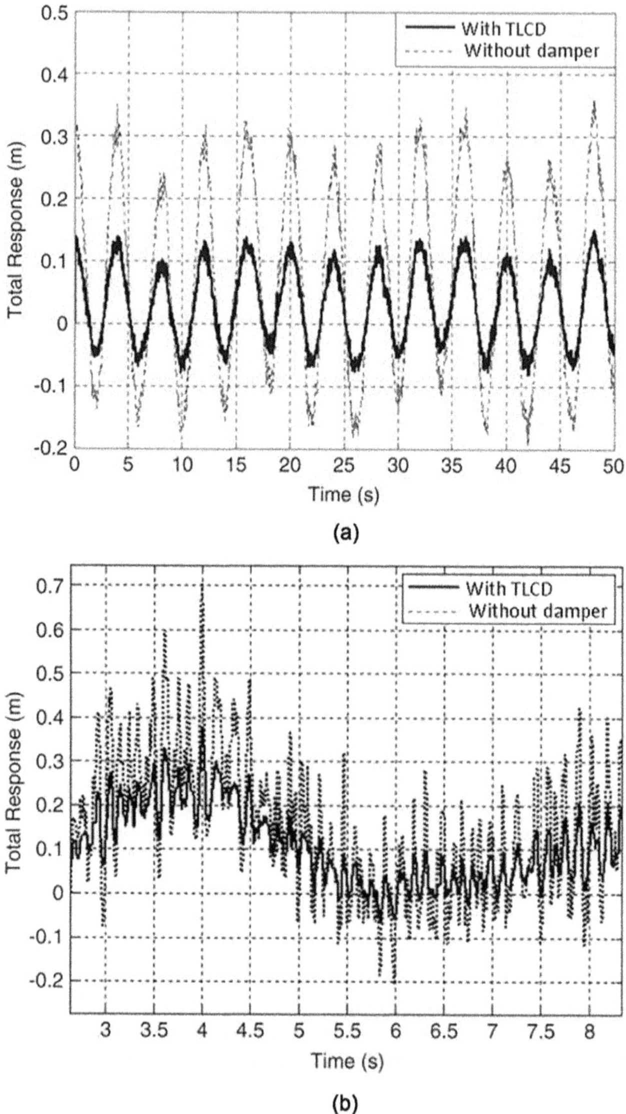

FIGURE 6.4 Time history of the displacement response of offshore wind turbine with rotating blades under the combined effect of wind and wave with and without tuned liquid column damper (a) "moderate" wind and wave excitation and (b) "strong" wind and wave excitation (Colwell and Basu, 2009).

approach of likelihood maximization. The installation of the TLCD was found to reduce the probability of failure of the wind turbine under multi-hazard conditions.

TLCDs have also been studied for wave-induced vibration control of different floating-type offshore wind turbines. Coudurier et al. (2015) conducted one of the earliest studies in this regard, considering barge-supported wind turbine. The wind turbine–barge system was modelled as a floating rigid body, and the study was carried out in the frequency domain using the JONSWAP wave loading spectrum. The TLCD was considered to be placed within the floater. It was reported that with a 2% mass ratio, the TLCD reduced the rms value of the pitch of the floater by 27%. In a further work, Coudurier et al. (2018) introduced a special type of TLCD, with multiple vertical limbs interconnected through different arrangements of horizontal limbs, for the control of the roll and pitch motions of the barge-supported wind turbine under wave loading. Although the authors have referred to the special TLCD proposed by them as the tuned liquid multi-column damper (TLMCD), it is essentially the forerunner of the omnidirectional TLCD (see Section 3.2.3 of Chapter 3). The damper was located in the barge, and under regular wave loading, a mass ratio of 4% could achieve significant and uniform response reductions for different wave incidences in comparison to that of a conventional TLCD. In a more recent study on the barge-supported wind turbine, Han et al. (2023) conducted a detailed parameter optimization for the design of a TLCD to reduce the tower-top displacement under both wind and wave loading, modelled by the Kaimal spectrum and the JONSWAP spectrum, respectively. In this work, a 3-DOF model of the structure-TLCD system was considered. The results indicated that the TLCD would be effective only with very large mass ratios (as high as 15%). The reason for this was possibly because the model of the barge-wind-turbine-TLCD system considered a hinge at the base of the wind turbine, and the TLCD was located within the barge.

Some experimental studies on offshore wind turbines with TLCD have also been carried out. Scaled models of wind turbine on tension leg platform and semi-submersible platform were investigated by Jaksic et al. (2015) and O'Donnell et al. (2017) (see Figure 6.5), respectively, under wave loading alone, generated in wave tanks. In each case, the TLCD was placed on the platform. Different values of mass ratios, some as high as 10%, were considered, against which the response reductions that were obtained were moderate, to the tune of 16% or less. In a more rigorous experimental study on the semi-submersible platform for a wind turbine (Xue et al., 2022), an omnidirectional TLCD (again termed as a TLMCD in the work) with a mass ratio of 2% was studied and found to reduce the pitch motion by 11%–19%. A numerical model in the OpenFOAM software was also developed, and a good agreement of the numerical results with the experimental data was obtained. A similar kind of TLCD for the floating offshore wind turbine was also

FIGURE 6.5 A model of semi-submersible platform having a wind turbine in a wave tank (O'Donnell et al., 2017).

investigated both numerically and experimentally by Yu et al. (2023), who stressed the importance of optimizing the individual subsystems and properly accounting for the interactions between them, in order to achieve an actual enhancement in the vibrational performance of the system.

A numerical investigation into the effect of liquid sloshing in the vertical limbs of the TLCD system studied by Xu et al. (2022) was conducted by Zhou et al. (2023) for suppression of the pitch motion of the semi-submersible platform supporting a wind turbine. The results indicated that the sloshing in vertical limbs does not always bring a positive impact on the damper performance.

In case of spar-type floating offshore wind turbines, Zhang and Høeg (2020) presented a detailed theoretical investigation on the use of the TLCD for mitigating the in-plane vibrations of the structure under wind and wave loading. A 5-DOF model of the spar-tower-TLCD system, with the damper placed in the nacelle, was analysed in time domain. The results indicated that a TLCD having a 3% mass ratio could achieve up to 30% reduction in the side–side vibrational response of the wind turbine tower.

Other variants of the TLCD have also been studied for the wave-induced vibration control of offshore wind turbines. For example, the sealed TLCD (Dezvareh, 2019; Dezvareh et al., 2016), toroidal TLCD (Ding et al., 2023,2022), bidirectional TLCD (Wei and Zhao, 2020), tuned liquid multi-column gas damper (Hokmabady et al., 2022) have been investigated and have been reported to be successful in controlling vibration.

Apart from offshore wind turbines, TLCDs have also been examined for wave-induced vibration of floating breakwaters. Shahrabi and Bargi (2019, 2020) conducted extensive analytical and numerical studies on the efficacy of using a TLCD embedded within a floating breakwater to mitigate responses of the latter (see Figure 6.6). The breakwater was modelled using the ANSYS AQWA software. Both regular and irregular waves, modelled by Airy's linear wave theory and JONSWAP wave spectrum, respectively, were considered. A TLCD with a mass ratio of 1.84% reduced the peak surge motion and the peak roll motion by 16% and 14%, respectively. The TLCD also significantly reduced the tensile stress in the mooring lines and improved the fatigue life of the mooring system by 47%. The control performance of the TLCD is indicated in Figure 6.7, which presents a comparison of the displacement of the fairlead of the breakwater, without and with a TLCD having a 1% mass ratio.

TLCDs are also found to be effective in reducing wave-induced vibration of offshore platforms. Chatterjee and Chakraborty (2014) examined a jacket platform subjected to the Pierson–Moskowitz wave. The study revealed that a TLCD having a mass ratio of 5% can reduce the maximum displacement response of the platform by 27%. A comparison of the TLCD performance with that of a tuned liquid column ball damper (TLCBD) indicates that the latter is slightly superior. Feizian et al. (2020) investigated response control of a semisubmersible drilling platform through the incorporation of a TLCD. The TLCD was considered to be embedded within the semisubmersible platform (see Figure 6.8). A numerical study was carried out using the ANSYS AQWA software. Irregular wave time histories were generated compatible with the JONSWAP wave spectrum. An optimally designed TLCD system with a 3.44% mass ratio reduced the pitching motion of the platform by 9.6%.

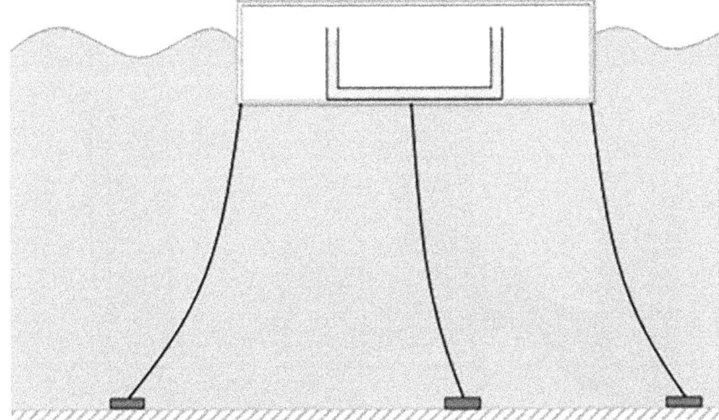

FIGURE 6.6   A floating breakwater with embedded tuned liquid column damper.

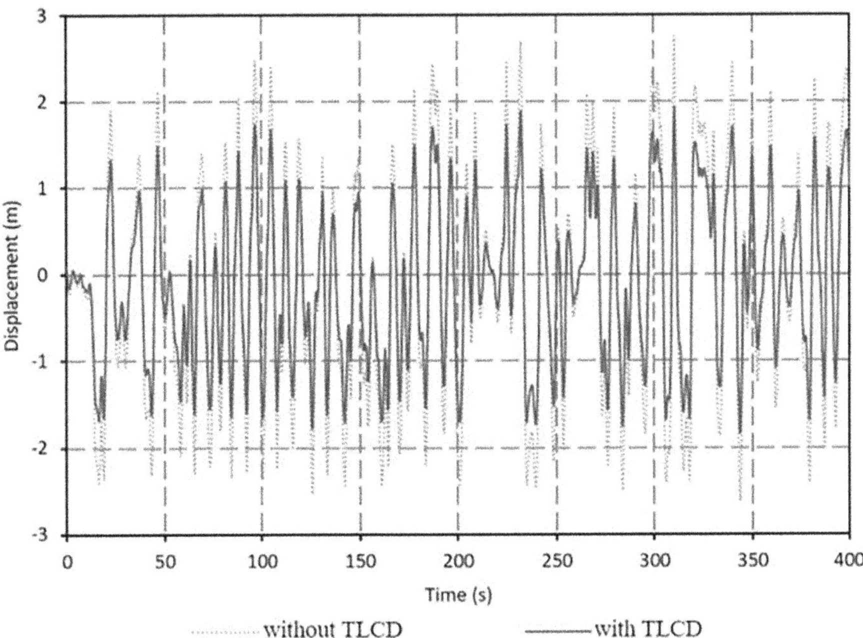

FIGURE 6.7   Displacement of fairlead of floating breakwater without and with a tuned liquid column damper having 1% mass ratio (Shahrabi and Bargi, 2019).

From the foregoing discussion in this section, it may be concluded that the current state of research in this area is not yet fully matured. More rigorous modelling of the structure-TLCD system is necessary, especially in case of complex structures such as offshore wind turbine systems. More in-depth investigations are required on the aspects of damper location, damper mass ratio, wave load modelling and combination of wave load with other loads to arrive at realistic estimates of the TLCD performance.

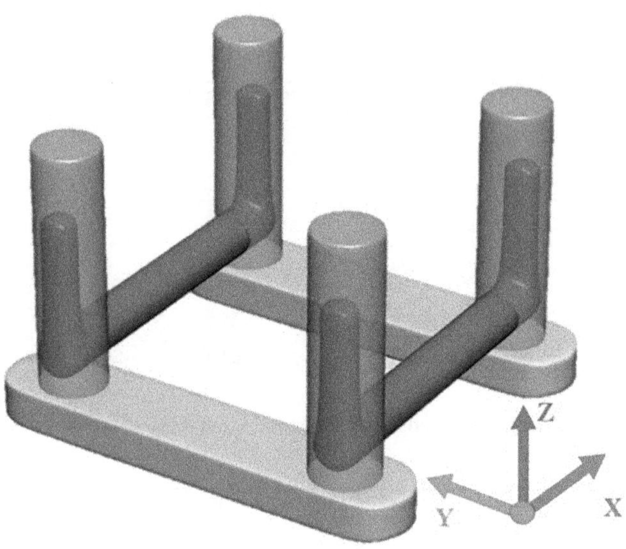

FIGURE 6.8   Tuned liquid column damper embedded within a semisubmersible platform (Feizian et al., 2020).

6.3   CONTROL OF PEDESTRIAN-INDUCED EXCITATION BY TLCD

Reports of bridge vibrations induced by pedestrian movement can be traced back to the late 1950s in China (Ingólfsson et al., 2012). The Toda Park Bridge in Japan, which was completed in 1989, suffered from excessive vibrations due to crowd movement immediately after it was opened to public (Nakamura and Fujino, 2002). However, pedestrian-induced vibration and its control have come to the focus of researchers only since the highly publicized closure of the London Millennium Footbridge at its opening on June 10, 2000, due to very large vibrations resulting from pedestrian movement (Dallard et al., 2001).

When a pedestrian walks, a ground reaction force arises as a result of the acceleration (and deceleration) of the centre of mass of the body of the pedestrian. This force can be represented by a three-dimensional vector and is a function of time and space, due to the forward progression of the pedestrian (Racic et al., 2009). Several simulation and experimental studies, including full-scale ones, have been conducted to develop mathematical models of pedestrian-induced dynamic loading (see Figure 6.9). Most of the mathematical models have approximated this pedestrian-induced force by the superposition of harmonics. The walking force, $F_w(t)$, may thus be expressed as (Kerr and Bishop, 2001)

$$F_w(t) = W + \sum_{i=1}^{n} W\lambda_i \sin\left(2\pi i f_p t - \varphi_i\right). \tag{6.17}$$

In Eq. (6.17), W is the static weight of the pedestrian, i the order number of the harmonic, n the total number of contributing harmonics, λ_i the dynamic loading factor

FIGURE 6.9   Simulation of pedestrian-induced vibration of London Millennium Footbridge (Belykh et al., 2021).

of the i-th harmonic, f_p the walking frequency in Hz and φ_i the phase angle of the i-th harmonic in radian. The Sétra (2006) guidelines suggest using $n = 3$ in above equation. The guidelines also suggest the values of the dynamic loading factors and phase angles to be taken as $\lambda_1 = 0.4$, $\lambda_2 = \lambda_3 \approx 0.1$, $\varphi_1 = 0$ and $\varphi_2 = \varphi_3 \approx \pi/2$ (in rad). The mean value of W for a single pedestrian may be taken as 700 N, and the mean value of f_p may be taken as 2 Hz.

The resolution of $F_w(t)$ into components in the vertical, longitudinal and transverse directions and consideration of only the first harmonic yield (Sétra, 2006)

$$F_{w,v}(t) = W + 0.4W \sin\left(2\pi f_p t\right), \tag{6.18}$$

$$F_{w,l}(t) = 0.2W \sin\left(2\pi f_p t\right) \tag{6.19}$$

and

$$F_{w,t}(t) = 0.05W \sin\left(\pi f_p t\right). \tag{6.20}$$

In Eqs. (6.18)–(6.20), $F_{w,v}(t)$, $F_{w,l}(t)$ and $F_{w,t}(t)$ denote the vertical, longitudinal and transverse components of $F_w(t)$, respectively. Further, it may be observed from Eq. (6.20) that the frequency of pedestrian-induced transverse loading is half the walking frequency. Another important aspect of pedestrian-induced vibration study is that the structures having natural frequencies up to 5 Hz are particularly vulnerable to this kind of vibration. More details about pedestrian-induced dynamic loading are available in Kerr and Bishop (2001), Racic et al. (2009), Sétra (2006) guidelines, and Belykh et al. (2021).

Amongst other vibration control techniques, TLCDs have also been examined as a potential device for pedestrian-induced vibration suppression. Probably, the earliest study in this regard was taken up by Reiterer (2004), who explored the possibility of using sealed TLCD for pedestrian-induced vibration control footbridges. In a subsequent work, Reiterer and Ziegler (2006) presented a detailed study on the design of a sealed TLCD for transverse and rotational vibration control of footbridges. The analytical formulation of the bridge-sealed TLCD system was developed and validated experimentally. The damper design was illustrated considering the example of the Toda Park Bridge and the London Millennium Footbridge. It was found that the effective damping ratio in the transverse direction of the main span of the London Millennium Footbridge could be enhanced six times by incorporating three TLCDs having a total mass ratio of 3%. In case of the Toda Park Bridge, a TLCD having a 0.6% mass ratio could increase the effective damping ratio of the main span in the transverse direction by three times. Ziegler and Amiri (2013) proposed an alternative design of the sealed TLCD for pedestrian-induced vertical vibration control of bridges. The proposed design consisted of a U-shaped damper container with vertical limbs close to each other and having unequal height. The vertical limbs of the sealed TLCD had unequal air pressure and unequal liquid levels at rest condition. A numerical study indicated that a sealed TLCD having the proposed configuration and mass ratio of 0.9% enhanced the damping ratio of the example bridge in the vertical vibration by as much as seven times.

Liu et al. (2023) studied the possibility of utilizing a vertical sealed TLCD (VSTLCD) (see Section 3.5.2 of Chapter 3) for the control of vertical vibration of a building floor system caused by human walking. An analytical model of the vertically excited floor-VSTLCD system was developed. The model was validated through shaking table tests. A numerical study was conducted considering a 64 m × 39 m floor supported on steel beams. A VSTLCD having a 2% mass ratio reduced the human walking-induced maximum floor acceleration by 43%. Although the maximum acceleration response of the floor increased with the increase in the number of pedestrians, the control percentage remained practically unchanged. A comparison of the Fourier spectra of the acceleration response of the floor, with and without the VSTLCD, for 20 people walking is presented in Figure 6.10.

6.4 MISCELLANEOUS

There has been some successful research on the use of TLCD for the control of bridge vibrations induced by a running train. Wendner et al. (2007) developed a variation of the sealed TLCD, namely the adaptive tuned liquid column damper (Ad-TLCD) for vibration mitigation of railway bridges under the action of moving trains. The Ad-TLCD has one end of the U-shaped container sealed, while the other end of the damper container is open to the atmosphere. An analytical formulation for the frequency of the Ad-TLCD was presented and was validated through laboratory experiments. Details on the Ad-TLCD are available in Section 3.4.1 of Chapter 3. In a subsequent work, Reiterer et al. (2008) designed a multi-Ad-TLCD system comprising eight units and a total mass ratio of 2.1% for the Bridge Typ. SHB 304/12 of the Austrian Federal Railways (in German Österreichische Bundesbahnen) (see Figure 6.11). The dampers were intended to control the vertical vibration of the bridge. A finite-element model of the bridge-damper system was developed in ANSYS Mechanical 11.0 environment, and the damper efficacy was

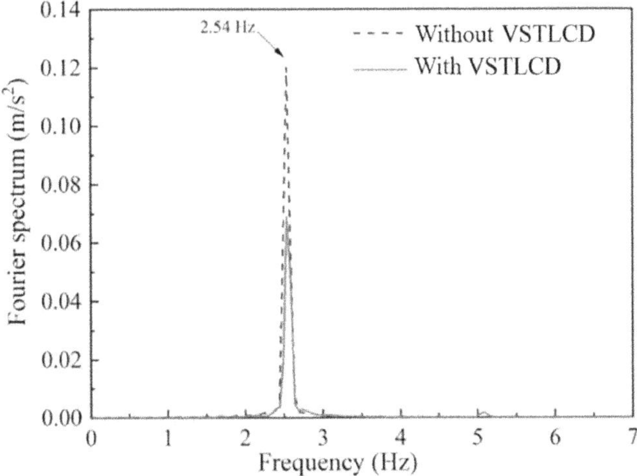

FIGURE 6.10 Comparison of Fourier spectra of acceleration response of the example floor, with and without vertical sealed tuned liquid column damper, for 20 people walking (Liu et al., 2023).

FIGURE 6.11 Ad–tuned liquid column damper installed in a bridge of the Austrian Federal Railways (Reiterer et al., 2008).

ascertained through a numerical study. The bridge was then fitted with the designed Ad-TLCD system, and the vertical acceleration of the bridge during the passage of a heavy freight train was measured. A comparison of the acceleration time histories revealed that the Ad-TLCD system reduced the decay time of the vibration by about 60%. The damper system also increased the effective damping ratio of the bridge in the vertical direction by 2.4 times.

In addition to the dynamic loading considered in Chapters 4 and 5, and in this chapter, so far, civil engineering structures may experience significant vibrations due to blast, drifting ice, vehicle movement, self-excitation, collision with moving or floating objects, etc. So far, literature on the control of structures from such types of dynamic loading by the use of the TLCD is absent. It may be noted that, the passive TLCD, as in the case of other passive inertial control systems, is not expected to provide effective control action for structures subjected to loadings, such as, blast and collision, since some initial time is required to mobilize the control force in these kinds of devices.

REFERENCES

Airy, G.B., 1841. Tides and waves, in: Rose, H.J., Smedley, E., Rose, H.J. (Eds.), *Encyclopaedia Metropolitana (1817–1845), Mixed Sciences*. B Fellop, London, p. 396.

Alujević, N., Ćatipović, I., Malenica, Š., Senjanović, I., Vladimir, N., 2019. Ship roll control and power absorption using a U-tube anti-roll tank. *Ocean Eng.* 172, 857–870. https://doi.org/10.1016/j.oceaneng.2018.12.007

Basu, B., 2020. A flow force reformulation for steady irrotational water waves. *J. Differ. Equ.* 268, 7417–7452. https://doi.org/10.1016/j.jde.2019.11.057

Basu, B., Kogelbauer, F., 2021. Global bifurcation of irrotational water waves using a flow force formulation. *J. Differ. Equ.* 301, 73–96. https://doi.org/10.1016/j.jde.2021.07.042

Belykh, I., Bocian, M., Champneys, A.R., Daley, K., Jeter, R., Macdonald, J.H.G., McRobie, A., 2021. Emergence of the London Millennium Bridge instability without synchronisation. *Nat. Commun.* 12, 7223. https://doi.org/10.1038/s41467-021-27568-y

Bretschneider, C.L., 1959. Wave variability and wave spectra for wind-generated gravity waves. *Technical Memorandum* 118. U.S. Army Beach Erosion Board, Washington, DC.

Chakrabarti, S.K., 2005. *Handbook of Offshore Engineering*. Elsevier Ltd, Amsterdam, The Netherlands.

Chatterjee, T., Chakraborty, S., 2014. Vibration mitigation of structures subjected to random wave forces by liquid column dampers. *Ocean Eng.* 87, 151–161. https://doi.org/10.1016/j.oceaneng.2014.05.004

Chen, L., Basu, B., 2021a. Numerical continuation method for large-amplitude steady water waves on depth-varying currents in flows with fixed mean water depth. *Appl. Ocean Res.* 111, 102631. https://doi.org/10.1016/j.apor.2021.102631

Chen, L., Basu, B., 2021b. Numerical investigations of two-dimensional irrotational water waves over finite depth with uniform current. *Appl. Anal.* 100, 1247–1255. https://doi.org/10.1080/00036811.2019.1636974

Colwell, S., Basu, B., 2009. Tuned liquid column dampers in offshore wind turbines for structural control. *Eng. Struct.* 31, 358–368. https://doi.org/10.1016/j.engstruct.2008.09.001

Constantin, A., Kalimeris, K., Scherzer, O., 2015. A penalization method for calculating the flow beneath traveling water waves of large amplitude. *SIAM J. Appl. Math.* 75, 1513–1535.

Constantin, A., Strauss, W., 2004. Exact steady periodic water waves with vorticity. *Commun. Pure Appl. Math.* 57, 481–527. https://doi.org/10.1002/cpa.3046

Coudurier, C., Lepreux, O., Petit, N., 2018. Modelling of a tuned liquid multi-column damper. Application to fl oating wind turbine for improved robustness against wave incidence. *Ocean Eng.* 165, 277–292. https://doi.org/10.1016/j.oceaneng.2018.03.033

Coudurier, C., Lepreux, O., Petit, N., 2015. Passive and semi-active control of an offshore floating wind turbine using a tuned liquid column damper. *IFAC-PapersOnLine* 48, 241–247. https://doi.org/10.1016/j.ifacol.2015.10.287

Crawford, K.C., 2022. Vibration Control in Bridges, in: Fischer, C., Náprstek, J. (Eds.), *Vibration Control of Structures.* pp. 1–21. https://doi.org/10.5772/intechopen.107501

Dallard, P., Fitzpatrick, T., Flint, A., Low, A., Smith, R.R., Willford, M., Roche, M., 2001. London millennium bridge: pedestrian-induced lateral vibration. *J. Bridg. Eng.* 6, 412–417.https://doi.org/10.1061/(ASCE)1084-0702(2001)6:6(412)

Dezvareh, R., 2019. Application of soft computing in the design and optimization of tuned liquid column–gas damper for use in offshore wind turbines. *Int. J. Coast. offshore Eng.* 2, 47–57. https://doi.org/10.29252/ijcoe.2.4.47

Dezvareh, R., Bargi, K., Mousavi, S.A., 2016. Control of wind/wave-induced vibrations of jacket-type offshore wind turbines through tuned liquid column gas dampers. *Struct. Infrastruct. Eng.* 12, 312–326. https://doi.org/10.1080/15732479.2015.1011169

Ding, H., Chen, Y.N., Wang, J.T., Altay, O., 2022. Numerical analysis of passive toroidal tuned liquid column dampers for the vibration control of monopile wind turbines using FVM and FEM. *Ocean Eng.* 247, 110637. https://doi.org/10.1016/j.oceaneng.2022.110637

Ding, H., Wang, W., Liu, J.F., Wang, J.T., Le, Z.J., Zhang, J., Yu, G.M., 2023. On the size effects of toroidal tuned liquid column dampers for mitigating wind- and wave-induced vibrations of monopile wind turbines. *Ocean Eng.* 273. https://doi.org/10.1016/j.oceaneng.2023.113988

Feizian, H., Shafieefar, M., Panahi, R., 2020. Application of tuned liquid column damper for motion reduction of semisubmersible platforms. *Int. J. Coast. Offshore Eng.* 5, 23–40. https://doi.org/10.22034/IJCOE.2020.149344

Fenton, J.D., 1990. Nonlinear wave theories, in: Le Méhauté, B., Hanes, D. M. (Eds.), *He Sea,* Vol. 9. Ocean Engineering Science, Wiley, New York City, USA.

Frahm, H., 1911. Results of trials of the anti-rolling tanks at sea. *Trans. Inst. Nav. Archit.* 53, 183–201.

Han, D., Li, X., Wang, W., Su, X., 2023. Dynamic modeling and vibration control of barge offshore wind turbine using tuned liquid column damper in floating platform. *Ocean Eng.* 276, 114299. https://doi.org/10.1016/j.oceaneng.2023.114299

Hasselmann, K., Barnett, T.P., Bouws, E., Carlson, H., Cartwright, D.E., Eake, K., Euring, J.A., Gicnapp, A., Hasselmann, D.E., Kruseman, P., Meerburg, A., Mullen, P., Olbers, D.J., Richren, K., Sell, W., Walden, H., 1973. *Measurements of Wind-Wave Growth and Swell Decay During the Joint North Sea Wave Project (JONSWAP).* Hamburg, Germany.

Hemmati, A., Oterkus, E., Barltrop, N., 2019. Fragility reduction of offshore wind turbines using tuned liquid column dampers. *Soil Dyn. Earthq. Eng.* 125, 105705. https://doi.org/10.1016/j.soildyn.2019.105705

Hogben, N., Dacunha, N.M.C.., Olliver, G.F., 1986. *Global Wave Statistics Online.* Unwin Brothers Ltd, Old Woking, UK.

Hokmabady, H., Mojtahedi, A., Mohammadyzadeh, S., 2022. Development of a novel tuned liquid multiple columns gas damper under wave and earthquake incidence. *Appl. Ocean Res.* 118, 102956. https://doi.org/10.1016/j.apor.2021.102956

Ingólfsson, E.T., Georgakis, C.T., Jönsson, J., 2012. Pedestrian-induced lateral vibrations of footbridges : a literature review. *Eng. Struct.* 45, 21–52. https://doi.org/10.1016/j.engstruct.2012.05.038

Jaksic, V., Wright, C.S., Murphy, J., Afeef, C., Ali, S.F., Mandic, D.P., Pakrashi, V., 2015. Dynamic response mitigation of floating wind turbine platforms using tuned liquid column dampers. *Philos. Trans. A* 373, 20140079. https://doi.org/10.1098/rsta.2014.0079

Kalimeris, K., 2018. Analytical approximation and numerical simulations for periodic travelling water waves. *Philos. Trans. R. Soc. A Math. Phys. Eng. Sci.* 376, 20170093. https://doi.org/10.1098/rsta.2017.0093

Kandasamy, R., Cui, F., Townsend, N., Foo, C.C., Guo, J., Shenoi, A., Xiong, Y., 2016. A review of vibration control methods for marine offshore structures. *Ocean Eng.* 127, 279–297. https://doi.org/10.1016/j.oceaneng.2016.10.001

Kerr, S.C., Bishop, N.W.M., 2001. Human induced loading on flexible staircases. *Eng. Struct.* 23, 37–45. https://doi.org/10.1016/S0141-0296(00)00020-1

Ko, J., Strauss, W., 2008. Effect of vorticity on steady water waves. *J. Fluid Mech.* 608, 197–215. https://doi.org/10.1017/S0022112008002371

Konar, T., Ghosh, A.D., 2023. A review on various configurations of the passive tuned liquid damper. *J. Vib. Control* 29, 1945–1980. https://doi.org/10.1177/10775463221074077

Le Méhauté, B., 1976. *An Introduction to Hydrodynamics and Water Waves.* Springer, Berlin, Heidelberg.

Liu, K., Shi, Q., Liu, Y., Liu, L., Zhou, F., 2023. Investigation of an improved tuned liquid column gas damper for the vertical vibration control. *Mech. Syst. Signal Process.* 196, 110340. https://doi.org/10.1016/j.ymssp.2023.110340

Morison, J.R., Johnson, J.W., Schaaf, S.A., 1950. The Force Exerted by Surface Waves on Piles. *J. Pet. Technol.* 2, 149–154. https://doi.org/10.2118/950149-g

Murtagh, P.J., Basu, B., Broderick, B.M., 2005. Along-wind response of a wind turbine tower with blade coupling subjected to rotationally sampled wind loading. *Eng. Struct.* 27, 1209–1219. https://doi.org/10.1016/j.engstruct.2005.03.004

Nakamura, S.I., Fujino, Y., 2002. Lateral vibration on a pedestrian cable-stayed bridge. *Struct. Eng. Int. J. Int. Assoc. Bridg. Struct. Eng.* 12, 295–300. https://doi.org/10.2749/101686602777965162

Nguyen, P.T.T., Manuel, L., 2024. A bi-fidelity surrogate model for extreme loads on offshore structures. *Ocean Eng.* 307, 118175. https://doi.org/10.1016/j.oceaneng.2024.118175

O'Donnell, D., Murphy, J., Desmond, C., Jaksic, V., Pakrashi, V., 2017. Tuned Liquid Column Damper based Reduction of Dynamic Responses of Scaled Offshore Platforms in Different Ocean Wave Basins. *J. Phys. Conf. Ser.* 842, 012043. https://doi.org/10.1088/1742-6596/842/1/012043

Paul, S., Ghosh, A. (Dey), Basu, B., 2024. A computationally efficient approach to estimate design wave force on seawalls under large amplitude irrotational waves. *Ocean Eng.* 311, 118745. https://doi.org/10.1016/j.oceaneng.2024.118745

Pierson, W.J., Moskowitz, L., 1964. A proposed spectral form for fully developed wind seas based on the similarity theory of S. A. Kitaigorodskii. *J. Geophys. Res.* 69, 5181–5190. https://doi.org/10.1029/jz069i024p05181

Racic, V.Ã., Pavic, A., Brownjohn, J.M.W., 2009. Experimental identification and analytical modelling of human walking forces : literature review. *J. Sound Vib.* 326, 1–49. https://doi.org/10.1016/j.jsv.2009.04.020

Reiterer, M., 2004. Control of pedestrian-induced vibrations of footbridges using tuned liquid column dampers, in: *Third European Conference on Structural Control (3ECSC).* Vienna, Austria.

Reiterer, M., Altay, O., Wendner, R., Hoffmann, S., Strauss, A., 2008. Adaptive Flüssigkeitstilger für Vertikalschwingungen von Ingenieurstrukturen, Teil 2 – Feldversuche. *Stahlbau* 77, 205–212. https://doi.org/10.1002/stab.200810022

Reiterer, M., Ziegler, F., 2006. Control of pedestrian-induced vibrations of long-span bridges. *Struct. Control Heal. Monit.* 13, 1003–1027. https://doi.org/10.1002/stc.91

Sétra, 2006. *Footbridges: Assessment of Vibrational Behaviour of Footbridges under Pedestrian Loading.* Service d'Etudes Techniques des Routes et Autoroutes, Paris, France.

Shahrabi, M., Bargi, K., 2020. Structural control of the floating breakwaters by implementation of tuned liquid column damper. *Adv. Struct. Eng.* 23, 1587–1600. https://doi.org/10.1177/1369433219898081

Shahrabi, M., Bargi, K., 2019. Enhancement the fatigue life of floating breakwater mooring system using tuned liquid column damper. *Lat. Am. J. Solids Struct.* 16, e220. https://doi.org/10.1590/1679-78255692

Stokes, G.G., 1847. On the theory of oscillatory waves. *Trans. Cambridge Philos. Soc.* 8, 441–455.

Wei, X., Zhao, X., 2020. Vibration suppression of a floating hydrostatic wind turbine model using bidirectional tuned liquid column mass damper. *Wind Energy* 23, 1887–1904. https://doi.org/10.1002/we.2524

Wendner, R., Reiterer, M., Hoffmann, S., Strauss, A., Bergmeister, K., 2007. Adaptive Flüssigkeitstilger für Vertikalschwingungen von Ingenieurstrukturen: teil 1 – Laborversuche. *Stahlbau* 76, 916–923. https://doi.org/10.1002/stab.200710096

Xue, M., Dou, P., Zheng, J., Lin, P., Yuan, X., 2022. Pitch motion reduction of semisubmersible floating offshore wind turbine substructure using a tuned liquid multicolumn damper. *Mar. Struct.* 84, 103237. https://doi.org/10.1016/j.marstruc.2022.103237

Yu, W., Lemmer, F., Wen, P., 2023. Modeling and validation of a tuned liquid multi-column damper stabilized floating offshore wind turbine coupled system. *Ocean Eng.* 280, 114442. https://doi.org/10.1016/j.oceaneng.2023.114442

Zhang, Z., Høeg, C., 2020. Dynamics and control of spar-type floating offshore wind turbines with tuned liquid column dampers. *Struct Control Heal. Monit.* 27, e2532. https://doi.org/10.1002/stc.2532

Zhou, Y., Qian, L., Bai, W., 2023. Sloshing dynamics of a tuned liquid multi-column damper for semi-submersible floating offshore wind turbines. *Ocean Eng.* 269, 113484. https://doi.org/10.1016/j.oceaneng.2022.113484

Ziegler, F., Amiri, A.K., 2013. Bridge vibrations effectively damped by means of tuned liquid column gas dampers. *Asian J. Civ. Eng.* 14, 1–16.

7 Active and Semi-Active Tuned Liquid Column Damper

7.1 INTRODUCTION

As discussed in Section 1.1 of Chapter 1, an active damper is a type of damper where the control force, necessary to alter the motion of a structure, is produced by a control actuator powered externally. To generate the control force, the actuation depends upon feedback from the structural response acquired through sensor(s). In other words, active control involves time-dependent control parameters that can be modified, based on the feedback from the structural response to achieve the desired behaviour of the structure. Active control systems have found extensive application in alleviating structural vibrations within aerospace, mechanical and automobile industries. Significant research and development have also been conducted to extend this approach to large civil engineering structures due to its notable advantages, including enhanced control performance, the ability to selectively target specific control objectives and adaptability to various types and levels of excitations. Active damper systems are also implemented in some real-life civil engineering structures with the first installation being in the form of an active mass driver to the Kyobashi Seiwa Building in Tokyo, Japan, in 1989 (Kobori, 1996). However, the number of installations of active control systems in civil engineering structures is very much limited as compared to the passive dampers (Ghosh and Konar, 2023; Kareem et al., 1999). A study by Lago et al. (2019) shows that only three of the top 50 tallest buildings with dampers have active damper systems.

Similar to some other variants of passive dampers, the tuned liquid column damper (TLCD) is also provided with active control strategies to develop its active form. There are different methods for the application of control forces and determination of control parameters through the application of various control algorithms. Given these diverse approaches, the active TLCDs (ATLCDs) studied so far can be classified into three major categories. These are ATLCD with a compliant mechanism (ATLCD-C), ATLCD with propellers (ATLCD-P) and sealed TLCD with an active pressure control mechanism (STLCD-APC). In the subsequent sections, a detailed discussion on these ATLCD varieties is presented.

A shift in research more recently has been taking place with the focus moving from active to semi-active control. As mentioned in Section 1.1 of Chapter 1, semi-active devices do not introduce any additional mechanical energy into the controlled system and, hence, have several advantages in terms of the system stability and implementation of the controller. Semi-active controllers modify the physical properties of the damper to optimally control the structural vibration according

DOI: 10.1201/9781003377894-7

to a prescribed control law. Hrovat et al. (1983) were the first to study the implementation of semi-active control systems in civil engineering structures. Several devices are currently being studied for use in semi-active structural control such as variable-stiffness devices, controllable fluid dampers, friction control devices and fluid viscous devices.

Semi-active versions of the passive TLCD have also been studied by the researchers. The devices of semi-active TLCDs can be of various types, so can be the control algorithms. Semi-active TLCDs have been classified into three major types: TLCD with semi-active orifice control, semi-active magneto-rheological tuned liquid column damper (MR-TLCD) and semi-active spring-connected TLCD. These are discussed in detail in this chapter. Also, damper systems consisting of combinations of passive, active and semi-active devices, of which one or more may be TLCD(s), termed as hybrid systems are briefly introduced.

7.2 ATLCD WITH A COMPLIANT MECHANISM

The ATLCD with a compliant mechanism (ATLCD-C) may be viewed as a compliant liquid column damper (CLCD) with an active control mechanism. As detailed in Section 3.4.2 of Chapter 3, a CLCD consists of a partially liquid-filled U-shaped container attached to the structure through a flexible system comprising a spring and a dashpot element (see Figure 3.29). A CLCD essentially combines the principles of the tuned mass damper (TMD) with those of the TLCD. It has two degrees-of-freedom (2-DOFs): one corresponds to the lateral movement of the damper container, and the other relates to the motion of the damper liquid in vertical direction (Ghosh and Basu, 2004; Roy et al., 2023).

The schematic of an ATLCD-C mounted on a single-degree-of-freedom (SDOF) structure is shown in Figure 7.1. In addition to the standard configuration of a CLCD, the ATLCD-C is provided with three strategically placed sensors, a controller and an actuator attached to the damper container. It may be noted that the combined structure-damper system has 3-DOFs, namely $x(t)$, $y_c(t)$ and $y_l(t)$ representing the displacements of the structure, the whole damper system and the damper liquid, respectively. These responses are recorded through the sensors. Utilizing the information received from the sensors, the controller produces a force signal, and the actuator then applies the necessary control force to the damper container.

7.2.1 Equations of Motion of ATLCD-C

Let us assume that the length of the liquid column, length of the horizontal limb of the damper container T and the cross-sectional area of the damper container are denoted by L, b and A, respectively. m_c, c_c and k_c, respectively, denote the mass of the damper container, the damping coefficient of the dashpot element and the stiffness of the spring element connecting the damper container with the structure. The mass, stiffness and damping of the SDOF structure are denoted by M_S, K_S and C_S, respectively. The structure is subjected to a base acceleration denoted by $\ddot{Z}_0$.

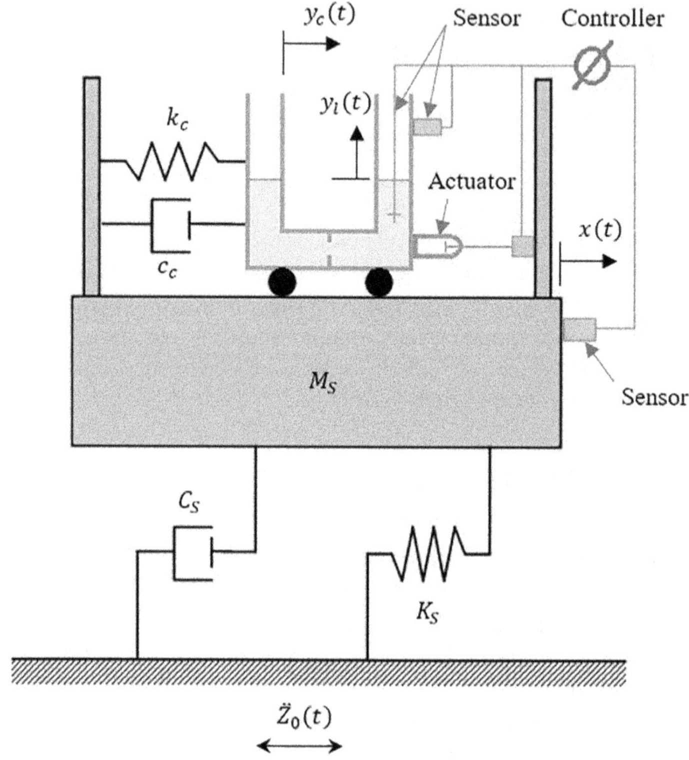

FIGURE 7.1   An active tuned liquid column damper with compliant mechanism mounted on a single-degree-of-freedom structural system.

Now, the mathematical model of the structure-damper system can be established by considering the dynamic equilibrium of the liquid in the damper container, the whole damper and the SDOF structure.

The consideration of the dynamic equilibrium of the liquid in the damper container and subsequent simplifications yield

$$\ddot{y}_l + \frac{1}{2}\frac{\xi}{L}|\dot{y}_l|\dot{y}_l + \omega_l^2 y_l = -\alpha\left(\ddot{y}_c + \ddot{x} + \ddot{Z}_0\right). \tag{7.1}$$

In Eq. (7.1), ξ is the orifice head-loss coefficient, α is the length ratio, given by b/L, and $\omega_l\left[=\sqrt{(2g/L)}\right]$ is the natural frequency of oscillation of the liquid column.

Let the actuator force be denoted by $f(t)$. Then, the consideration of the dynamic equilibrium of the whole damper system leads to

$$\ddot{y}_c + 2\zeta_c\omega_d\dot{y}_c + \omega_d^2 y_c + \frac{\alpha}{(1+\tau)}\ddot{y}_l + \ddot{x} + \ddot{Z}_0 = \frac{f(t)}{(\rho AL + m_c)}. \tag{7.2}$$

In Eq. (7.2), $\omega_d \left[= \sqrt{\left(k_c/\left(\rho AL + m_c \right) \right)} \right]$ is the natural frequency of the whole damper system, $\zeta_c = c_c / \left[2\omega_d \left(\rho AL + m_c \right) \right]$ is the damping ratio of the element connecting the damper container with the structure, $\tau \left[= m_c/\rho AL \right]$ is the ratio of the damper container mass to the liquid mass, and ρ is the mass density of the damper liquid.

The equation of motion for the SDOF structural system, normalized with respect to its mass, M_S is given by

$$\ddot{x} + 2\zeta_s\omega_s\dot{x} + \omega_s^2 x + \ddot{Z}_0 = \mu_c\omega_d^2 y_c + 2\mu_c\zeta_c\omega_d\dot{y}_c - \frac{f(t)}{M_S}. \tag{7.3}$$

In Eq. (7.3), ω_s is the natural frequency of the structure, ζ_s is the damping ratio of the structure and μ_c is the mass ratio of the ATLCD-C given by the mass of the whole damper (the mass of the damper container as well as the damper liquid) to that of the structure.

Now, by adopting an equivalent linearization procedure, as detailed in Section 2.2.2 of Chapter 2, Eq. (7.1) may be written as

$$\ddot{y}_l + 2\frac{C_p}{L}\dot{y}_l + \omega_l^2 y_l = -\alpha\left(\ddot{y}_c + \ddot{x} + \ddot{Z}_0 \right). \tag{7.4}$$

In Eq. (7.4), C_p represents the equivalent linearized damping coefficient, the expression for which is given by (Konar and Ghosh, 2023; Xu et al., 1992)

$$C_p = \frac{\xi\sigma_{\dot{y}_l}}{\sqrt{2\pi}} \tag{7.5}$$

where $\sigma_{\dot{y}_l}$ is the standard deviation of the liquid velocity $\dot{y}_l(t)$.

7.2.2 Controller Design Algorithms

7.2.2.1 Linear Quadratic Regulator Controller

Different techniques can be employed for the application of control force to the ATLCD-C. One of the ways to obtain the optimal control force is by using the linear quadratic regulator (LQR) algorithm, which is discussed here in detail. Here, it is pertinent to mention that the geometrical constraints of the damper system are required to be checked explicitly and cannot be handled in the derivation of the optimal control formulation. In case of the ATLCD-C, the geometrical constraints are the maximum liquid displacement in the damper that must be restricted to $y_l|_{max} = (L - b)/2$ to prevent loss of any liquid from the damper container and the maximum displacement of the damper container that may be determined from practical considerations.

The LQR control formulation represents a dynamical system in state-space matrix form as (Agrawal and Yang, 1999)

$$\{\dot{X}\} = [A]\{X\} + [B]\{U\} + \{F\}. \tag{7.6}$$

In Eq. (7.6), $[A]$, $[B]$, $\{X\}$, $\{U\}$ and $\{F\}$, respectively, denote the state matrix, the control influence matrix, the state vector, the input vector and the external excitation vector. Based on Eqs. (7.2)–(7.4), these matrices and vectors can be determined as

$$[A] = \begin{bmatrix} 0 & 1 & 0 & 0 & 0 & 0 \\ -\omega_s^2 & -2\zeta_s\omega_s & \mu_c\omega_d^2 & 0 & 0 & 0 \\ 0 & 0 & 0 & 1 & 0 & 0 \\ \omega_s^2 & 2\zeta_s\omega_s & -\omega_d^2\dfrac{1+\tau+\delta\mu_c}{\delta} & 0 & \dfrac{\alpha\omega_l^2}{\delta} & \dfrac{\alpha}{\delta}\dfrac{2C_p}{L} \\ 0 & 0 & 0 & 0 & 0 & 1 \\ 0 & 0 & \alpha\omega_d^2\dfrac{1+\tau}{\delta} & 0 & -\omega_l^2\dfrac{1+\tau}{\delta} & -\dfrac{1+\tau}{\delta}\dfrac{2C_p}{L} \end{bmatrix},$$

$$\tag{7.7}$$

$$[B] = \begin{bmatrix} 0 & \dfrac{1}{M_S} & 0 & \dfrac{1+\tau+\delta\mu_c}{\mu_c\delta M_S} & 0 & -\dfrac{\alpha(1+\tau)}{\mu_c\delta M_S} \end{bmatrix}^T \tag{7.8}$$

and

$$\{X\} = \begin{Bmatrix} x & \dot{x} & y_c & \dot{y}_c & y_l & \dot{y}_l \end{Bmatrix}^T \tag{7.9}$$

where $\delta = 1 + \tau - \alpha^2$.

Since the LQR operates as an optimal state-feedback controller, its primary goal is to ensure the proximity of the state to equilibrium while simultaneously optimizing the control effort based on the minimization of the cost function, J. The state-feedback law is articulated as

$$\{U\} = -[G]\{X\}. \tag{7.10}$$

In Eq. (7.10), $[G]$ is the optimal gain matrix, which is evaluated in such a way that the state-feedback law minimizes the quadratic cost function defined as

$$J = \int_0^\infty \left(\{X\}^T[Q]\{X\} + \{U\}^T[R]\{U\} \right) dt. \tag{7.11}$$

In Eq. (7.11), $[Q]_{6\times6}$ is a symmetric positive semi-definite weighting matrix associated with the state vector and $[R]_{1\times1}$ is a strictly positive-definite symmetric weighting matrix associated with the control force.

For the structure-ATLCD-C system, the weighting matrix $[Q]$ may be expressed as

$$[Q] = \begin{bmatrix} q_{11}^{(S)} & 0 & 0 & 0 & 0 & 0 \\ 0 & q_{22}^{(S)} & 0 & 0 & 0 & 0 \\ 0 & 0 & q_{33}^{(c)} & 0 & 0 & 0 \\ 0 & 0 & 0 & q_{44}^{(c)} & 0 & 0 \\ 0 & 0 & 0 & 0 & q_{55}^{(l)} & 0 \\ 0 & 0 & 0 & 0 & 0 & q_{66}^{(l)} \end{bmatrix}. \tag{7.12}$$

In Eq. (7.12), the elements $q_{11}^{(S)}$ and $q_{22}^{(S)}$, respectively, are the weights associated with the structural displacement and velocity responses. Further, $q_{33}^{(c)}$, $q_{44}^{(c)}$, $q_{55}^{(l)}$ and $q_{66}^{(l)}$, respectively, denote the weights for the displacement and velocity responses of the damper container and the liquid column.

For any given excitation, the value of C_p is first obtained from Eq. (7.5) and by using the system matrix $[A]$ as given by Eq. (7.7), the corresponding optimal $[G]$ is determined. Then, by using the optimal $[G]$ and the state-feedback of the nonlinear system represented by Eq. (7.6), the control force is obtained from Eq. (7.10).

To introduce the constraints on the displacement of the liquid and the damper container, the weighting matrix $[R]_{1 \times 1}$ applied on the control force relative to the weighting matrix $[Q]$, which acts on the state $\{X\}$, is appropriately determined for use in the optimal LQR algorithm. The minimum value of the scalar R (i.e. $[R]_{1 \times 1}$), which achieves a targeted maximum response reduction is denoted by R_{min}. Higher values of R would lead to diminished response reductions corresponding to lower values of control force requirement, which provides design options with lower energy cost. The value of R that corresponds to the maximum permissible energy cost is designated as R_{opt} and provides the design solution. The matrix $[Q]$ controls the emphasis on the structural response, and higher weights corresponding to this matrix are given if the minimization of the structural response is the main priority. The optimal $[Q]$ is determined for each value of R considered. The optimal combination of R and $[Q]$ thereby achieves the highest reduction in structural response corresponding to the minimum control force with the applied constraints as discussed earlier. This algorithm was developed and used by Bhattacharyya et al. (2018) in their study on ATLCD-C device. No damping element between the damper container and the structure was considered in their study. The values for $q_{11}^{(S)}$ and $q_{22}^{(S)}$ considered in the study were in the order of 10^{10}. On the other hand, the values of other weights were taken as $q_{33}^{(c)} = q_{44}^{(c)} = 10^9$, and $q_{55}^{(l)} = q_{66}^{(l)} = 10^{-10}$. The study revealed that the ATLCD-C is able to reduce the structural displacement response by 1.4–3 times under seismic excitations as compared to an equivalent passive CLCD.

7.2.2.2 Multiresolution-Based Wavelet Linear Quadratic Regulator Controller

Multiresolution-based wavelet linear quadratic regulator (WLQR) controller is a new class of controller, originally conceptualized and developed by the researchers Basu and Nagarajaiah (2010, 2008). They had synthesized two types of controllers (Basu and Nagarajaiah, 2010, 2008), among which the system level one proposed in Basu and Nagarajaiah (2010) was used in the study by Bhattacharyya et al. (2018). The basis for the design of the controller was LQR, and hence, this is an optimal controller, although the use of multiscale wavelet analysis modifies the optimization problem and its solution. For solving the optimization problem, the weights normally used in a conventional LQR controller are frequency-dependent in this case. The weights are chosen such that response in certain frequency bands can be suppressed than others, if and as desired. Since multiresolution analysis (MRA) resolves the response signal at different stages or levels, an independent set of weights at each of these levels may be assigned to cast the optimization formulation. For details on how the signals are split (filtered), the reader may refer to Basu and Nagarajaiah (2010). The filtered signals in different bands are then used to synthesize the control action with the help of discrete wavelet transform and are expressed in the wavelet domain as (Basu and Nagarajaiah, 2010)

$$\left\{ W_{\psi_a} U \right\} = -\left[G \right]_a \left\{ W_{\psi_a} X \right\}. \tag{7.13}$$

In Eq. (7.13), $W_{\psi_a}(\cdot)$ is the wavelet transform of $(\cdot)$ with respect to the basis ψ for a particular value of the dilation parameter a, and $\left[G \right]_a$ is the control gain matrix, dependent on a, which controls the frequency content of $\left\{ W_{\psi_a} U \right\}$.

The multiresolution LQR as a modified quadratic cost function, which corresponding to the frequency band with dilation parameter a, yields

$$J_a = \int_{t_0}^{t_c} \left(\left\{ W_{\psi_a} X \right\}^T \left[Q \right]_a \left\{ W_{\psi_a} X \right\} + \left\{ W_{\psi_a} U \right\}^T \left[R \right]_a \left\{ W_{\psi_a} U \right\} \right) dt. \tag{7.14}$$

The weighting matrices $\left[Q \right]_a$ and $\left[R \right]_a$ in Eq. (7.14) are dependent on the dilation parameter a. The frequency-dependent optimal control gains are obtained by minimizing the cost function in Eq. (7.14) and solving the optimization problem. These gains are then used to synthesize the control signal in time domain, which is given by

$$\left\{ U \right\} = -\left[G \right]_{a_l} \left\{ X \right\}_l - \sum_{j=l}^{p-1} \left[G \right]_{a_j} \left\{ X \right\}_j. \tag{7.15}$$

In Eq. (7.15), the subscripts l and p, respectively, denote the level below which the signal can be represented by a low frequency approximation and the level above which the frequency bands can be neglected without significant approximation. The expression for equivalent time-varying gain may be written as

$$\left[G_e(t) \right] = -\left[G \right]_{a_l} \left\{ X \right\}_l \left\{ X \right\}^T \left(\left\{ X \right\} \left\{ X \right\}^T \right)^{-1} - \sum_{j=l}^{p-1} \left[G \right]_{a_j} \left\{ X \right\}_j \left\{ X \right\}^T \left(\left\{ X \right\} \left\{ X \right\}^T \right)^{-1} \tag{7.16}$$

which leads to the control force vector

$$\left\{ U \right\} = \left[G_e(t) \right] \left\{ X \right\}. \tag{7.17}$$

In addition to the LQR controller, Bhattacharyya et al. (2018) considered a separate case when the ATLCD-C is provided with multiresolution-based WLQR. The study revealed that the use of multiresolution-based WLQR controller can result in structural displacement response reduction comparable to what is achieved by a conventional LQR controller, but with a lower amount of control effort.

7.2.2.3 H_∞ Controller

The symbol "H_∞" represents the Hardy space, comprising all complex-valued functions of a complex variable that are analytic and bounded in the open right-half complex plane. In the context of a linear system, the H_∞ norm of the transfer matrix refers to the maximum of the largest singular value of the output across all frequencies. Thus, H_∞ control minimizes the H_∞ norm of the transfer function from the disturbance to the controlled output, subject to various system constraints (Basar and Bernhard, 1995; Glover, 2013).

The transfer function matrix of the structure-damper system $\left[H(\omega)\right]$ may be obtained from the frequency domain representation and may be given by the relation

$$\{Y(\omega)\} = \left[H(\omega)\right]\{F(\omega)\}. \tag{7.18}$$

In Eq. (7.18), $\{Y(\omega)\}$ is a column vector containing the ordinates of the Fourier transform of the displacement responses of all the DOFs at ω, and $\{F(\omega)\}$ is the Fourier transform of the input loading.

Balendra et al. (2001) adopted the H_∞ algorithm to evaluate the control force for the ATLCD designed for wind-induced vibration control of towers modelled as an SDOF system. The ATLCD considered in the study had a partially filled U-shaped damper container fixed on a movable platform attached to the structure through a spring and a dashpot. They neglected the mass of the damper container. This system is similar to the ATLCD-C discussed in this chapter and can be represented using Eqs. (7.2)–(7.4) only by replacing the mass of the damper container with the mass of the movable platform.

Other types of controller used for ATLCD-C include hedge-algebra (HA)-based controller which is used by Bui et al. (2023) in a recent work. HA-based controller is a type of fuzzy control algorithm, and the detailed theoretical background on this topic is available in Ho et al. (2008). In the work of Bui et al. (2023), the study on ATLCD-C was extended to multi-degree-of-freedom (MDOF) structural systems under seismic excitation. Additionally, they considered a case where $m_c = \tau = 0$ (in Eq. (7.2)). However, in such a situation, the performance of the damper deteriorates.

7.3 ATLCD WITH PROPELLERS

As the name suggests, this type of ATLCD is provided with two propellers in the horizontal limb of the damper container (see Figure 7.2). The propellers are controlled by a servo-motor actuator. Based on the feedback from the sensors, the actuator operates the propellers and thereby applies the control force on the liquid column. As the propellers are provided near the centre of the horizontal limb of the damper, the orifices of the damper may be provided close to the ends of the horizontal limb of

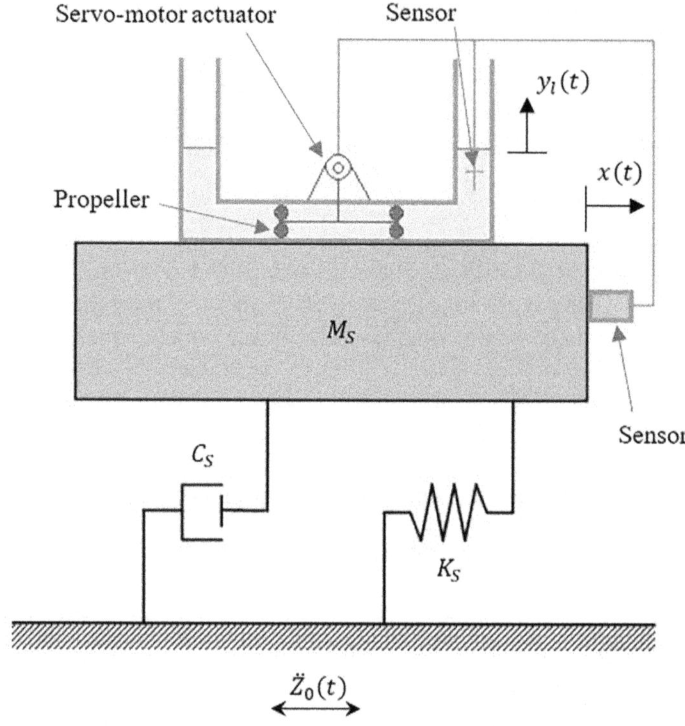

FIGURE 7.2 An active tuned liquid column damper with propellers mounted on a single-degree-of-freedom structural system.

the damper. This kind of arrangement has already been used in a real-world implementation of a passive TLCD system in the Random House Tower at New York City, USA (Konar and Ghosh, 2021; Tamboli et al., 2005).

7.3.1 EQUATIONS OF MOTION OF ATLCD-P

As shown in Figure 7.2, the structure is modelled as an SDOF system. The structure has the same parameters as described in Section 7.2.1.

The consideration of the dynamic equilibrium of the liquid in the ATLCD-P and of the mass of the structural system, with subsequent simplifications, yields

$$\ddot{y}_l + \frac{1}{2}\frac{\xi}{L}|\dot{y}_l|\dot{y}_l + \omega_l^2 y_l = -\alpha\left(\ddot{x} + \ddot{Z}_0\right) + \frac{f(t)}{\rho A L} \tag{7.19}$$

and

$$\ddot{x} + \frac{2\zeta_s\omega_s}{(1+\mu)}\dot{x} + \frac{\omega_s^2}{(1+\mu)}x = -\ddot{Z}_0 - \frac{\alpha\mu}{(1+\mu)}\ddot{y}. \tag{7.20}$$

Here, the parameters have the same meaning as described in Eqs. (7.1)–(7.3).

The linearization of the damping term in Eq. (7.19) leads to

$$\ddot{y}_l + 2\frac{C_p}{L}\dot{y}_l + \omega_l^2 y_l = -\alpha\left(\ddot{x} + \ddot{Z}_0\right) + \frac{f(t)}{\rho AL}.$$

(7.21)

7.3.2 Controller Design and Implementation

An appropriate control algorithm needs to be designed to achieve optimum performance of the ATLCD-P. In their study with ATLCD-P, Chen and Ko (2003) adopted the LQR algorithm–based feedback control technique. The numerical study considered a 210.8 kg structure subjected to a 1/100 scaled Kobe earthquake ground acceleration. The damper mass ratio was 0.95%. The study was conducted for two structural damping ratios, namely 0.2% and 2%. A comparison of the reductions in peak structural displacement response achieved by the ATLCD-P and an equivalent passive TLCD is presented in Table 7.1, which highlights the superior performance of the active variant of TLCD. In a subsequent work, Chen and Ding (2006) designed an ATLCD-P system for a five-storey building and reported that the ATLCD-P outperformed both passive TLCD and TMD.

7.4 GAS-SEALED TLCD WITH ACTIVE PRESSURE CONTROL

Another approach to generate an active control force for controlling structural vibration is by actively controlling the air-spring effect in a sealed passive TLCD. This mechanism extends the passive TLCD to an active TLCD (ATLCD), and the damper can be classified as a gas sealed TLCD with active pressure control. This ATLCD was proposed by Hochrainer (2005) and is based on a more practical method of adjusting the gas pressure at either of the two vertical limbs of the TLCD by injecting or releasing gas (generally air), which can be stored in an external high-pressure reservoir. This STLCD-APC has the advantage that it does not need any moving parts for its operation unlike a propellor or an actuator (as was required in other ATLCD configurations).

7.4.1 Equations of Motion of STLCD-APC

An STLCD-APC mounted on an SDOF structural system is shown in Figure 7.3. This device can also act in a passive mode by utilizing the passive air spring effects (when the limbs are sealed at the top), but no external pressure is applied. If in addition to

---

TABLE 7.1

Comparison of Peak Displacement Reductions Achieved by ATLCD-P and Passive TLCD

	Reduction in Peak Structural Displacement Response by	
Structural Damping Ratio (%)	Passive TLCD (%)	ATLCD-P (%)
0.2	16.4	24.6
2	1.3	9.6

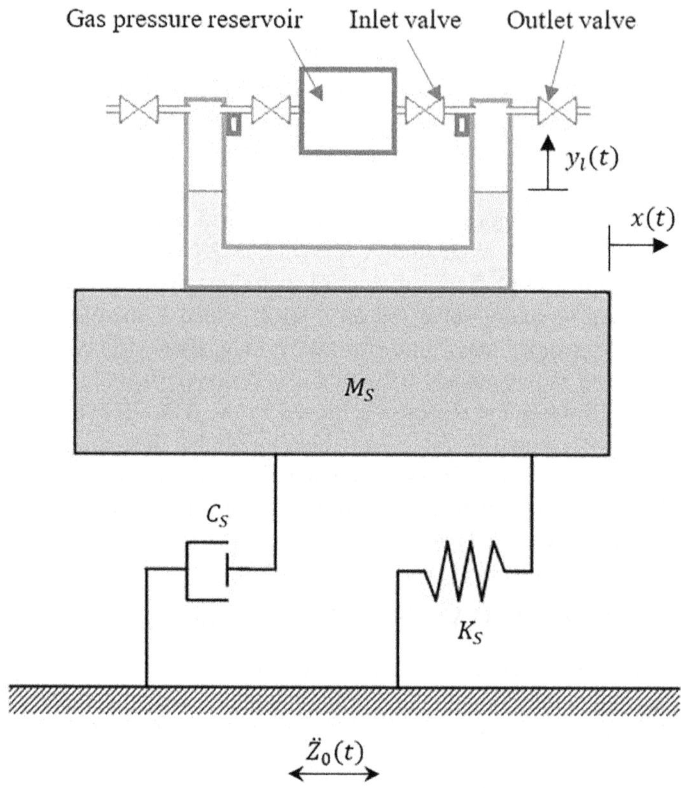

FIGURE 7.3 An active tuned liquid column gas damper mounted on a single-degree-of-freedom structural system.

the passive air spring effect, an active pressure difference Δp_a is applied by controlling the gas flow, the linearized equation of the fluid flow in the device is expressed as

$$\ddot{y}_l + 2\zeta_d\omega_{\text{sealed TLCD}}\dot{y}_l + \omega_{\text{sealed TLCD}}{}^2 y_l = -\alpha\left(\ddot{x} + \ddot{Z}_0\right) - \frac{\Delta p_a}{\rho L}. \qquad (7.22)$$

Eq. (7.22) has been derived by Hochrainer (2005) with an appropriate extension of nonstationary Bernoulli's equation, see, e.g. Hochrainer (2001), Ziegler (1998) (as Bernoulli's equation for ideal fluid flow is not applicable here). In Eq. (7.22), Δp_a denotes the quasi-static time-dependent pressure input which can be actively controlled to achieve a desired dynamic behaviour of the STLCD-APC. The inertia of the gas is neglected in deriving Eq. (7.22). Applying control signals to the valves, both positive and negative pressures can be generated by pressurizing either side of the damper container. The natural frequency of a passive sealed TLCD $\omega_{\text{sealed TLCD}}$ is given in Eq. (3.107), and ζ_d is the equivalent viscous damping ratio of the STLCD-APC. All other parameters are the same as described in Section 7.2.

7.4.2 Controller Design and Implementation

To control the air pressure for generating the active pressure difference in the STLCD-APC, a suitable algorithm needs to be designed. It is important to synthesize a control strategy which minimizes the structural vibration while limiting the air consumption. The air consumption in turn can be controlled by putting constraints on the number of pressurizers, operating pressure required and air volume to be pressurized. An important aspect of the STLCD-APC is its independence from any energy supply for generating the control action, which is entirely based on storage of pressurized air in an external high-pressure reservoir. In this respect, this control device has similarities with semi-active devices as the air pressure effectively modifies the dynamics of the liquid motion in the liquid column by changing the properties of the damper, although an active force is used in the process. Hence, we could also loosely classify this device to be at the borderline between active and semi-active controllers.

A numerical illustration of the implementation of the proposed STLCD-APC was presented by Hochrainer (2005) to demonstrate the effectiveness of the active damper on a 3-DOF system subjected to El Centro earthquake–based acceleration. The study was based on the benchmark definition paper (Spencer and Sain, 1997) where a steel frame structure considered by Wu et al. (1995) was equipped with a passive and an ATLCD for comparing their performances. For the study, the active tendon control was replaced by the STLCD-APC installed at the top of the building. Even with the application of the passive damper, a significant vibration reduction for response quantities such as inter storey draft and base shear was observed. To investigate the additional capability of the STLCD-APC, the vibration response during the initial transient period was focussed on. For facilitating the practical implementation of the active controller, a switching control strategy was applied, in which measurement values available from cheap sensors were used to control the active pressurization. The specific strategy that was implemented was a bang-bang control based on LQR algorithm with the active pressure input $\Delta p_a = \Delta p_{max} \cdot \text{sign}(\Delta p_{a,\text{LQR}})$, where Δp_{max} and $\Delta p_{a,\text{LQR}}$ are the maximum pressure that can be applied to the STLCD-APC and the optimal control pressure input obtained by solving the LQR problem, respectively. The controller is activated if the sum of the relative kinetic and strain energy of the structure exceeds a threshold value, since the active system is not required when the passive system performs satisfactorily. The superiority of the STLCD-APC over the passive system was well established, particularly during the initial transient phase. While the passive TLCD took several cycles of oscillation before it could start reducing the structural vibrational response, the beneficial effect of the STLCD-APC was almost immediate.

7.5 TLCD WITH SEMI-ACTIVE ORIFICE CONTROL

A variable damping TLCD can be achieved by semi-actively controlling the orifice diameter in a TLCD. The orifice damping has a significant impact on the response of a structure equipped with a TLCD, as has been observed from the studies on passive TLCD in Chapter 2. Hence, efficient control may be achieved by semi-actively or adaptively controlling the orifice opening ratio with a small amount of energy requirement (Haroun and Pires, 1996; Abe et al. 1996, Yalla and Kareem, 2000). Conceptually, this has led to the development of semi-active TLCD with damping controlled by orifice head-loss control (Yalla et al., 2001; Yalla and Kareem, 2003).

7.5.1   Equations of Motion of TLCD with Semi-Active Orifice Control

The equations of motion of an SDOF system equipped with a semi-active orifice controller (shown in Figure 7.4) subjected to a base acceleration $\ddot{Z}_0(t)$ can be written as

$$
\begin{bmatrix} M_s + m_f & \alpha m_f \\ \alpha m_f & m_f \end{bmatrix} \begin{Bmatrix} \ddot{x} \\ \ddot{y}_l \end{Bmatrix} + \begin{bmatrix} C_s & 0 \\ 0 & \dfrac{\rho A \xi}{2}|\dot{y}_l| \end{bmatrix} \begin{Bmatrix} \dot{x} \\ \dot{y}_l \end{Bmatrix}
$$
$$
+ \begin{bmatrix} K_s & 0 \\ 0 & 2\rho A g \end{bmatrix} \begin{Bmatrix} x \\ y_l \end{Bmatrix} = - \begin{Bmatrix} M_s + m_f \\ \alpha m_f \end{Bmatrix} \ddot{Z}_0(t)
$$

$$(7.23)$$

which can be rewritten as

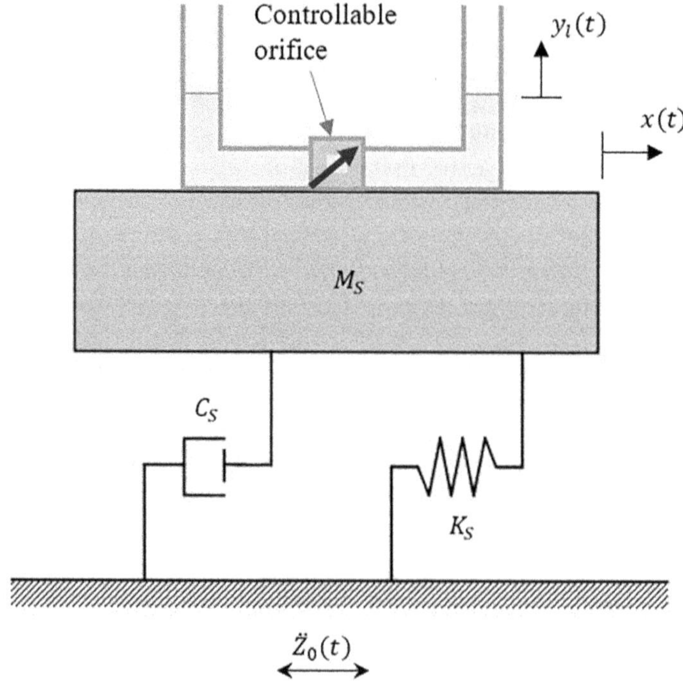

FIGURE 7.4   Semi-active tuned liquid column damper with controllable orifice mounted on a single-degree-of-freedom structural system.

$$\begin{bmatrix} M_s + m_f & \alpha m_f \\ \alpha m_f & m_f \end{bmatrix} \begin{Bmatrix} \ddot{x} \\ \ddot{y}_l \end{Bmatrix} + \begin{bmatrix} C_s & 0 \\ 0 & 0 \end{bmatrix} \begin{Bmatrix} \dot{x} \\ \dot{y}_l \end{Bmatrix}$$

$$+ \begin{bmatrix} K_s & 0 \\ 0 & 2\rho A g \end{bmatrix} \begin{Bmatrix} x \\ y_l \end{Bmatrix} = -\begin{Bmatrix} M_s + m_f \\ \alpha m_f \end{Bmatrix} \ddot{Z}_0(t) + \begin{Bmatrix} 0 \\ 1 \end{Bmatrix} u(t)$$

$$(7.24)$$

when an equal amount of active control force $u(t)$ is used to replace the semi-active force generated by the damping from the orifice head-loss. In Eqs. (7.23) and (7.24), m_f denotes the total mass of liquid in the damper container.

An external constraint has to be imposed in (7.23) to maintain the presence of the liquid in the vertical limbs for the TLCD to be effective. Also, a control mechanism such as a controllable valve is required to control the coefficient of head-loss by changing the orifice opening by using a command signal.

If $u(t)$ represents the semi-active control force instead, then $u(t)$ (as a function of coefficient of head-loss) and head-loss through the orifice h_l are given by

$$u(t) = -\frac{\rho A \xi(\Lambda, t)}{2}|\dot{y}_l|\dot{y}_l \tag{7.25}$$

and

$$h_l = \frac{\xi(\Lambda, t)}{2g}\dot{y}_l^{\,2} \tag{7.26}$$

where $\xi(\Lambda, t)$ is the head-loss coefficient, which is a function of the applied voltage Λ at a given time t.

7.5.2 Controller Design and Implementation

Two different semi-active control algorithms: (a) clipped-optimal continuously varying and (b) on-off strategies have been proposed and extensively studied by Yalla et al. (2001) with numerical examples presented for MDOF systems subjected to harmonic and random excitations. Controllers were designed with (a) full state feedback, (b) observer-based feedback and (c) fuzzy control.

The control force for the first strategy is based on LQR/LQG type of optimal control theory and is expressed as

$$\xi(t) = -2u(t)/\left(\rho A|\dot{y}_l|\dot{y}_l\right) \le \xi_{\max} \text{ if } u(t)\dot{y}_l(t) < 0 \tag{7.27a}$$

$$\xi(t) = \xi_{\min} \text{ if } u(t)\dot{y}_l(t) \ge 0 \tag{7.27b}$$

where $\xi_{\max}$ and $\xi_{\min}$ are the feasible maximum and minimum head-loss coefficients, respectively.

The control law in Eqs. (7.27a) and (7.27b) is a clipped-optimal controller as this emulates a fully active controller when the control force is dissipative (Dyke et al., 1996;

Karnopp et al., 1974). Also, the control force is bounded by the following inequalities as the force is dependent on the physical limitation of the valve and the maximum coefficient of head-loss feasible:

$$\left(-\frac{\rho A \xi_{\min}}{2}|\dot{y}_l|\dot{y}_l\right) \le u(t) \le \left(-\frac{\rho A \xi_{\max}}{2}|\dot{y}_l|\dot{y}_l\right) \tag{7.28}$$

The second strategy is a variation of the first one where the controller is turned on or off instead of supplying a continuously varying control signal. The on-off strategy can be represented as

$$\xi(t) = \xi_{\max} \text{ if } u(t)\dot{y}_l(t) < 0 \tag{7.29a}$$

$$\xi(t) = \xi_{\min} \text{ if } u(t)\dot{y}_l(t) \ge 0. \tag{7.29b}$$

In order to design the controller, a state space representation of the equations with the controller can be formulated, and the control force may be obtained by using the control gain acting on the states. The gain can be computed using the Ricatti matrix obtained by solving the Ricatti equation (Friedland, 1986; Meirovitch, 1990; Yalla et al., 2001).

Yalla et al. (2001) had considered an MDOF-TLCD system from Soong (1991) for numerical demonstration of the application of their proposed control strategies. The 5-DOF system under consideration had a mass of 131,338.6 ton on each floor, and a damping ratio of 3% was assumed for each mode. The natural frequencies of the 5-DOF were 0.23, 0.35, 0.42, 0.49 and 0.56 Hz. Figure 7.5 presents the rms values of the floor displacements and accelerations, and maximum values of storey shear and inter-storey displacements for the different control strategies with full state feedback, when the 5-DOF system is subjected to superposition of harmonic lateral forces at the floors. The numerical simulations indicate that the reduction in response achieved by the semi-active controllers is larger than the passive system. This is possible with a negligible amount of power requirement which can even be supplied by a battery. It was also concluded that the continuously varying semi-active control does not provide any significant benefit as compared to the simple on–off control strategy.

An early experimental study on an SDOF-semi-active TLCD system was conducted by Yalla and Kareem (2003) to verify a gain-scheduling based control law, providing optimum damping using a look-up table. Figure 7.6 is a schematic representation of the experimental setup consisting of an SDOF system with a TLCD fixed on a shaking table. The U-shaped tube of the TLCD had an electromagnetically controlled ball valve situated at the centre of the tube. By controlling the electromagnetic ball valve, optimal damping in a wide range of structural motion amplitudes could be provided. The effect of tuning ratio, damping and excitation amplitude and the equivalent damping provided by the TLCD as a function of excitation amplitude and the valve opening were studied. A general conclusion that was drawn from the studies was the fact that semi-active TLCDs can enhance the performance of passive TLCDs with fixed orifices by 15%–25% typically.

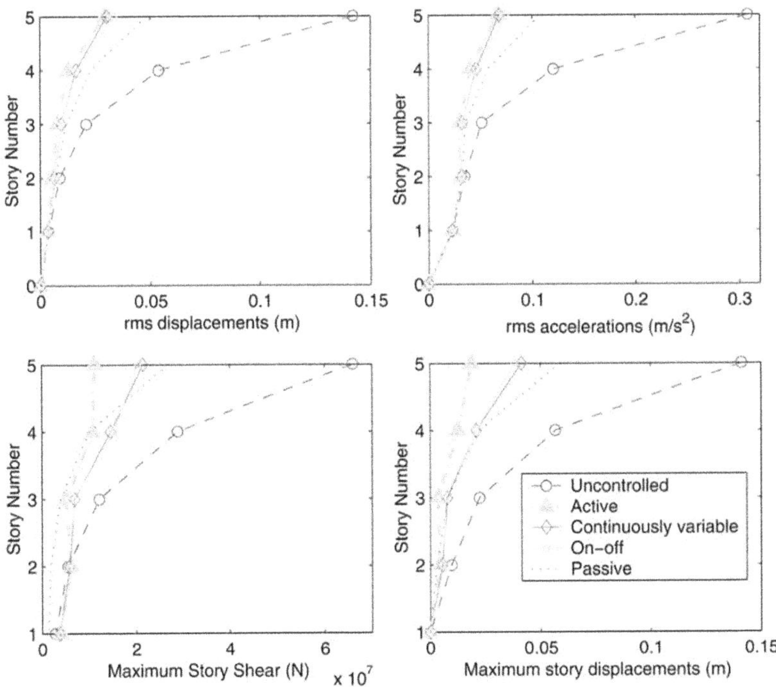

FIGURE 7.5 Variation of root-mean-square (rms) displacements, rms accelerations, maximum story shear and maximum inter-storey displacements (Yalla et al., 2001).

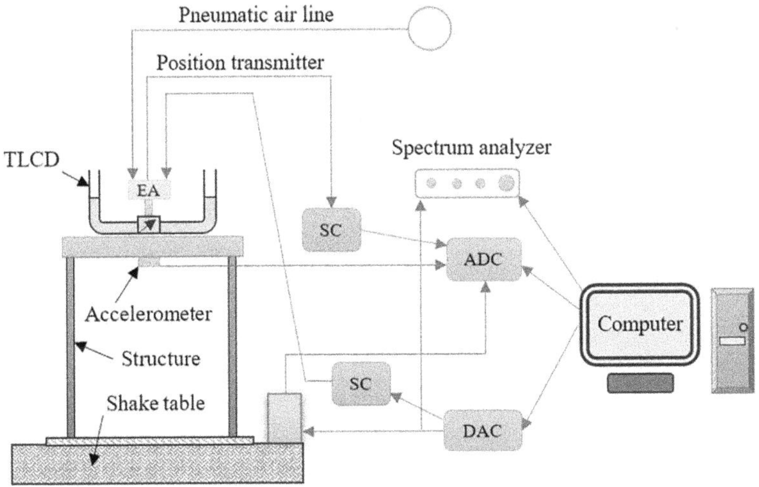

FIGURE 7.6 Schematic of experimental setup used by Yalla and Kareem (2003) for testing semi-active tuned liquid column damper with controllable orifice mounted on a single-degree-of-freedom structural system.

7.5.3 Irregular Structures and Controllers Based on Neural Networks

TLCD with semi-active orifice control has also been used for response control of irregular structures under seismic excitations (Bigdeli and Kim, 2016; Li and Huo, 2003). Both the investigations used neural networks but for different applications. Li and Huo (2003) used a back propagation artificial neural network (ANN) to predict the response of the structure subjected to two-dimensional seismic excitations, whereas Bigdeli and Kim (2016) had trained a neurocontroller based on the theory of multilayer neural network avoiding the requirement to design a controller based on optimal control theory.

A schematic of the 3D building model used by Bigdeli and Kim (2016) is shown in Figure 7.7. Material and geometric nonlinearities were considered in the 3D irregular system. The material nonlinearity considered the hysteresis behaviour suggested by Barber and Wen (1981). The semi-active TLCD used in this work had an orifice located in the middle of the U-shaped tube which had an on-off controller. The controller was designed based on the theory of multilayer neural network, and the head-loss was controlled by controlling the opening area of the valve. A cost function was used to train the neuro-controller, and hence, specific application of optimal control theory to compute the optimal control input was not required. The 1940 El Centro earthquake excitation record was used to train the algorithm, and two other earthquakes were then utilized to test the performance of the proposed algorithm. The results demonstrate significant reduction in rms displacement and acceleration responses including appreciable reduction in hysteretic response of the controlled system.

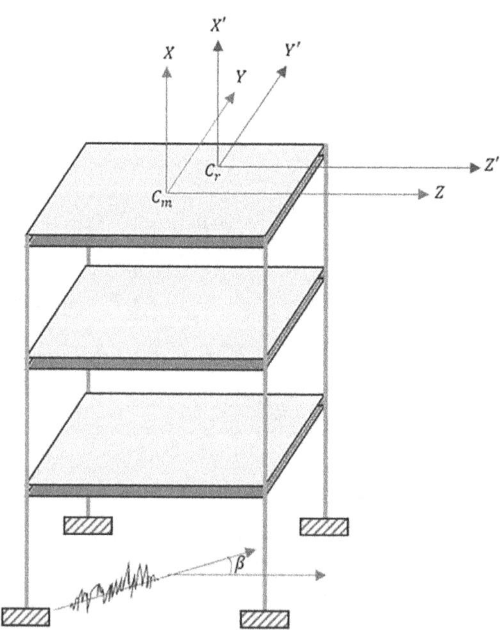

FIGURE 7.7 Multi degree-of-freedom structure model used by Bigdeli and Kim (2016).

The approach followed by Li and Huo (2003) was applied to an irregular structure subjected to multi-dimensional ground motions, where two semi-active TLCDs were employed in two orthogonal directions to control the translational and torsional responses of the structure, eccentric in plan. The strategy of Abe et al. (1996) was employed for semi-actively controlling the variable orifice opening and the associated hydraulic head-loss to suppress the translational-torsional coupled vibrations. To compensate for the time delay in the process of control, structural responses were predicted using the ANN.

In another research investigation on irregular structures subjected to multi-dimensional earthquake ground motions, Li et al. (2003) explored and confirmed the effectiveness of semi-active TLCDs in controlling the translational and rotational vibrations using the principle of fuzzy neural network (FNN) based on the Takagi-Sugeno model. FNN was used to train the controller for semi-actively controlling the variable orifice damping (as proposed in Abe et al. (1996)).

7.6 SEMI-ACTIVE MR-TLCD

An alternative way to modify and control the damping due to the motion of the liquid within the liquid column in a TLCD is to use different types of fluids (instead of water) and vary the damping properties of the fluid itself. This can have the same effect of semi-actively changing the head-loss by controlling the orifice opening. Magnetorheological (MR) fluids have the characteristic of reversibly changing the inherent viscous property from a free-flowing linear viscous one to a semi-solid nature with very high viscosity when subjected to a magnetic field. This functionality to change the viscosity of the MR fluid and to be able to do so in a short time span of the order of milliseconds (Jolly et al., 1999) makes the MR fluids an attractive choice as a resident fluid in a semi-active TLCD with alterable fluid viscosity. Experimental investigations into the passive damping properties of MR fluids (Colwell and Basu, 2008) have also confirmed the capability of the MR fluids in dissipating energy by head-loss in an adjustable orifice diameter TLCD to control structural vibration and that low viscous MR fluids can be implemented in semi-active MR-TLCDs.

Utilizing the sharply alterable fluid viscosity property of the MR fluids, adjustable and controllable fluid viscosity (or damping) in a TLCD can be used to control structural vibration over a wide range of excitations. Exploiting this concept, semi-active MR-TLCDs have been developed (Wang et al., 2002), and their effectiveness in controlling the vibrations in wind-excited tall buildings has been demonstrated. This research work was further developed for real-time semi-active vibration suppression of tall buildings incorporating nonlinear MR-TLCDs under random excitation using optimal LQ control strategy (Ni et al., 2004).

7.6.1 EQUATIONS OF MOTION WITH SEMI-ACTIVE MR-TLCD

A schematic diagram of a semi-active MR-TLCD is shown in Figure 7.8. An MR-TLCD consists of a U-tube filled with MR fluid. An orifice is placed at the middle of the horizontal part of the U-tube.

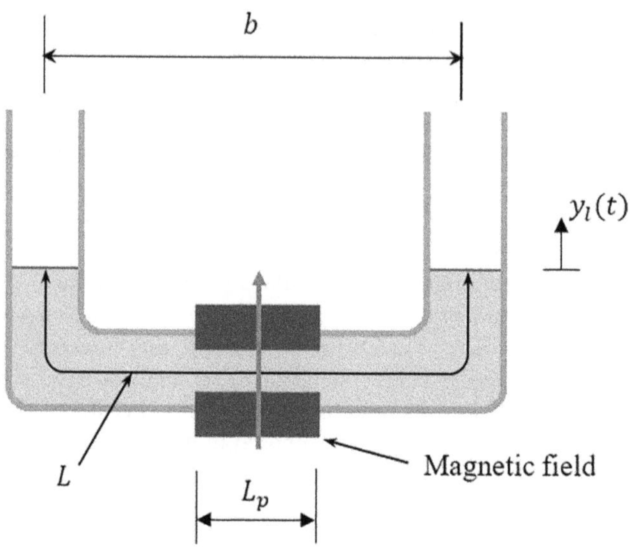

FIGURE 7.8   Schematic of MR-TLCD.

Using the parallel plate theory (Yang et al., 2002), the equation of motion of the MR fluid between fixed poles can be written as

$$\ddot{y}_l + \frac{1}{2}\frac{\delta}{L}|\dot{y}_l|\dot{y}_l + \frac{c\tau_y L_p}{\rho L h_p}\frac{\dot{y}_l}{|\dot{y}_l|} + \omega_l^2 y_l = -\alpha \ddot{x}_0 \tag{7.30}$$

where L_p and h_p are the length and depth of the flow between the fixed poles. τ_y is the yield stress developed by the applied magnetic field. c is a coefficient relying on the flow velocity and has a value ranging from 2.07 to 3.07. $\ddot{x}_0$ is the acceleration at the base of the damper. In Eq. (7.30), the overall head-loss coefficient δ is given by

$$\delta = \frac{48}{\text{Re}\left(1+h_b/w_p\right)^2}\frac{L}{h_b} + \sum_j \xi_j \tag{7.31}$$

where w_p denotes the width of the flow between the fixed poles, h_b denotes the height of the bottom limb of the damper, and ξ_j denotes the coefficient of minor head-losses, including those of elbows and orifices. The Reynolds number Re is defined by

$$\text{Re} = \frac{2\rho V_f\left(w_p + h_p\right)}{\eta_i w_p h_p}. \tag{7.32}$$

In Eq. (7.32), V_f and η_i denote the fluid velocity and the field-independent viscosity, respectively.

The damping term in (7.30) is nonlinear both due to the presence of orifice and the applied magnetic field. By adjusting the applied external voltage, the yield stress of the MR fluid can be adjusted to produce a controllable damping force. This controllable damping force always acts opposite to the relative motion, which ensures that energy is always dissipated, and hence, MR-TLCD always exerts dissipative control force.

7.6.2 CONTROLLER DESIGN AND IMPLEMENTATION

An appropriate control algorithm needs to be designed to control the damping in the semi-active MR-TLCD. For example, Ni et al. (2004) designed an optimal control law to generate the command signal to the semi-active MR-TLCD by minimization of a chosen performance index computed over an infinite time interval. A linear quadratic (LQ) optimal control strategy was implemented to obtain the optimal control force. A numerical demonstration of the semi-active MR-TLCD on a 50-storey residential building under construction in Hong Kong was presented. The initial design of the building was not satisfactory according to the wind resistant design of the Hong Kong code. Hence, the use of several supplemental damping devices in this building was studied. The total height of the building is 161.65 m, and the mass of the building is $2{,}774 \times 10^7$ kg. It was observed that the semi-active MR-TLCD with the optimal control strategy can achieve significant displacement, inter-storey drift, and acceleration response reduction as compared to a passive TLCD. An increase in the intensity of wind loading further increased the ability of the semi-active MR-TLCD to reduce the vibrational responses of the structure. It may be concluded that the semi-active MR-TLCD has the advantage of utilizing the dual effects of a smart fluid (with adjustable viscosity) and TLCD (with head-loss from orifice opening control).

7.7 SEMI-ACTIVE SPRING-CONNECTED TLCD

While semi-active TLCD with variable orifice or with MR fluids possessing adjustable damping properties are more effective than their passive counterpart, the effectiveness of such semi-active devices in tuning the frequency to the frequency of the primary structure is restricted to long-period structures only. In the context of passive dampers (as discussed in Section 3.4.2 of Chapter 3), as a development to cater for short-period (or stiff) structures, Ghosh and Basu (2004) proposed a new type of TLCD where the TLCD or CLCD was connected by a spring to the stiff (or short period) structure. However, this spring-connected TLCD or CLCD is still based on a passive control strategy and might lose its effectiveness when the properties of the primary structure or that of the excitations vary. To overcome this limitation of the passive spring-connected TLCD, Sonmez et al. (2016) proposed a semi-active TLCD (sTLCD) where the sTLCD was connected to the primary structure using an adaptive spring. This type of damper is also effective when damage occurs in the primary structure leading to a change in its natural frequency.

7.7.1 Equations of Motion with Semi-Active Spring-Connected TLCD

An MDOF system with an installed sTLCD is shown in Figure 7.9a. The reduced SDOF system from the MDOF system with a CLCD or a sTLCD has three DOFs (see Figure 7.9b). The equations of motion of the 3-DOF system in a matrix form yields

$$
\begin{bmatrix}
M_s & 0 & 0 \\
(\rho AL + m_c) & (\rho AL + m_c) & \rho AB \\
\rho AB & \rho AB & \rho AL
\end{bmatrix}
\begin{Bmatrix}
\ddot{x} \\
\ddot{y}_c \\
\ddot{y}_l
\end{Bmatrix}
+
\begin{bmatrix}
C_s & -c_c & 0 \\
0 & c_c & 0 \\
0 & 0 & 2\rho A C_p
\end{bmatrix}
$$

$$
\times
\begin{Bmatrix}
\dot{x} \\
\dot{y}_c \\
\dot{y}_l
\end{Bmatrix}
+
\begin{bmatrix}
K_s & -k_c & 0 \\
0 & k_c & 0 \\
0 & 0 & 2\rho A g
\end{bmatrix}
\begin{Bmatrix}
x \\
y_c \\
y_l
\end{Bmatrix}
= -
\begin{Bmatrix}
M_s \\
(\rho AL + m_c) \\
\rho AB
\end{Bmatrix}
\ddot{Z}_0(t).
$$

$$(7.33)$$

In Eq. (7.33), the parameters have the same meaning as defined in Section 7.2.1. Recasting in frequency domain gives

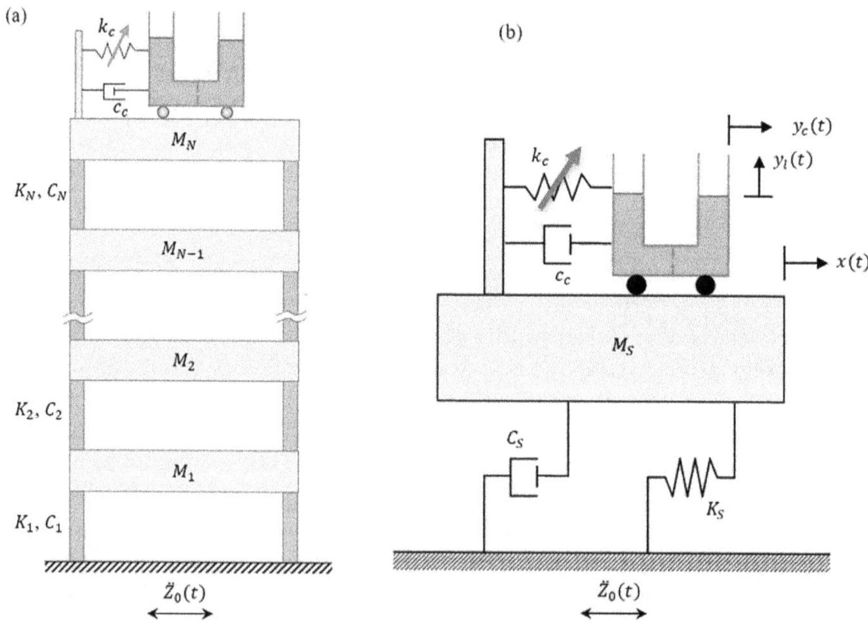

FIGURE 7.9 (a) A multi-degree-of-freedom (MDOF) structure with an sealed tuned liquid column damper (sTLCD), (b) reduced single-degree-of-freedom system from the MDOF system with sTLCD.

$$
\begin{bmatrix}
\dfrac{1}{H_1(\omega)} & -\mu_c\left(2\zeta_c\omega_d\omega i + \omega_d^2\right) & 0 \\[3mm]
-\omega^2 & \dfrac{1}{H_2(\omega)} & -\dfrac{\alpha\omega^2}{1+\tau} \\[3mm]
-\dfrac{1+\tau}{\alpha} & -\dfrac{1+\tau}{\alpha} & \dfrac{1}{\beta(\omega)}
\end{bmatrix}
\begin{Bmatrix}
H_X(\omega) \\[2mm]
H_{YC}(\omega) \\[2mm]
H_{YL}(\omega)
\end{Bmatrix}
=
\begin{Bmatrix}
(-1)^q \\[2mm]
-q \\[2mm]
-q\left(\dfrac{1+\tau}{\alpha\omega^2}\right)
\end{Bmatrix}
$$

$$(7.34)$$

where $H_X(\omega)$, $H_{YC}(\omega)$ and $H_{YL}(\omega)$ denote the transfer functions for the motions of the SDOF primary mass, the damper container and the liquid column elevation, respectively. Further, μ_c, ζ_c, ω_d, τ and α have the same meaning as defined in Section 7.2.1. The other parameters in Eq. (7.34) are defined as

$$H_1(\omega) = \frac{1}{-\omega^2 + i2\zeta_s\omega_s\omega + \omega_s^2}, \tag{7.35}$$

$$H_2(\omega) = \frac{1}{-\omega^2 + i2\zeta_c\omega_d\omega + \omega_d^2}, \tag{7.36}$$

$$\beta(\omega) = \frac{\alpha^2\omega^2}{(1+\tau)\left(-\omega^2 + i2\dfrac{C_p}{L}\omega + \omega_l^2\right)} \tag{7.37}$$

and

$$q = \begin{cases} 0 & \text{for forced excitaton} \\ 1 & \text{for base excitation} \end{cases}. \tag{7.38}$$

In Eqs. (7.35)–(7.37), ω_s, ζ_s and ω_l have the same meaning as defined in Section 7.2.1.

The terms $H_1(\omega)$ and $H_2(\omega)$ are the transfer functions of the primary SDOF system and the TLCD, respectively, treated as individual SDOF systems excited by the base acceleration. From Eq. (7.34) we arrive at

$$H_X(\omega)$$

$$= \frac{H_1(\omega)\left\{(-1)^q\left[1-\omega^2\beta(\omega)H_2(\omega)\right]-q\mu_c\left[1+\omega^2H_2(\omega)\right]\left[1+\beta(\omega)\right]\right\}}{1-\omega^2\beta(\omega)H_2(\omega)-\mu_c\omega^2H_1(\omega)\left[1+\omega^2H_2(\omega)\right]\left[1+\beta(\omega)\right]} \tag{7.39}$$

and

$$H_{YL}(\omega) = \frac{(1+\tau)\beta(\omega)}{\alpha\omega^2}\left[\frac{\omega^2 H_2(\omega)\left[1+\beta(\omega)\right]\left[\omega^2 H_X(\omega)-q\right]}{1-\omega^2\beta(\omega)H_2(\omega)} + \omega^2 H_X(\omega)-q\right].$$

(7.40)

7.7.2 Controller Design and Implementation

Two control algorithms have been proposed by Somnez et al. (2016), which are (i) feedforward and (ii) feedback. In the feedforward algorithm, the tuning of the semi-active stiffness and the choice of the optimal head-loss or damping are based on the estimated evolutionary spectrum of the excitation, while in the case of feedback algorithm, the same are obtained from the evolutionary spectrum of the structural response. The block diagrams for the two control algorithms are presented in Figure 7.10.

Once the semi-active tuning is decided for the sTLCD, the value of linear equivalent damping coefficient, C_p needs to be estimated. One approach is estimating C_p iteratively using the transfer function of the liquid velocity and a target spectral intensity of the input excitation (see Ghosh and Basu (2004)). If $S(t,\omega)$ is the evolutionary input excitation spectrum, the power spectral density function (PSD) of the liquid velocity in the sTLCD and the PSD of the primary structural displacement response are, respectively, estimated as

$$S_{\dot{y}_l}(t,\omega) \approx \omega^2\left|H_{YL}(t,\omega)\right|^2 S(t,\omega)$$

(7.41)

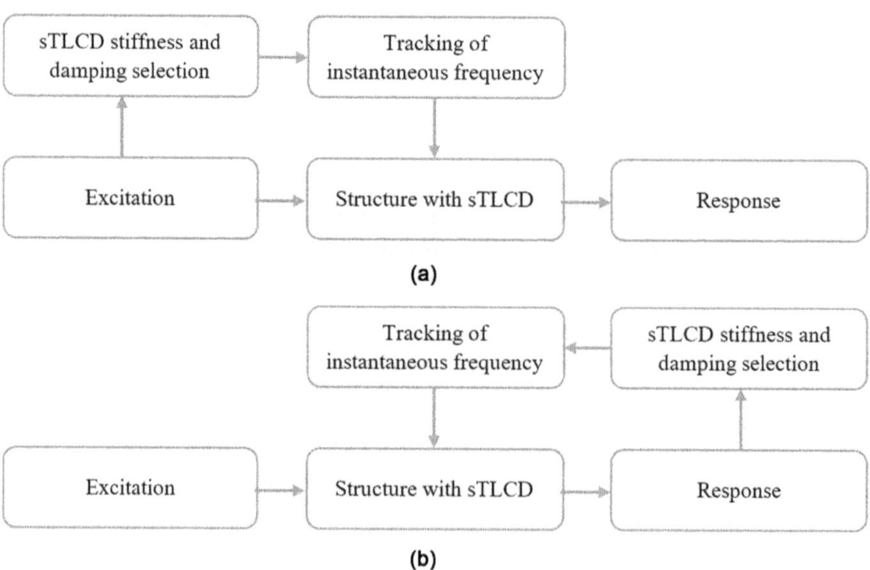

FIGURE 7.10 (a) Flowchart of control algorithm proposed by Somnez et al. (2016) (a) feedforward and (b) feedback.

and

$$S_x(t, \omega) \approx |H_X(t,\omega)|^2 \, S(t,\omega), \tag{7.42}$$

leading to the following expressions for the rms of the liquid velocity and of the primary structural displacement response,

$$\sigma_{\dot{y}_l} \approx \sqrt{\int_{-\infty}^{\infty} S_{\dot{y}_l}(t,\omega)\,d\omega} \tag{7.43}$$

and

$$\sigma_x \approx \sqrt{\int_{-\infty}^{\infty} S_x(t,\omega)\,d\omega} \tag{7.44}$$

respectively. The optimized head-loss coefficient is obtained iteratively by solving Eqs. (7.43) and (7.5) for a design input spectral intensity. This intensity can be assumed as the mean spectral intensity of a design evolutionary PSD. Hence, time-varying C_p can be iteratively obtained for each window over which the evolutionary PSD is estimated at a given instant of time.

There are five steps in the semi-active control algorithm, which are summarized as follows.

Step 1: At the initial time $t = 0$, the stiffness of the sTLCD is set to the default optimum for passive spring connected TLCD.

Step 2: A moving window is chosen of an appropriate length for time-frequency analysis to obtain the spectra (see Somnez et al. (2016) for details) and is incremented by a number of time steps to proceed to the following window with a new time array.

Step 3: Instantaneous PSD of the excitation (in case of feedforward) or of the response (in case of feedback) are estimated using short time Fourier transform (STFT) (see Cohen (1995), Nagarajaiah and Varadarajan (2005), and Nagarajaiah (2009)). The instantaneous dominant frequency is tracked by averaging the frequencies of the highest energy detected over an averaging time duration.

Step 4: Stiffness and damping variation start after a certain time instant $t = t_0$ to allow for sufficient data to be collected to perform the estimation.

Step 5: For feedforward control, if the estimated instantaneous frequency $f_{\text{ins}}(t)$ is within a prescribed lower and upper limit, f_{low} and f_{up}, respectively, then the sTLCD is tuned to the instantaneous frequency. Otherwise, the sTLCD is tuned to the nearest limit of the instantaneous frequency. Default limits prescribed are $f_{\text{low}} = 0.5 f_n$ and $f_{\text{up}} = 2 f_n$, where f_n is the fundamental frequency of the structure.

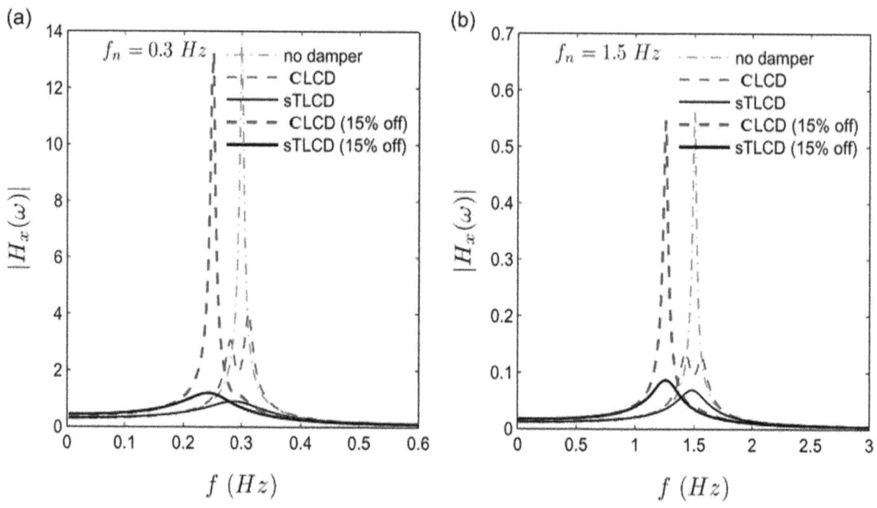

FIGURE 7.11 Transfer functions for single-degree-of-freedom system with sTLCD and CLCD: (a) $f_n = 0.3$ Hz (forced excited) and (b) $f_n = 1.5$ Hz (base excited) (Sonmez et al., 2016).

Some parametric results on sTLCDs were reported by Somnez et al. (2016) for feedforward control. The feedback control performance of the sTLCD will be same as that of a CLCD without any off tuning. The structural damping ratio was taken as 1% of the critical and the mass ratio, μ_c was assumed to be 0.01. The ratio of the horizontal length to the total length of the liquid column was set to 0.9, and the tuning ratio of the damper was obtained from $1/(1+\mu_c)$. The liquid column mass was assumed to be the same as the liquid container mass. Two different primary structures, one with $f_n = 0.3\,$Hz and the other with $f_n = 1.5\,$Hz, were considered.

Figure 7.11 shows the transfer functions of the two different SDOF primary structures for five different cases: (i) primary structure without any damper, (ii) primary structure with CLCD, (iii) primary structure with sTLCD, (iv) damaged primary structure with mistuned CLCD and (v) damaged primary structure with sTLCD. The performance of sTLCD is spectacular with a much smaller transfer function as compared to the cases corresponding to the structure without any damper or with a CLCD. An interesting point to note is that the transfer function of the structure with sTLCD has only one peak as opposed to the optimized two peaks for the case with CLCD. This might be attributed to the fact that the sTLCD with feedforward control tunes to the excitation frequency.

The 1940 El Centro earthquake time history has also been used, to study the performance of the sTLCD under complex realistic non-stationary ground motions. Time history response of an SDOF primary structure (damaged at $t=2.6\,$s and f_n changing from 2.0 Hz to 1.4 Hz) and the corresponding variable tuning frequency of the sTLCD with feedforward control and feedback control are shown in Figures 7.12 and 7.13, respectively.

Overall, the performance of the sTLCD is superior to the passive CLCD for harmonic excitations and similar when subjected to more complex random excitations. The main advantage of the sTLCD is its robustness against changes in the frequency

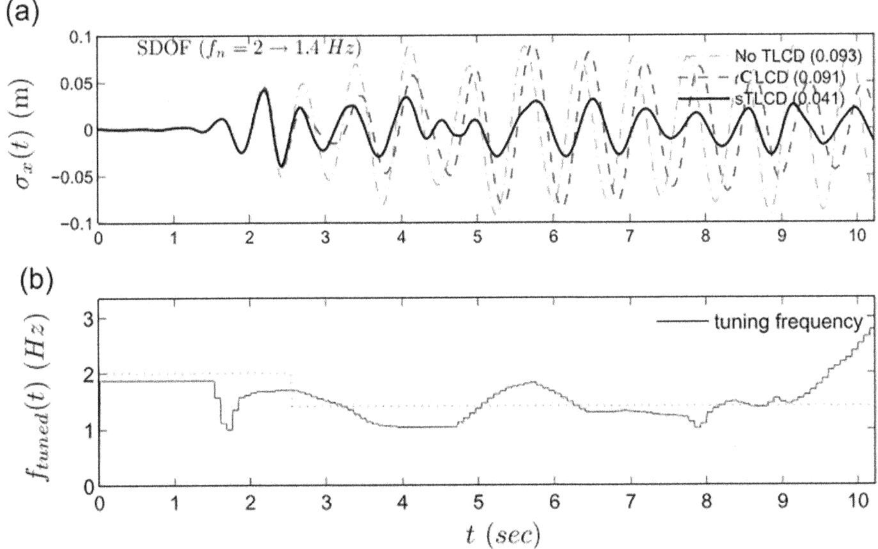

FIGURE 7.12   1940   El Centro Earthquake excitation of single-degree-of-freedom structural system (feedforward): (a) time history response and (b) variable spring frequency (Sonmez et al., 2016).

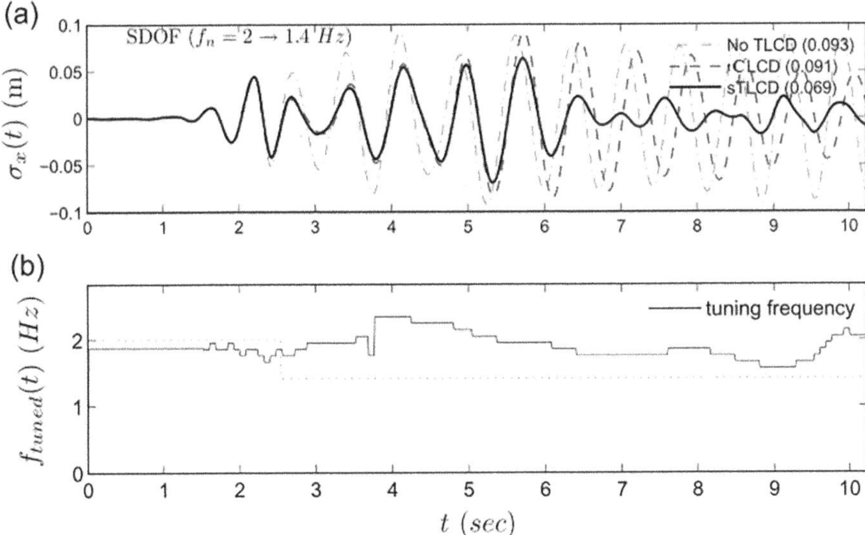

FIGURE 7.13   1940   El Centro Earthquake excitation of single-degree-of-freedom structural system (damaged at $t = 2.6$ s) along (feedback): (a) time history response and (b) variable spring frequency (Sonmez et al., 2016).

of the primary structure (e.g., due to some damage or change in other environmental conditions) which may lead to mistuning in case of a passive damper.

7.8   HYBRID DAMPERS WITH SEMI-ACTIVE TLCD

A damper system, consisting of multiple dampers, where one or more are semi-active TLCDs, can be classified as a hybrid damper with semi-active TLCD. The concept of a hybrid liquid column damper was proposed by Haroun et al. (1996). In the hybrid system proposed, instantaneous optimal control was used to control the orifice damping semi-actively while the liquid column pressure was controlled actively, leading to a dual control strategy. Numerical implementation of the hybrid controller demonstrated the effectiveness of the hybrid damper with dual control in suppressing earthquake and wind-induced vibrations in tall buildings in an energy efficient way along with low maintenance.

By definition, hybrid dampers are non-unique and so are hybrid semi-active TLCDs. Another variation of the hybrid damper with TLCDs was proposed by Kim and Adeli (2005) in which a passive viscous fluid supplementary damping system was combined with a semi-active TLCD. Figure 7.14 illustrates a schematic of the hybrid damper system. A robust wavelet-hybrid feedback least mean square (LMS) control scheme developed by the same researchers (Adeli and Kim, 2004) was used to obtain the optimal control parameters. The effectiveness of the hybrid semi-active TLCD was numerically evaluated and confirmed by assessing the reduction in vibration of an eight-storey frame with the hybrid semi-active TLCD, subjected to

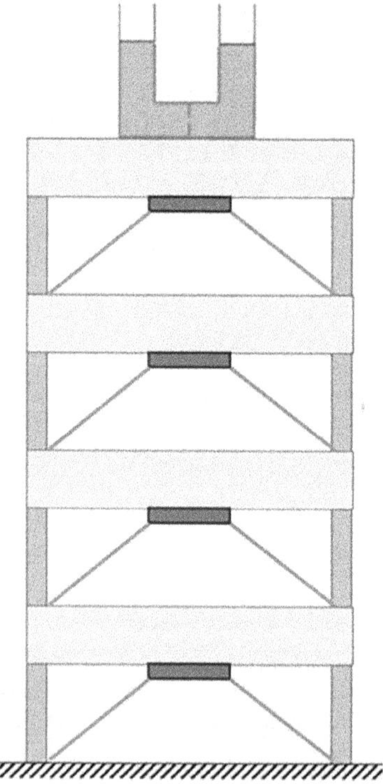

FIGURE 7.14   A hybrid vibration control system consisting of tuned liquid column damper and viscous fluid damper as studied by Kim and Adeli (2005).

various earthquake excitations. The hybrid damper possesses the advantage of both the passive (viscous fluid) and the semi-active (tuned liquid column) dampers, thus improving the overall performance, reliability and operability of the controller during normal as well as emergency situations when power failure or computer system malfunctioning might occur.

REFERENCES

Abe, M., Kimura, S., Fujino, Y., 1996. Control laws far semi-active tuned liquid column damper with variable orifice openings, in: *Second International Workshop on Structural Control*. Hong Kong, pp. 5–10.

Adeli, H., Kim, H., 2004. Wavelet-hybrid feedback-least mean square algorithm for robust control of structures. *J. Struct. Eng.* 130, 128–137. https://doi.org/10.1061/(ASCE)0733-9445(2004)130:1(128)

Agrawal, A.K., Yang, J.N., 1999. Design of passive energy dissipation systems based on LQR control methods. *J. Intell. Mater. Syst. Struct.* 10, 933–1014. https://doi.org/10.1106/FB58-N1DG-ECJT-B8H4

Baber, T.T., Wen, Y.K., 1981. Random vibration of hysteretic, degrading structures. *J. Eng. Mech. Div.* 107, 1069–1087. https://doi.org/https://doi.org/10.1061/JMCEA3.0002768

Balendra, T., Wang, C.M., Yan, N., 2001. Control of wind-excited towers by active tuned liquid column damper. *Eng. Struct.* 23, 1054–1067. https://doi.org/10.1016/S0141-0296(01)00015-3

Basar, T., Bernhard, P., 1995. *H/sup ∞/-Optimal Control and Related Minimax Design Problems: A Dynamic Game Approach*, 2nd ed. Birkhäuser, Boston. https://doi.org/10.1109/tac.1996.536519

Basu, B., Nagarajaiah, S., 2008. A wavelet-based time-varying adaptive LQR algorithm for structural control. *Eng. Struct.* 30, 2470–2477. https://doi.org/10.1016/j.engstruct.2008.01.011

Basu, B., Nagarajaiah, S., 2010. Multiscale wavelet-LQR controller for linear time varying systems. *J. Eng. Mech.* 136, 1143–1151. https://doi.org/10.1061/(asce)em.1943-7889.0000162

Bhattacharyya, S., Ghosh, A.D., Basu, B., 2018. Design of an active compliant liquid column damper by LQR and wavelet linear quadratic regulator control strategies. *Struct Control Heal. Monit.* 25, e2265. https://doi.org/10.1002/stc.2265

Bigdeli, Y., Kim, D., 2016. Damping effects of the passive control devices on structural vibration control: TMD, TLC and TLCD for varying total masses. *KSCE J. Civ. Eng.* 20, 301–308. https://doi.org/10.1007/s12205-015-0365-5

Bui, H., Tran, N., Cao, H.Q., 2023. Active control based on hedge-algebras theory of seismic-excited buildings with upgraded tuned liquid column damper. *J. Eng. Mech.* 149, 04022091. https://doi.org/10.1061/JENMDT.EMENG-6821

Chen, Y.-H., Ding, Y.-J., 2006. Passive, semi-active, and active tuned-liquid-column dampers, in: *ASME 2006 Pressure Vessels and Piping/ICPVT-11 Conference*. ASME, Vancouver, Canada, pp. 99–106. https://doi.org/https://doi.org/10.1115/PVP2006-ICPVT-11-93834

Chen, Y.H., Ko, C.H., 2003. Active tuned liquid column damper with propellers. *Earthq. Eng. Struct. Dyn.* 32, 1627–1638. https://doi.org/10.1002/eqe.295

Cohen, L., 1995. *Time-Frequency Analysis: Theory and Applications*. Prentice-Hall, Englewood Clifs, NJ.

Colwell, S., Basu, B., 2008. Experimental and theoretical investigations of equivalent viscous damping of structures with TLCD for different fluids. *J. Struct. Eng.* 134, 154–163. https://doi.org/10.1061/(ASCE)0733-9445(2008)134:1(154)

Dyke, S.J., Spencer, B.F., Sain, M.K., Carlson, J.D., 1996. Modeling and control of magnetorheological dampers for seismic response reduction. *Smart Mater. Struct.* 5, 565–575. https://doi.org/10.1088/0964-1726/5/5/006

Friedland, B., 1986. *Control System Design: An Introduction to State Space Methods.* McGraw-Hill Book, New York.

Ghosh, A.D., Basu, B., 2004. Seismic vibration control of short period structures using the liquid column damper. *Eng. Struct.* 26, 1905–1913. https://doi.org/10.1016/j.engstruct.2004.07.001

Ghosh, A.D., Konar, T., 2023. Popular passive dampers for structural control: a review. *J. Struct. Eng.* 50, 24–38.

Glover, K., 2013. H-infinity control. *Encycl. Syst. Control* 1–9. https://doi.org/10.1007/978-1-4471-5102-9

Haroun, M.A., Pires, J.A., Won, A.Y.J., 1996. Suppression of environmentally-induced vibrations in tall buildings by hybrid liquid column dampers. *Struct. Des. Tall Build.* 5, 45–54. https://doi.org/10.1002/(SICI)1099-1794(199603)5:1<45::AID-TAL58>3.0.CO;2-F

Ho, N.C., Nhu Lan, V., Xuan Viet, L., 2008. Optimal hedge-algebras-based controller: design and application. *Fuzzy Sets Syst.* 159, 968–989. https://doi.org/10.1016/j.fss.2007.11.001

Hochrainer, M.J., 2001. *Control of Vibrations of Civil Engineering Structures with Special Emphasis on Tall Buildings.* Technische Universität Wien, Wien.

Hochrainer, M.J., 2005. Tuned liquid column damper for structural control. *Acta Mech.* 175, 57–76. https://doi.org/10.1007/s00707-004-0193-z

Hrovat, D., Barak, P., Rabins, M., 1983. Semi-active versus passive or active tuned mass dampers for structural control. *J. Eng. Mech.* 109, 691–705. https://doi.org/10.1061/(asce)0733-9399(1983)109:3(691)

Jolly, M.R., Bender, J.W., Carlson, J.D., 1999. Properties and applications of commercial magnetorheological fluids. *J. Intell. Mater. Syst. Struct.* 10, 5–13. https://doi.org/10.1177/1045389x9901000102

Kareem, A., Kijewski, T., Tamura, Y., 1999. Mitigation of motions of tall buildings with specific examples of recent applications. *Wind Struct.* 2, 201–251. https://doi.org/10.12989/was.1999.2.3.201

Karnopp, D., Crosby, M.J., Harwood, R.A., 1974. Karnopp, D., Crosby, M. J., & Harwood, R. A. (1974). Vibration control using semi-active force generators.. *J. Eng. Ind.* 96, 619–626.

Kim, H., Adeli, H., 2005. Hybrid control of smart structures using a novel wavelet-based algorithm. *Comput. Civ. Infrastruct. Eng.* 20, 7–22. https://doi.org/10.1111/j.1467-8667.2005.00373.x

Kobori, T., 1996. Future direction on research and development of seismic-response-controlled structures. *Comput. Civ. Infrastruct. Eng.* 11, 297–304. https://doi.org/10.1111/j.1467-8667.1996.tb00444.x

Konar, T., Ghosh, A.D., 2021. Flow damping devices in tuned liquid damper for structural vibration control: a review. *Arch Comput Methods Eng* 28, 2195–2207. https://doi.org/10.1007/s11831-020-09450-0

Konar, T., Ghosh, A.D., 2023. A review on various configurations of the passive tuned liquid damper. *J. Vib. Control* 29, 1945–1980. https://doi.org/10.1177/10775463221074077

Lago, A., Trabucco, D., Wood, A., 2019. *Damping Technologies for Tall Buildings.* Butterworth-Heinemann, Oxford. https://doi.org/10.1016/b978-0-12-815963-7.00008-7

Li, H.-N., Huo, L.-S., 2003. Seismic control of eccentric structures using TLCD semi-active neural networks, in: *Proceedings of 2003 IEEE Conference on Control Applications.* IEEE, Istanbul, Turkey, pp. 336–340.

Li, H.N., Jin, Q., Song, G., Wang, G.X., 2003. Fuzzy neural network based TLCD semi-active control methodology for irregular buildings, in: *International Symposium on Intelligent Control*. IEEE, Houston, Texas, USA, pp. 176–181. https://doi.org/10.1109/isic.2003.1253934

Meirovitch, L., 1990. *Dynamics and Control of Structures*. Wiley, New York.

Nagarajaiah, S., 2009. Adaptive passive, semiactive, smart tuned mass dampers: identification and control using empirical mode decomposition, hilbert transform, and short-term Fourier transform. *Struct. Control Heal. Monit.* 16, 800–841. https://doi.org/10.1002/stc.349

Nagarajaiah, S., Varadarajan, N., 2005. Short time Fourier transform algorithm for wind response control of buildings with variable stiffness TMD. *Eng. Struct.* 27, 431–441. https://doi.org/10.1016/j.engstruct.2004.10.015

Ni, Y.Q., Ying, Z.G., Wang, J.Y., Ko, J.M., Spencer, B.F., 2004. Stochastic optimal control of wind-excited tall buildings using semi-active MR-TLCDs. *Probabilistic Eng. Mech.* 19, 269–277. https://doi.org/10.1016/j.probengmech.2004.02.010

Roy, A.K., Konar, T., Ghosh, A., 2023. Mitigation of structural vibrations due to pulse-type-near-fault earthquake by the compliant liquid column damper. *J. Earthq. Tsunami* 17, 2350004. https://doi.org/10.1142/S1793431123500045

Sonmez, E., Nagarajaiah, S., Sun, C., Basu, B., 2016. A study on semi-active Tuned Liquid Column Dampers (sTLCDs) for structural response reduction under random excitations. *J. Sound Vib.* 362, 1–15. https://doi.org/10.1016/j.jsv.2015.09.020

Soong, T.T., 1991. *Active Structural Control: Theory and Practice*. Longman and Wiley, London.

Spencer, B.F., Sain, M.K., 1997. Controlling buildings: a new frontier in feedback. *IEEE Control Syst. Mag.* 17, 19–35. https://doi.org/10.1109/37.642972

Tamboli, A., Christoforou, C., Brazil, A., Joseph, L., Vadnere, U., Malmsten, B., 2005. Manhattan's mixed construction skyscrapers with tuned liquid and mass dampers, in: *CTBUH 2005–7th World Congress Renewing the Urban Landscape, Proceedings*. New York City, USA.

Wang, J.Y., Ni, Y.Q., Ko, J.M., Spencer, B.F., 2002. Semi-active TLCDS using magneto-rheological fluids for vibration mitigation of tall buildings, in: Anson, M., Ko, J. M., Lam, E. S. S. (Eds.), *Advances in Building Technology (Proceedings of the International Conference on Advances in Building Technology)*. Elsevier, Oxford, pp. 537–544. https://doi.org/10.1016/b978-008044100-9/50069-3

Wu, Z., Soong, T., Gattulli, V., Lin, R.C., 1995. *Nonlinear Control Algorithms for Peak Response Reduction*. University of Buffalo, New York.

Xu, Y.L., Kwok, K.C.S., Samali, B., 1992. The effect of tuned mass dampers and liquid dampers on cross-wind response of tall/slender structures. *J. Wind Eng. Ind. Aerodyn.* 40, 33–54. https://doi.org/10.1016/0167-6105(92)90519-G

Yalla, S.K., Kareem, A., 2000. Optimum absorber parameters for tuned liquid column dampers. *J. Struct. Eng.* 126, 906–915. https://doi.org/10.1061/(ASCE)0733-9445(2000)126:8(906)

Yalla, S.K., Kareem, A., 2003. Semiactive tuned liquid column dampers: experimental study. *J. Struct. Eng.* 129, 960–971. https://doi.org/https://doi.org/10.1061/(ASCE)0733-9445(2003)129:7(960)

Yalla, S.K., Kareem, A., Kantor, J.C., 2001. Semi-active tuned liquid column dampers for vibration control of structures. *Eng. Struct.* 23, 1469–1479. https://doi.org/10.1016/S0141-0296(01)00047-5

Yang, G., Spencer, B.F., Carlson, J.D., Sain, M.K., 2002. Large-scale MR fluid dampers: modeling and dynamic performance considerations. *Eng. Struct.* 24, 309–323. https://doi.org/10.1016/S0141-0296(01)00097-9

Ziegler, F., 1998. *Mechanics of Solids and Structures*, 2nd. ed. Springer, New York. https://doi.org/10.1016/b978-0-12-818283-3.00002-6

8 Practical Implementation of TLCD

8.1 INTRODUCTION

One of the most important benchmarks in the development of any new technology is the extent of its practical implementation and subsequent performance. The early stages include those of conceptual development, mathematical modelling and full-scale experimental verification. In case of tuned liquid column damper (TLCD), due to the sheer simplicity of its design, its practical implementation commenced within a few years from the conceptualization of the device by Saoka et al. (1988). The first installation of TLCD in a real-world structure was in the Higashi-Kobe Bridge, Japan, which was opened to public in 1992 (Kitazawa et al., 1992). Ever since, several structures, including bridges, buildings, towers, etc., have been built with TLCDs installed for improved performance under dynamic loading. However, among the several configurational variations of TLCDs, only the conventional TLCD, bi-directional TLCD, TLCD with air-spring, conventional liquid column vibration absorber (LCVA) and bi-directional LCVA have been installed in structures. The other TLCD varieties have not yet reached the installation stage.

Kareem et al. (1999) have provided a list of installations of inertia-based dampers in buildings that contain a few instances of installations of TLCDs. Further information on the practical implementations of TLCDs is also available in the work of Konar and Ghosh (2023). Konar et al. (2024) presented six case studies on real-world implementations of TLCDs in landmark buildings for wind-induced vibration control. A comprehensive list of significant practical applications of TLCDs is presented in Table 8.1, all of which utilize water as the damper liquid. Additionally, there is a large number of patents related to TLCDs. The details of some of the notable patents on TLCDs are listed in Table 8.2. It may be observed from Table 8.2 that a large number of patents have been awarded in recent times. This indicates that the field is a fast-expanding one, and it is expected that a large number of TLCDs will be installed in buildings, towers, bridges and wind turbines in the coming days. In the subsequent sections, some specific cases of practical implementations of TLCDs are elaborated upon. The aspects of damper design, modelling and performance assessment are discussed.

DOI: 10.1201/9781003377894-8

TABLE 8.1

Application of TLCDs in Real-World Structures

Name of Structure	Location	Type of Structure	Height/Number of Storeys/Total Floor Area	Year of installation	Type of TLCD	Approximate Volume of Water in TLCD(s) (L)
Higashi-Kobe bridge	Kobe, Japan	Bridge pylons[a]	146.5 m/–/–	1992	Single conventional TLCD on each column of the pylons	0.27×10^5
Hotel Cosima (later known as Hotel Sofitel)	Tokyo, Japan	Commercial building	106 m/ 26+3 basement/9,797 m²	1994[b]	Single bi-directional TLCD with period adjustment equipment	0.58×10^5
Hyatt Hotel	Osaka, Japan	Commercial building	112 m/28/78,417 m²	1995	Single bi-directional TLCD with period adjustment equipment	1.04×10^5
Ichida Building[c]	Osaka, Japan	–	–	–	–	–
Prospect Communications Tower	Sydney, Australia	Tower	67 m/–/–	Between 1996 and 1999	20 bi-directional LCVAs	280
One Wall Centre	Vancouver, Canada	Commercial-residential building	149.8 m/48/42,955 m²	2001	2 conventional TLCDs	4.546×10^5
Random House Tower	New York City, USA	Residential-commercial building	208 m/52 + 2 basement/79,900 m²	2003	2 conventional TLCDs in mutually perpendicular direction	11×10^5
Bridge Typ SHB 304/12 of ÖBB[d]	Austria	Railway Bridge	Span 29.9 m	2007	8 TLCDs with air spring (Ad-TLCD)	–
Comcast Center	Philadelphia, USA	Commercial building	297 m/58 + 3 basement/130,064 m²	2008	Single conventional TLCD	11.36×10^5
New Songdo City First World Towers (4 buildings)	Incheon, Republic of Korea	Residential building	237 m/67 + 3 basement/344,000 m²	2009	2 conventional LCVAs in mutually perpendicular direction on each building	6.62×10^5

(*Continued*)

TABLE 8.1 (*Continued*)
Application of TLCDs in Real-World Structures

Name of Structure	Location	Type of Structure	Height/Number of Storeys/Total Floor Area	Year of installation	Type of TLCD	Approximate Volume of Water in TLCD(s) (L)
Sky Gate	Beirut, Lebanon	Residential building	180 m/42 + 6 basement/58,000 m^2	2014	2 conventional TLCDs	0.8×10^5
Gama Tower[e]	Jakarta, Indonesia	Commercial building	285.5 m/64 + 4 basement/163,466 m^2	2016	2 conventional TLCDs	Data not available
B2 tower	New York, USA	Residential building	109.4 m/32 + 2 basement/32,163 m^2	2016	4 TLCDs with air spring (Air tuned damper)	Data not available
Australia 108[f]	Melbourne, Australia	Residential building	318.7 m/100 + 1 basement/2,828 m^2	2021	Single conventional TLCD	3×10^5

[a] TLCDs are used to control the vibration of free-standing pylons at the construction stage.

[b] The building has been demolished in 2008.

[c] Details not available.

[d] ÖBB - Österreichische Bundesbahnen (in German, the English meaning of which is Austrian Federal Railways)

[e] Earlier known as Rasuna Tower and Cemindo Tower.

[f] Tallest building to have a TLCD.

TABLE 8.2
Patents on TLCDs

Patent Title	Patent Publication Number	Inventor	Publication Date (yyyy-mm-dd)
Damping device for tower-like structure	US5070663A	Fujikazu Sakai, Shingo Takaeda, Toshihiro Tamaki	1991-12-10
Propeller-controlled active tuned-liquid-column damper	US6857231B2	Yung-Hsiang Chen	2005-02-22
Damping of wind turbine	WO2006062390A1	Marcel Van Duijvendijk, Engbert Wilmink, Anton Herrius De Roest	2006-06-15
Fluid damper for damping oscillations in constructions	WO2006099650A1	Michael Reiterer, Franz Ziegler	2006-09-28
Magnetic-fluid change type regulation-liquid column damper	CN101070893A	Xiuyong Wang, Hongxin Sun, Zhengqing Chen	2007-11-14
Pendulum-type liquid column damper (PLCD) for controlling the vibration of a building structure	WO2014046549A1	Arunjyoti Sarkar, Ove Tobias Gudmestad	2014-03-27
Liquid column damping system	WO2014206507A1	Okyay Altay	2014-12-31
Novel tuning gas-liquid column damper with damping and frequency dual regulation function and structural vibration control system	CN103669631B	Yuanfeng Duan, Ming Zhao, Hongmei Zhang, Yuanchang Duan	2017-04-12
Annular tuning liquid column damper	CN210342303U	Jin-Ting Wang, Hao Ding, Li-Qiao Lu	2020-04-17
Stabilization system, in particular for a floating support, comprising at least three interconnected liquid reserves	US10683065B2	Olivier Lepreux, Christophe Coudurier	2020-06-16
Vibration damper for a wind turbine, method for installing a vibration damper in a tower of a wind energy system and wind energy system	EP3048326B1	Marc Seidel, Andreas Kluibenschedl, Michael Reiterer	2021-07-07
Stabilization system, in particular for a floating support, comprising multiple u-shaped damping devices	US11267543B2	Olivier Lepreux, Christophe Coudurier	2022-03-08
Multilayer circular ring-shaped tuned liquid column damper	WO2022151580A1	Jin-Ting Wang, Hao Ding, Okyay Altay, Jian-Wen Pan	2022-07-21
Floating fan with intelligent tuning multi-liquid-column damper system	CN115013249A	Peng Dou, Zhidong Wang, Hongjie Ling, Xiaosen Xu, Daiyu Zhang, Wenchao Cong	2022-09-06
A mass damper system with tuned liquid column	WO2023107047A2	Muaz Kemerli, Tahsin Engin, Ahmet Aydin	2023-06-15

8.2 HIGASHI-KOBE BRIDGE

8.2.1 OVERVIEW

The Higashi-Kobe Bridge connects two artificially created islands in the Kobe-Osaka Bay. It forms a vital part of the Hanshin Expressway, which is an important transportation artery in Japan (Ganev et al., 1998). It is a three-span double-decker cable-stayed bridge (see Figure 8.1). There are three-lane roads on each deck. The bridge has a total length of 885 m, and the centre span is 485 m. Some important information related to the bridge is provided in Table 8.3.

The bridge is located in one of the most seismically active zones in the world and also beside the sea. Thus, in addition to the expected traffic load, both earthquake and wind loading played vital roles in the design of the bridge (Kitazawa et al., 1993). The Higashi-Kobe Bridge is provided with a special type of viscous damper, namely the vane-type oil damper, at the ends of the girders to prevent excessive vibration under strong earthquake. Aerodynamic modifications of the stay cables were carried out to improve the performance of the bridge under wind-induced excitation (Kitazawa et al., 1993). Further, at the construction stage, when the deck of the bridge was not integrated with the pylons, it was necessary to put measures to reduce wind-induced vibration of the free-standing pylons. Free-standing pylons are, in general, extremely slender and long-period structural systems. This makes them highly susceptible to wind-induced vibration. The Higashi-Kobe Bridge has 146.5 m tall H-shaped pylons (see Figure 8.2) (Kitazawa et al., 1993). The mass of each pylon of the bridge is 40.27×10^5 kg (Ganev et al., 1998). The width of the columns of the pylons along the carriageway is 6.5 m at the base, and it reduces to 4.5 m at the top (Yamada et al., 1991). Thus, the height-to-base-width ratio of the pylons along the carriageway is as high as 22.54.

FIGURE 8.1   The Higashi-Kobe Bridge (Kitazawa et al., 1993).

TABLE 8.3

Salient Points on the Higashi-Kobe Bridge

Type of Bridge	:	Three-Span Double-Decker Cable-Stayed Steel Bridge
Total span	:	885 m
Width	:	17 m
Pylons	:	H-shaped pylons (146.5 m tall)
Cable arrangement	:	Double-plain multi-cable system with harp pattern (a total of 96 cables)
Deck detail		Upper and lower decks together form a Warren-type truss with no vertical members. This forms the 9 m high main girder of the bridge.
	:	The deck is of continuous type.
Material used in main structural system	:	Steel
Substructure	:	Caisson foundation for pylons. Pile foundation for piers.
Built by (Owner)	:	Hanshin Expressway Public Corp.
Main engineering consultant	:	Sogo Engineering Inc.
Main contractors	:	Mitsubishi Heavy Industries, Kawasaki Heavy Industries
Year of completion	:	1992

During the construction phase of the bridge in the early 1990s, Japanese engineers chose to use liquid dampers, both tuned sloshing dampers (TSDs) and TLCDs, mainly due to the ease of installation, tuning and dismantling, to mitigate wind-induced vibrations in the free-standing pylons of the bridge. It is of interest to note that prior to the Higashi-Kobe Bridge, TSDs were used in the Sakitama Bridge (Kaneko and Ishikawa, 1999) and Ikuchi Bridge (Ueda et al., 1992) for the same purpose of wind-induced vibrational control of the free-standing bridge pylons.

8.2.2 Description of Damper and Performance

As is evident from the foregoing sub-section, although the Higashi-Kobe Bridge was the first real-world structure to be fitted with TLCDs, it was only a temporary control measure for the free-standing pylons (Sakai et al., 1991). The TLCDs were of conventional type. The U-shaped damper container had a rectangular cross section of $1.412\,m \times 0.632\,m$ (see Figure 8.3). Water was used as the damper liquid. There was a single orifice plate in each TLCD with a rectangular opening of $1.12\,m \times 0.48\,m$. The TLCDs were tuned to the fundamental frequency of the pylons along the axis of the carriageway. The dampers were placed near the top of the columns.

To assess the performance of the TLCDs, experimental studies were conducted on the actual pylons (Kitazawa et al., 1992). A pair of exciters were fixed near the top of a pylon. From the experiments, the fundamental natural period of the free-standing pylon was obtained as 3.906 seconds. By applying the logarithmic decrement approach to the vibrational response of the pylon, the damping ratio of the fundamental mode was estimated to be around 0.5%. This was much less than the value assumed during the design process (Kitazawa et al., 1992). The vibration experiment was repeated after

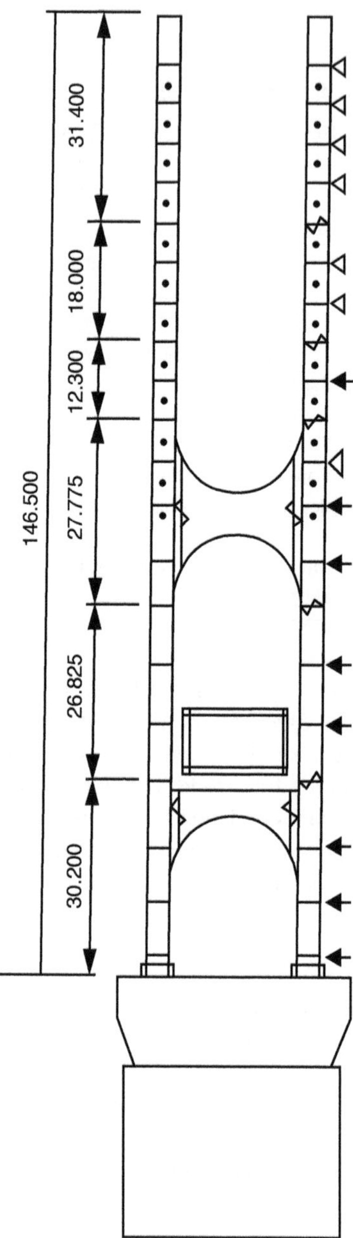

FIGURE 8.2 Front view of a pylon of the Higashi-Kobe Bridge (dimensions are in m) (Kitazawa et al., 1993).

installation of the TLCDs, and it was observed that the damping ratio of the free-standing pylon increased to 1.751%, which was within the acceptable range as per the design to control the wind-induced vibrations of the structure. The frequency response functions obtained from the experimental data showed a 67% response reduction in the TLCD-controlled structure as compared to the uncontrolled structure.

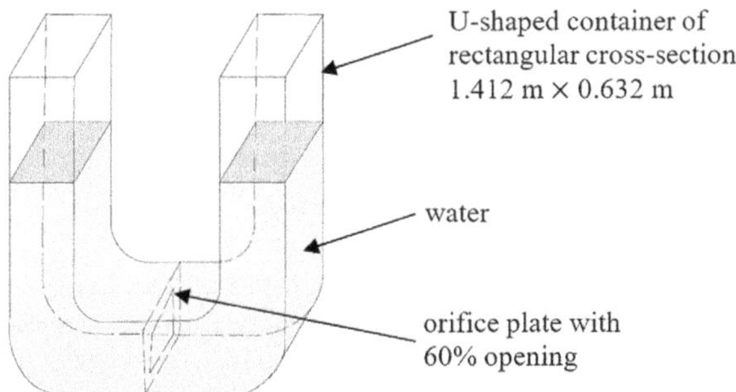

FIGURE 8.3   Schematic of tuned liquid column damper used in the Higashi-Kobe Bridge.

8.3 HOTEL COSIMA

8.3.1 OVERVIEW

Hotel Cosima (see Figure 8.4) was designed by the renowned Japanese architect Kiyonori Kikutake. Its architecture combined the modern metabolism style with traditional Shinto values. Later, it was renamed as Hotel Sofitel. Due to a financial crisis, the property was sold to Accor Group in 1999, within 5 years of the opening of the hotel (Pettafor, 1999). After refurbishing, Accor Group ran the hotel for another 7 years, and ultimately, the building was demolished in 2008. The building had 26 storeys above grade with three basements and stood like a massive chest of drawers. The 106 m tall building had a weight of about 4600 ton. Table 8.4 provides some of the salient points of the building data. The height-to-base-width ratio of the building along the longitudinal and transverse directions was about 4:1 and 5:1, respectively. Thus, the building was slender along both its principal axes and was therefore

FIGURE 8.4   Hotel Sofitel (left of the two tall buildings) along with other surrounding buildings (Murayama, 2005).

TABLE 8.4

Salient Points on the Hotel Cosima

Type of Building	:	Commercial (4-star luxury hotel)
Height/number of storeys	:	106 m/26 above grade + 3 basements
Total floor area	:	9,797 m²
Location	:	1–48, 2 Ikenohata, Taito-ku, Tokyo, Japan
Architectural style	:	Metabolism, with close resemblance to Japanese pagoda and pine trees.
Material used in main structural system	:	Steel
Developer	:	Hokke Club Co., later taken over by Accor Group in March 1998
Architect	:	Kiyonori Kikutake
Year of completion	:	1994
Year of demolition	:	2008 (the hotel closed service in December 2006)

sensitive to vibration. The fundamental natural frequencies of the building along the transverse and longitudinal directions were 0.48 and 0.5 Hz, respectively. The building was lightly damped, with damping ratios of 0.55% and 0.68% along the transverse and longitudinal directions, respectively. Thus, the building required supplemental damping to improve serviceability conditions.

8.3.2 DESCRIPTION OF DAMPER AND PERFORMANCE

Although many years have elapsed since the demolition of the hotel, the engineering community is still curious about the building due to its unique architecture and the fact that this was the first prominent building in the world to have a TLCD. The damper used in the building was a bi-directional TLCD with period adjustment equipment (BTLCD-PA) (Shimizu and Teramura, 1994). As mentioned earlier, the damper was installed to improve occupant comfort by reducing vibration under environmental excitation. Further, the TLCD was connected to the emergency sprinkler system of the building for possible utilization of the water in the damper tank during an outbreak of fire. This is interesting, as there is current research that focusses on the dual functionality of tank dampers (Das et al., 2023; Konar, 2024a,b; Konar and Ghosh, 2023b), while this had already been explored and implemented nearly 30 years ago.

The BTLCD-PA had a base dimension of 6 m × 6 m with four 1.5 m × 1.5 m vertical limbs at the corners (see Figure 8.5). The horizontal limbs of the damper had varying heights between 1 and 1.3 m. The total weight of water in the damper was 58 ton, which was about 1.26% of the weight of the building. The period adjustment equipment was also in the form of a U-shaped tank of smaller size and partly filled with water (Shimizu and Teramura, 1994). The number of period adjustment equipment was four, and the vertical limbs of each period adjustment equipment were connected to two vertical limbs of the damper through air ducts. When laterally excited, there was liquid movement in the horizontal and vertical limbs of the TLCD. As a result, a difference in air pressure in the vertical limbs

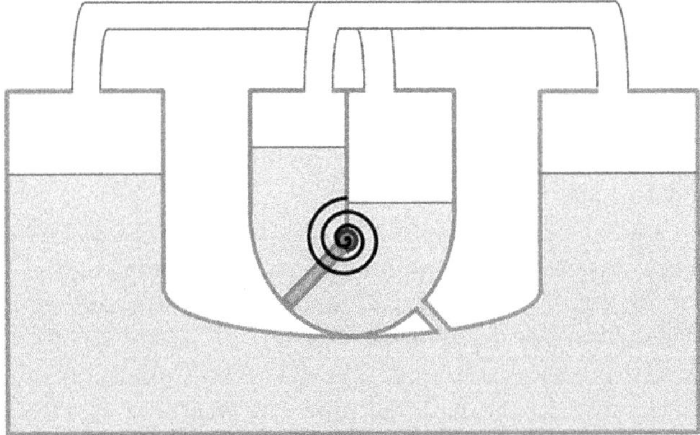

Elevation

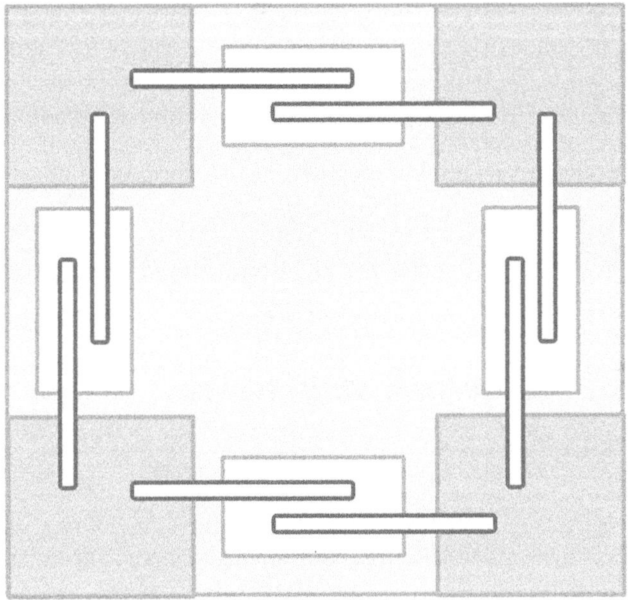

Plan

FIGURE 8.5   Schematic of bi-directional TLCD with period adjustment equipment used in Hotel Sofitel (Teramura and Yoshida, 1996).

of the period adjustment equipment was created. This led to the movement of a valve and its shaft. By adjusting the stiffness of the springs connected to the shaft of the valve, the period of oscillation of the liquid in the TLCD could be controlled. This was used for tuning the damper frequencies with the frequencies of the building along the two principal directions. The details of the analytical formulation behind the functioning of the period adjustment equipment are available in Teramura and Yoshida (1996).

To evaluate the performance of the BTLCD-PA, a vibration experiment was conducted on the building post installation of the damper. From the free vibration response of the building with the empty BTLCD-PA and with the BTLCD-PA having the designed volume of water, it was found that the damper was able to increase the damping ratio by about 10 times along both the principal axes of the building (Teramura and Yoshida, 1996).

To further assess the control effectiveness of the BTLCD-PA, the vibration of the building was continuously monitored after the completion of the building. On December 28, 1994, there was a 7.7 magnitude earthquake with the epicentre at far-off Sanriku. The earthquake was felt in Tokyo, 632.9 km away from the epicentre. During the earthquake, the BTLCD-PA was functional, and the maximum acceleration of the building roof was measured as 17.9 cm/s^2. It was calculated numerically that without the damper, the peak acceleration of the building roof would have been 30.8 cm/s^2 (Teramura and Yoshida, 1996). Thus, the damper achieved a 42% reduction in peak roof acceleration under seismic excitation. Again, on September 17, 1995, the building was exposed to a strong storm designated as "typhoon9512." The maximum wind velocity measured during the storm at the roof of the building was 78.12 km/hr. During the storm, the damper was functional, and the root-mean-square (rms) and peak acceleration of the roof of the building were recorded as 1.18 and 4.54 cm/s^2, respectively. The acceleration response of the uncontrolled building in the storm condition was obtained numerically, and it was calculated that the damper reduced the rms and peak roof acceleration response of the building by 56% and 42%, respectively (Teramura and Yoshida, 1996). Thus, the damper was highly effective in the control of both wind and seismic excitations.

8.4 PROSPECT COMMUNICATION TOWER

8.4.1 Overview

This is a microwave communications tower constructed in 1992 at Prospect, situated in the western suburbs of Sydney, Australia. The 67 m tall free-standing tower was designed to meet strict deflection criteria under dynamic loads, especially due to wind, to minimize signal transmission interference and loss. Similar to most lattice structures, the predominant wind action on the tower is the along-wind turbulence buffeting (Glanville and Kwok, 1995). The general arrangement of the tower is shown in Figure 8.6. Table 8.5 summarizes some data relevant to the tower.

After the construction of the tower, several full-scale dynamic tests were conducted to identify the dynamic characteristics of the tower (Glanville et al., 1996; Glanville and Kwok, 1997, 1995). The frequencies of the first three modes along the east-west direction were determined as 1.08, 1.41 and 3.22 Hz; while the frequencies of the first three modes along the north-south direction were 1.07, 1.39 and 3.28 Hz (Glanville and Kwok, 1995). The damping ratio of the first mode along the east-west and the north-south directions was estimated to be 0.5% and 0.8%, respectively (Kwok, 2013). The generalized modal mass for the first mode along both principal axes was found to be 24 ton, which is less than 20% of the total mass of the tower.

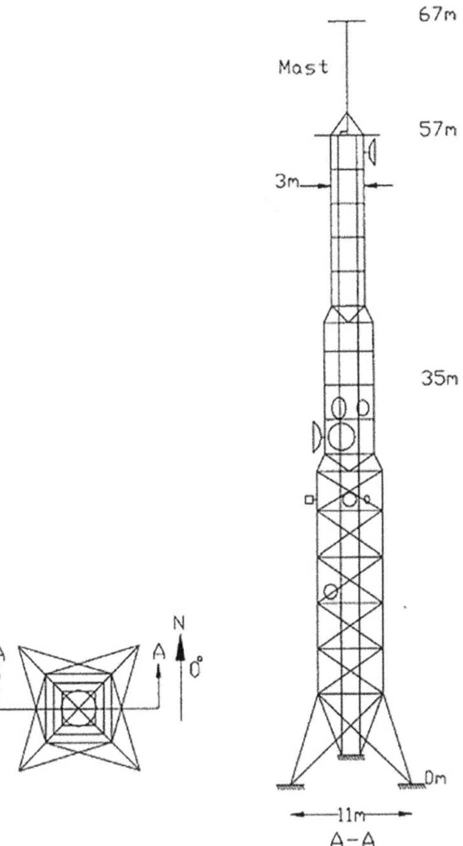

FIGURE 8.6 General arrangement of Prospect Combination Tower, top view on the left and sectional view on the right (Hitchcock et al., 1999).

8.4.2 DESCRIPTION OF DAMPER AND PERFORMANCE

Sometime between 1996 and 1999, a bi-directional LCVA system was installed on the Prospect Communication Tower (Hitchcock et al., 1999). A total of 20 damper units were fixed at the 57 m level of the tower. Bi-directional LCVAs were considered as the appropriate choice of TLCD configuration as the tower has nearly equal modal frequencies along the principal axes. Initially, several sets of LCVAs were examined. In the first set, all 20 damper units had equal frequency with a tuning ratio of 0.99. The total mass of liquid (water) in the damper was 280 kg, which was equal to 1.17% of the first generalized modal mass Tf the tower (Hitchcock et al., 1999). The performance of the damper system was investigated through free vibration experiments. From an analysis of the decay of the structural response along the east-west direction recorded during the free vibration experiment, it was found that the bi-directional LCVA system improved the damping ratio of the fundamental mode by five times. In the second set of experiments, the dampers were identical to the first set except for the tuning ratio, which was now 1.03. A reduced effectiveness of the damper was reported for

TABLE 8.5

Salient Points on the Prospect Communication Tower

Type of Structure	:	Free-Standing Lattice Tower
Height	:	67 m including the 10 m tall mast
Base dimensions	:	11 m × 11 m
Location	:	Prospect (a suburb of Sydney), New South Wales, Australia
Weight of the structure	:	126 ton
Material used in main structural system	:	Steel frame with rigid joints. Columns and bracings have hollow circular sections and intermediate platforms are made of rolled section.
Built by (Owner)	:	Prospect Electricity. On 1 March 1996, Prospect Electricity was merged with Illawarra Electricity to form Integral Energy.
Year of completion	:	1992

this set. In the third set of experiments, the bi-directional LCVA units had natural frequencies distributed in a range about the fundamental natural frequency of the tower. Here, the performance of the damper system was comparable to that of the first set. Ultimately, the configuration of the bi-directional LCVA system used in the first set of experiment was chosen for wind-induced vibration control of the tower.

After the installation of the bi-directional LCVA system, long-term monitoring of the wind-induced response of the tower was carried out. The standard deviation of the along-wind acceleration response of the tower for winds having similar magnitude and direction, before and after the installation of the bi-directional LCVA system, was compared for a large set of data (see Figure 8.7). The results indicated that the damper had little effectiveness when the mean wind speed was less than 10 m/s. With the increase in mean wind speed, the effectiveness of the damper increased. At a mean wind speed of 20 m/s, about 50% reduction in the rms acceleration response of the tower could be achieved (Hitchcock et al., 1999; Kwok, 2013).

8.5 ONE WALL CENTRE

8.5.1 Overview

The Sheraton Vancouver Wall Centre Hotel, popularly known as the One Wall Centre, is located at the southern edge of the business district of Vancouver, Canada (see Figure 8.8). When it was completed in the year 2001, it was the tallest building in Vancouver (Fortner, 2001). The horizontal cross section of the building has the shape of an ellipse with pointed vertices. The maximum width of the building along the north-south and the east-west directions is 21.3 and 39.6 m, respectively. The tower has a height-to-width ratio of 7:1, making it one of the most slender buildings in the skyline of Vancouver. The fundamental natural frequency of the building along the north-south direction was determined as 0.28 Hz from ambient vibration testing (Lord and Ventura, 2002). The building has an outrigger structural system made of reinforced cement concrete resting on a raft situated at a depth of

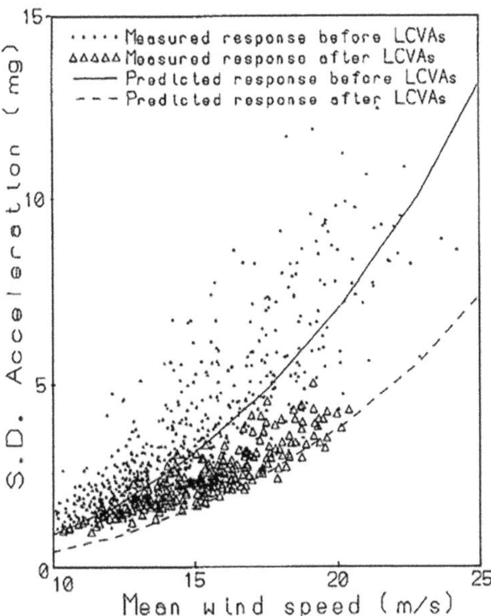

FIGURE 8.7 Standard deviation of acceleration response of the tower versus mean wind speed before and after the installation of bi-directional LCVA (Hitchcock et al., 1999).

23 m below the ground level (https://glotmansimpson.com/project/one-wall-centre/). The central concrete core of the building has a wall thickness of up to 0.9 m and provides space for two stairways and six elevators. The outrigger system consists of a 1.5 m wide and 6.4 m deep beam at the fifth storey level and two 1.8 m × 1.8 m beams at the 21st and 31st storey level. The TLCDs are installed on the roof of the building, and the container walls of concrete of height 8 m and width 460 mm also act as outrigger beams (Fortner, 2001; Simpson, 2001). Some general information on the building is provided in Table 8.6.

During the design of the tower, different structural systems were considered. Due to the high height-to-width ratio of the building, special attention had to be devoted to ensure occupant comfort under wind-induced vibration. Considering the unique shape of the building, wind tunnel tests on scaled-down models were conducted. The results predicted that the maximum acceleration under wind loading (with a 10-year return period) was in the range of 28–40 milli-g along the north-south direction, depending on the structural system used (Irwin and Breukelman, 2001). To ensure occupant comfort, the designers decided to limit the maximum acceleration to within 15 milli-g (Hesson, 2023). For this, it was decided to enhance the damping of the structure by the installation of supplemental damping devices.

8.5.2 Description of Damper and Performance

Two identical unidirectional TLCDs were installed on the roof of the building to control the lateral vibration along the north-south direction. There was no requirement

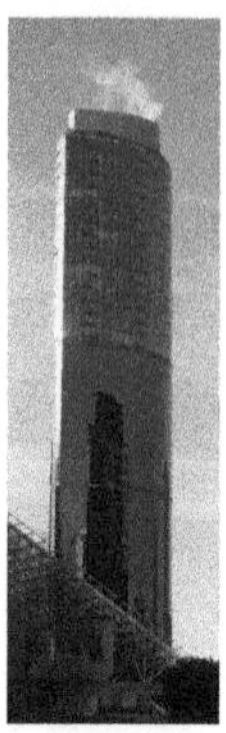

FIGURE 8.8   One Wall Centre (https://commons.wikimedia.org/wiki/File:Onewallcentre3. jpg?uselang=de#Lizenz).

for supplemental damping along the east-west direction. As the building is made of concrete, concrete tanks for the TLCDs were preferred. This was the first real-world installation of TLCD in a building outside Japan.

Initially, the designers of the damper system determined its size based on an analytical model. A scaled-down model of the TLCDs was also examined experimentally and found to provide comparable results (Irwin and Breukelman, 2001). Further, full-scale field vibration tests were conducted after completion of the structure up to 41 storeys (Simpson, 2001). The natural frequency of the analytical model of the building was compared with that obtained from field measurements, and the design of the TLCDs was finalized accordingly. The TLCDs have a base dimension of 16 m by 4.5 m (Simpson, 2001). The vertical limbs of the TLCDs are 8 m tall. Each TLCD contains about 227,300 litre of water.

From an analytical study, the predicted maximum acceleration of the building with the TLCDs under wind having a 10-year return period was 16 milli-g, which marginally exceeded the target value by 1 milli-g, and was deemed to be acceptable (Irwin and Breukelman, 2001). Apart from vibration control, the TLCD system is utilized to meet other functional requirements. The water in the damper system can be used for the sprinkler system of the tower in case of a fire outbreak. This eliminates the requirements of a high-speed pump system and a backup generator at the base of the building, initially required for firefighting (Fortner, 2001). The TLCDs also act as a heat sink for the high-efficiency heat pump of the building (Fortner, 2001). Further, as already mentioned, the walls of the damper tanks are integrated with the outrigger structural system of the building (Simpson, 2001). It has been estimated that the installation of TLCDs has saved at least $2 million in construction costs (Fortner, 2001).

8.6   RANDOM HOUSE TOWER

8.6.1   Overview

The Random House Tower stands as a 52-storey mixed-use skyscraper located in Manhattan, New York City, USA (see Figure 8.9). On the ground floor of the

TABLE 8.6
Salient Points on the One Wall Centre

Type of Building	:	Commercial (Hotel) — Residential
Height/number of storeys	:	149.8 m/48
Total floor area	:	42,955 m²
Location	:	1000 Burrard Street, Vancouver, B.C., Canada
Material used in main structural system	:	Outrigger structural system made of reinforced cement concrete
Developer	:	Wall Financial Corporation
Architect	:	Busby & Associates Architects
Structural designer	:	Glotman Simpson Consulting Engineers
Damper designed by	:	Rowan, Williams, Davis, and Irwin, Inc.
Main contractors	:	Siemens Development
Year of completion	:	2001

FIGURE 8.9   Random House Tower (Henderson, 2010).

building, there is 3,000 m² of retail space. The office floors span from the second to the 26th storeys, while the 27th to the 51st storeys comprise the residential floors. Within the residential portion, there are a total of 130 apartments. The building also has two storeys of basement for parking.

The Random House Tower is structurally divided into two distinct sections. The first 25 storeys are steel frame structures to provide clear spans for office spaces, while the top 25 storeys are of reinforced cement concrete frame construction that allows for variations in residential arrangements (Tamboli et al., 2005). Notably, the two sections do not perfectly align with each other, necessitating the construction of trusses on the 26th and 27th floors to efficiently transfer the structural load from the upper section to the lower section. The steel transfer truss also engages the lower perimeter columns as an outrigger (Brazil et al., 2006). Following the principle of sustainable structural design, the steel frame and transition trusses located halfway up the structure are made of recycled steel. Further, flyash concrete was utilized in the columns and floors of the top 25 storeys (Tamboli et al., 2008). Some relevant information about the building is listed in Table 8.7.

The dynamic analysis of the tower yielded the fundamental natural period of the building along the east-west and the north-south directions as 6.811 and 5.956 s, respectively. With such high fundamental natural periods, the building was susceptible to wind-induced vibrations.

From an architectural standpoint, to ensure that the residents of most of the condominiums can enjoy the view of a nearby park, the residential floor plan has a sawtooth design (see Figure 8.9). While the irregular shape of the building may assist in mitigating wind-induced motion, stringent residential comfort standards, a higher mass concentration in the upper floors and a steel framing support with limited inherent damping in the bottom half of the building necessitated the incorporation of supplementary damping measures.

TABLE 8.7

Salient Points on the Random House Tower

Type of Building	:	Commercial (office) — Residential
Height /Number of storeys	:	208 m/52 + 2 basements
Total floor area	:	79,900 m²
Location	:	1739 Broadway, Manhattan, New York City, USA
Material used in main structural system	:	Steel frame in lower portion housing offices and concrete frame in upper portion having residential apartments
Owner	:	SL Green Realty, Ivanhoé Cambridge, Witkoff, Lehman Brothers
Architect	:	Skidmore, Owings & Merrill LLP (office portion) Ismael Leyva Architects and Adam D. Tihany (residential portion)
Structural designer	:	Cosentini Associates; Thornton Tomasetti
Damper designed by	:	Motioneering Inc.
Year of completion	:	2003

8.6.2 Description of Damper and Performance

It was decided that two conventional TLCDs, one along the east-west and the other along the north-south direction, would be installed for wind-induced vibration control of the structure. This is the first instance of a TLCD installation in a building in the USA. The TLCDs are made of 406.4 mm thick concrete walls and are placed on the 50th floor of the building. The TLCD along the east-west direction has a base length of 20.295 m and a width of 5.588 m, while the base length and width of the TLCD along the north-south direction are 23.876 and 5.309 m, respectively. The height of both the TLCDs is 5.245 m (Tamboli et al., 2005). Each TLCD contains about 550 ton of water, which is equal to 0.33% of the total mass of the building (Brazil et al., 2006). Instead of an orifice, louvre-type blades have been used in the TLCDs to induce damping.

After the completion of the building, in the month of February 2004, a number of tests were conducted on the building and its TLCDs by the engineers of Motioneering Inc. (Tamboli et al., 2005). During each experiment, one vertical limb of a TLCD was sealed and pressurized, leading to the displacement of water from the pressurized vertical limb to the non-pressurized vertical limb. Once the water level in the non-pressurized vertical limb was raised by about 0.3 m from its equilibrium position, the sealed end was rapidly depressurized. Consequently, the water exhibited oscillatory motion, which gradually subsided over time. This motion of water within the tank induced vibrations in the building. Precise measurements were taken of both the water level in the tank and the acceleration of the building. These measurements were systematically repeated for various configurations of louvre blade positions and liquid heights in the TLCDs. Data obtained from the tests were analysed to obtain the actual periods of the structure along both the principal directions and the optimum louvred blade angle. The periods along the east-west and the north-south directions were obtained as 7.008 and 6.161 s, respectively, which were slightly higher than the analytical values. The TLCDs were designed with a tuning ratio of unity.

It was determined from an analytical study that after installation of the TLCDs, the inter storey drift of the building would be restricted to within the acceptable limit of storey height/350. Further, it was estimated that the TLCDs would reduce the peak acceleration response of the structure by about 40% (Tamboli et al., 2008, 2005), which is significantly high.

8.7 COMCAST CENTER

8.7.1 Overview

The Comcast Center stands as a prominent skyscraper in the heart of Center City Philadelphia, USA (see Figure 8.10). The building is also recognized as the Comcast Tower. Soaring to a height of 297 m with 58 storeys, it is the second tallest building in both Philadelphia and in the entire state of Pennsylvania. Out of the 58 floors of the building, 56 are designated for occupancy. Originally conceived as "One Pennsylvania Plaza" when its development plans were first unveiled in 2001, the Comcast Center underwent two significant redesigns before construction commenced in 2005 (Stephens, 2009). The building has the form of a faceted obelisk, clad in high-performance non-reflective low-emissivity silvery glazing, with ultra-clear low-iron glass at the corners of the tower and at the crown (https://www.ramsa.com/projects/project/comcast-center). The Comcast Center showcases two distinct cutouts near its pinnacle, located on both the north and the south sides. This choice of design not only lends a modern aesthetic to the building but also contributes to energy efficiency. The building is a recipient of the prestigious Urban Land Institute Award for Excellence in the Americas. Key details about the building can be found in Table 8.8.

The structural system of the building consists of a central concrete core housing the elevators and the stairs and a steel outer frame made up of floor beams and peripheral columns (Stephens, 2009). To have a small footprint, the tower has a high height-to-core width ratio. This necessitates an unusually thick core wall. The thickness of the core wall is 1.372 m till the 20th floor of the building (https://www.thorntontomasetti.com/project/comcast-center). The building is more slender along the north-south direction. Wind tunnel tests on the scaled-down model of the tower revealed that the building would experience a significant cross-wind oscillation under design wind load (https://rwdi.com/en_ca/projects/comcast-center/). The structural designers, by means of an analytical study, ensured that the tower met the design requirements for strength, but the wind-induced oscillation exceeded

FIGURE 8.10 Comcast Center (https://en.wikipedia.org/wiki/Comcast_Center#/media/File:Comcast_Philly.JPG).

TABLE 8.8
Salient Points on the Comcast Center

Type of Building	:	Commercial
Height /Number of storeys	:	297 m/58 +3 basements
Total floor area	:	130,064 m²
Location	:	1701 John F. Kennedy Blvd., Philadelphia, Pennsylvania, USA
Material used in main structural system	:	Concrete-Steel Composite
Owner	:	Liberty Property Trust
Architect	:	Robert A.M. Stern Architects (with Kendall /Heaton Associates as Architect of Record)
Structural designer	:	Thornton Tomasetti
Damper designed by	:	Motioneering Inc.; and Rowan, Williams, Davis, and Irwin, Inc.
Main contractors	:	L. F. Driscoll Company
Year of completion	:	2008

the acceptable limit for occupant comfort under design wind load. The decision was then taken to adopt a supplemental damping device to reduce the tower vibrations under wind load.

8.7.2 DESCRIPTION OF DAMPER AND PERFORMANCE

The preliminary study conducted by the designers indicated that a conventional TLCD system would be the most efficient way to mitigate the wind-induced vibration of the building (https://rwdi.com/en_ca/projects/comcast-center/). A double chamber conventional TLCD that could hold about 1300 ton of water was selected to suppress the lateral vibration of the building along the north-south direction, that is along the building's most slender axis. Dampers were not required along the east-west direction. The TLCD used in the Comcast Center is made of concrete and has a self-weight of 900 ton. It is divided into two chambers to avoid the off-axial flow of water and the resultant reduction in damper performance (Lago et al., 2019). Each chamber of the damper container is about 17 m long, 10 m wide and 6 m tall (D'Huy Engineering, 2023). It is recognized as the largest TLCD in the world. Until 2020, the Comcast Center was also considered as the tallest building to have a TLCD. For energy dissipation, the TLCD is provided with two sets of vertical steel vanes or louvres in the vicinity of the junction of the horizontal and vertical limbs of the damper (Stephens, 2009).

As it was decided that the TLCD would be made of concrete, cast-in-situ concrete was the natural option. However, considering the weather during the time of the year when the structure was planned to be built, the contractor for the work, L. F. Driscoll Company, decided to use precast concrete panels connected with steel plates and anchorage cast into the panels. D'Huy Engineering conducted a finite element analysis on a model consisting of more than 10,000 plate elements to arrive at the actual scheme of the panel system along with the joints between the panels (D'Huy Engineering, 2023). After assembling the TLCD container, waterproofing was done carefully as the precast system had a greater chance of leakage. Besides the hydro-dynamic load due to liquid motion, the TLCD container was designed to take the load of three large cooling towers that were placed on the damper container (D'Huy Engineering, 2023).

A full-scale experiment on the TLCD, after completion of its installation, was carried out to obtain the optimum opening of the steel louvres. For this, a similar procedure as described in Section 8.6.2 for the Random House Tower was followed (Lago et al., 2019).

It was estimated that the TLCD reduced the acceleration in the upper storeys of the building by about one-third during high wind. Thereby, the damper achieved the serviceability criteria in terms of occupant comfort and lateral drift (Lago et al., 2019). The total cost of construction of the TLCD was to the tune of $2 million (D'Huy Engineering, 2023). However, it saved much more as the overall structural cost would have gone up significantly if the stiffness of the tower had to be enhanced to reduce the wind-induced oscillation (https://rwdi.com/en_ca/projects/comcast-center/; Lago et al., 2019).

8.8 NEW SONGDO CITY FIRST WORLD TOWERS

8.8.1 Overview

New Songdo City First World Towers are one of the most attractive landmarks of Songdo International Business District (IBD), Incheon, Republic of Korea. The complex, also known as Songdo, The Sharp First World, is the first completed project in New Songdo City (https://www.kunwon.com) with four 67-storey almost identical towers that contain 2,654 apartments and associated amenities (see Figure 8.11). The architecture of the complex was inspired by the Korean social hierarchy which comprises four courtyard communities (https://www.kpf.com/project/songdo-first-world-towers). The buildings are 237 m tall and have three basement storeys designated for parking. The top three storeys of the buildings serve as mechanical penthouses (Cho et al., 2016). The buildings are constructed on reclaimed land. In fact, the entire

FIGURE 8.11 New Songdo City First World Towers (Cho et al., 2016).

TABLE 8.9
Salient Points on the New Songdo City First World Towers

Type of Building	:	Residential
Height /Number of storeys	:	237 m/67 + 3 basements
Total floor area	:	344,000 m²
Location	:	991-88 Dong Choon-Dong Yeon Soo-Gu, Incheon, South Korea
Material used in main structural system	:	Concrete
Owner	:	Gale International
Architect	:	Kohn Pedersen Fox Associates (Associate Architect: Kunwon Architects)
Structural designer	:	Thornton Tomasetti
Main contractors	:	POSCO E&C
Year of completion	:	2009

Songdo IBD is spread over 1,500 acres of reclaimed land (Henry, 2011). A general information summary on the building is provided in Table 8.9.

The towers are made of concrete and have a central core with a wall thickness of 1.2 m in the lower storeys. The thickness of the core wall is reduced to 0.6 m in the upper storeys. There are concrete columns along the outer perimeter of the buildings, and the floors comprise flat slabs. On the 33rd and 60th floors, there are two concrete outrigger walls (Cho et al., 2016).

The towers are almost square in plan and have an aspect ratio of about 7. The natural periods of each tower as determined from its finite element model were 5.525 and 6.098 s along the stronger and weaker axis, respectively. This made the flexible towers susceptible to wind-induced vibration along both principal axes. The wind-induced motion in the towers is resisted chiefly by the concrete core and outrigger system. Wind tunnel tests on the scaled-down model of the towers revealed that the peak acceleration in the upper residential storeys of all four buildings would be in the range from H-70 to H-90 as per the AIJ-GBV (2004) code for wind speed having a 1-year return period (Cho et al., 2016). The acceleration level H-70 implies that 70% of residents can perceive the vibration. The design acceleration level for the towers was set to H-50 for a wind speed with a 1-year return period, and as a consequence, the lateral loading resisting system of the towers was found to be insufficient to meet the design standard (Cho et al., 2016). It was thus decided to use supplemental damping measures to reduce the level of vibration within the design limit.

8.8.2 Description of Damper and Performance

To meet the design target, two conventional LCVAs, one along each principal axis, were installed in each tower on the 66th floor (see Figure 8.12). From a finite element model of the tower, the generalized mass corresponding to the first mode of vibration along the stronger and the weaker axis was obtained as 34,908 ton and 36,365 ton,

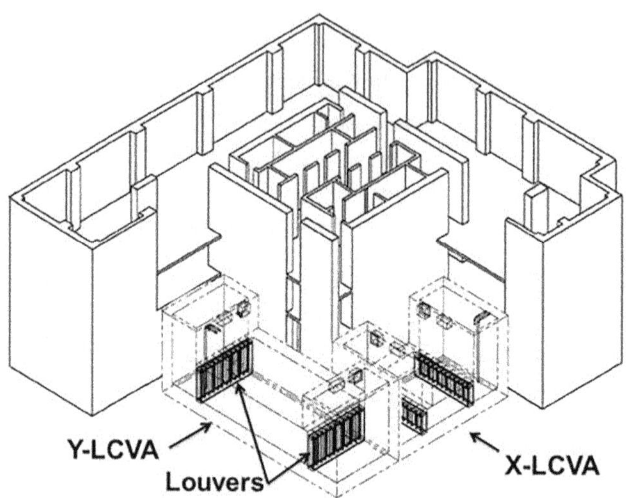

FIGURE 8.12 Placement of LCVA in New Songdo City First World Towers (X-LCVA and Y-LCVA, respectively, represent the damper along the stronger and weaker axis) (Cho et al., 2016).

respectively (Cho et al., 2016). In accordance with the preliminary design, the required mass ratios of the LCVAs along the stronger and weaker axis were 0.59% and 1.18%, respectively.

The LCVA along the weaker axis is 20.2 m long, 7.2 m wide and 7.3 m tall while the LCVA along the stronger axis is slightly smaller in size. In both the dampers, the horizontal limb of the LCVA has a cross section of 6 m × 3.5 m, while each vertical limb has a cross section of 6 m × 2.31 m (Cho et al., 2016). The LCVAs are made of concrete and are provided with two sets of adjustable louvres in their horizontal limbs.

For the final tuning of the dampers, ambient vibration experiments on the completed towers with empty LVCAs were carried out. During the experiments, the towers were exposed to light wind. The structural response along both axes was measured. Then, using the stochastic subspace identification method (Peeters and De Roeck, 1999), the actual natural periods of the towers along the stronger and the weaker axis were determined as 4.651 and 5.291 s, respectively (Cho et al., 2016). It may be noted that the measured natural periods of all four towers are very close to each other. To achieve the designed tuning ratio of 0.968, the water level in the LCVA along the stronger and the weaker axis was estimated as 2.5 and 3.9 m, respectively (Cho et al., 2016).

To examine the effectiveness of the LCVAs, first, a set of free vibration experiments along the weaker axis was conducted with the LCVAs filled with water up to the required level. The results indicated that the achieved tuning ratio is 0.964, which is only 0.4% below the designed tuning ratio (Cho et al., 2011). The equivalent damping ratio of the LCVA was determined as 3.65% (Cho et al., 2011).

Further, full-scale ambient vibration experiments were conducted on the towers with functional LCVAs. The results of these experiments indicated that the installation of the LCVAs had increased the damping ratio of the towers by 3.89 and 3.13 times along the stronger and the weaker axis, respectively (Cho et al., 2016).

8.9 SKY GATE

8.9.1 Overview

Over the years, tall buildings have been constructed in Ashrafieh, a hilly area by the sea in Beirut, Lebanon (https://www.nabilgholam.com/project.8), progressively forming a kind of an urban wall. There was a high and narrow gap framed by two residential towers, Tilal Beirut and Atomium Twin Towers, standing at 24 and 27 floors, respectively (Khalil and Ruta, 2023), in which the Sky Gate was constructed (see Figure 8.13). The 180 m high building seems to be consisting of four blocks, placed one over another. The 42-storey superstructure comprises 55 apartments along with elevated gardens and water bodies, and there are six basement storeys (https://new.manenterprise.com/our-work/skygate). General information on the building is provided in Table 8.10.

FIGURE 8.13 Sky Gate (https://commons.wikimedia.org/wiki/File:Skygate_-_Beirut_-_Nabil_Gholam_Architects_%28cropped%29.jpg).

TABLE 8.10

Salient Points on the Sky Gate

Type of Building	:	Residential
Height /Number of storeys	:	180 m/42 + 6 basements
Total floor area	:	58,000 m²
Location	:	Plots 5233-5240, Achrafieh, Beirut, Lebanon
Material used in main structural system	:	Concrete
Developer	:	Sky Gate Project
Architect	:	Nabil Gholam Architects
Structural designer	:	Bureau d'Études Rodolphe Mattar
Damper designed by	:	BMT Fluid Mechanics Ltd.
Main contractors	:	MAN Entreprises
Year of completion	:	2014

The structure is made of cast-in-situ reinforced cement concrete and has a central core with outer beams and columns. One of the major challenges in the execution of the project was the construction of the cantilevered volumes, consisting of stacked and shifted multi-storey blocks (Hajjar, 2017). For this, three large post-tensioned concrete cantilever beams were used at the 10th, 17th and 22nd storey level (https://www.cclint.com/case-studies/skygate-beirut-lebanon/). The base dimension of the building is approximately 22 m × 44 m, due to which the slenderness ratio of the building about its weaker axis is 1:8.2. From the analysis of a finite element model of the building, the fundamental natural period of the structure about the weaker axis was determined to be as high as 5.263 s (Cammelli et al., 2016). Further, the building is located only about 1 km from the shoreline of the Mediterranean Sea. Due to all these factors, the building was susceptible to wind-induced vibration about the weaker axis. To assess the possible impact of wind on the tower, a wind tunnel test was performed on a model of the building. The test results indicated that to restrict the vibration of the building about the weaker axis within the acceptable limits for a 1-year return wind and a 10-year return wind speed, the building requires a damping ratio of 2% (Cammelli et al., 2016). A full-scale on-site measurement when the building was nearing completion indicated that the tower had an intrinsic damping of 1%. Thus, installation of a supplemental damper was contemplated.

8.9.2 Description of Damper and Performance

Due to higher cost involvement, a vibration control solution by using a tuned mass damper (TMD) was ruled out. Further, among the liquid dampers, the TLCD was favourably considered due to its high volumetric efficiency and ease of estimation of damping (Cammelli et al., 2016). The design requirements indicated the need for a TLCD system with a total effective mass of approximately 80,000 kg, which corresponded to about 0.5% of the mass participating in the first mode of vibration of the structure about the weaker axis. The designed tuning ratio was 0.995 (Cammelli et al., 2016). Given these considerations, it was decided to install two

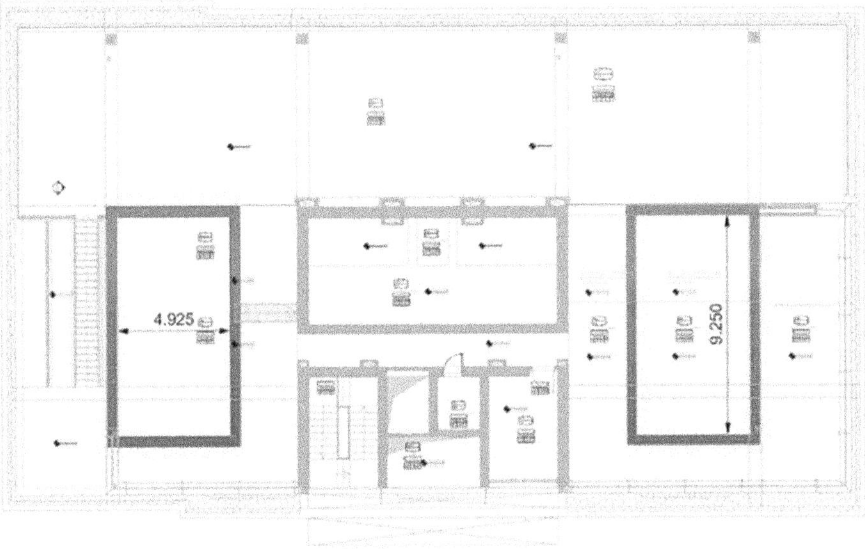

FIGURE 8.14 Footprint of the tuned liquid column dampers on the building plan (dimensions are in m) (Cammelli et al., 2016).

identical TLCDs just below the roof level of the tower. The footprint of each TLCD excluding the wall thickness is 9.25 m × 4.925 m (see Figure 8.14). In the TLCDs, instead of orifices, porous screens were recommended.

Shake table tests were conducted on a 1:20 scaled-down model of the TLCD, the results of which indicated that a TLCD system with three screens having a porosity of 75% would be able to provide a supplemental damping of 1.2%. Further, to obtain greater insight into the actual behaviour of the TLCD system, a full-scale numerical model of the TLCD was studied. It was found that in this numerical model, the added damping due to the TLCD under wind loading having a 10-year return period was 10% lower than what was obtained from the shake table test. Despite this, the pair of TLCDs would still be capable of reducing the wind-induced response of the building about the weaker axis by approximately 30%, and would limit the vibration of the building to an acceptable level from the perspective of occupant comfort (Cammelli et al., 2016).

8.10 B2 TOWER

8.10.1 OVERVIEW

B2 tower, also known as the 461 Dean, is a 109 m tall, 32-storeyed building comprising 363 apartments of different sizes and various amenity spaces. This is the first residential building to come up in the Atlantic Yards of Brooklyn, New York City, where the construction of 15 new residential buildings was planned to provide affordable housing. The building comprises 930 modules, some of which are apartment-size, that were fabricated and assembled off-site at a factory and then transported to the site where they were stacked around a braced core and connected to each other to make up the structure (McKnight, 2016). The elevator core and stairwells were likewise built using a modular construction approach, with the elevator cabs and rails being installed and aligned on-site as the final step. This construction technique, known as volumetric modular prefabricated construction, offers the advantage of fast construction, reduced noise and air pollution and minimum wastage of material. Further, as reported by Mafi (2016), the use of the volumetric modular prefabricated technique saved about 20% of the cost of construction for the B2 tower. When the building was topped up in the year 2016, it was the tallest residential tower on earth to be constructed using this method (https://www.shoparc.com/projects/b2/). More facts about the building are provided in Table 8.11.

Architecturally, the massing of the building is broken into three major blocks, with a recess reglet among them to provide scale to the tower and avoid a monolithic appearance (Farnsworth, 2014). The reglets are emphasized through a blend of cantilevers, setbacks and variations in the façade (see Figure 8.15).

As the modules are mostly made of steel, the building as a whole is significantly lighter compared to typical multi-storey construction. This necessitated a detailed assessment of the sway that the building may undergo under wind loading. Further, the architectural massing and slenderness of the building were also required to be considered in the estimation of the sway. Thus, it was decided to carry out wind tunnel tests on a model of the structure. Due to the unprecedented scale of modular construction, and the difficulty in the estimation of the contribution of the modules to the overall damping and stiffness of the building, wind tunnel tests were conducted under conservative assumptions. Based on the wind tunnel test results, the installation of a supplemental damping device was suggested to reduce the wind-induced acceleration of the building.

TABLE 8.11
Salient points on the B2

Type of Building	:	Residential
Height /Number of storeys	:	109.4 m/32 + 2 basements
Total floor area	:	32,163 m²
Location	:	461 Dean Street, Brooklyn, New York City, USA
Material used in main structural system	:	Steel - volumetric modular prefabricated construction
Developer	:	Forest City Ratner Companies
Architect	:	SHoP Architects
Structural designer	:	Arup
Damper designed by	:	Thornton Tomasetti
Main contractors	:	Turner Construction Company
Year of completion	:	2016

FIGURE 8.15 The B2 tower (https://upload.wikimedia.org/wikipedia/commons/b/b6/461 Dean.jpg).

8.10.2 Description of Damper and Performance

The structural consultant for the project, Arup, had initially proposed a bi-directional TMD solution involving two 100 ton dampers to restrict the lateral acceleration of the building to a permissible limit of 15 milli-g under a wind speed having a 10-year return period (Farnsworth, 2014). However, the developer of the building desired a different solution. One of the reasons for that could be the capacity of the tower crane available at the site, which was 26.5 ton (Oder, 2015), and it was difficult to work on the TMD with that crane. Further, as already mentioned, the precise influence of the modules on the overall stiffness and damping of the structure remained uncertain until the completion of the construction. Thus, the damper was required to be designed for a broad range of structural frequencies and damping ratios. Thornton Tomasetti, a structural engineering company, was engaged to find an appropriate alternative to the proposed TMDs. They explored the possibility of using a larger variety of fluid harmonic disruptors, initially developed by NASA as a solution to protect Ares I rocket passengers from intense vibrations during launch (Brenzel, 2015). Ultimately, they came up with the air-tuned damper (ATD), a special form of the TLCD with an air spring, that was easy to install and tune to the structure (Ghisbain et al., 2021).

Similar to a conventional TLCD, an ATD consists of a partially filled U-shaped container. In the ATD, the additional feature is the air spring arrangement at one end of the U-shaped container (https://hummingbirdkinetics.com/experience/).

More details on ATD are provided in Section 3.4.1 of Chapter 3. When the ATD is laterally excited, liquid oscillates in the U-shaped container. This causes the air to either compress or expand, resulting in a reactive force acting on the damper liquid. As a result, the oscillation frequency of the liquid in the damper increases. In the case of the B2 tower, a decision was made to fabricate the U-shaped container of the damper using readily available corrugated PVC pipes as a more cost-effective substitute for the conventional concrete or steel tanks commonly employed in TLCDs. The diameter of the PVC pipes is 0.914 m. Another PVC pipe of a smaller diameter was used to create the air spring. One end of the smaller PVC pipe was plugged and made airtight, while the other end was connected to one end of the U-shaped container. The stiffness of the air spring is dependent on the enclosed volume of air in the PVC pipe of smaller diameter, which can be regulated by adjusting the plug of the air spring (Ghisbain et al., 2021).

After the completion of construction, the natural period and the damping ratio of the building were measured through field experiments. It was found that the innate damping of the building along the long direction was sufficient to limit the wind-induced vibration to the permissible limit. However, the damper was required to control the first translation mode along the shorter direction and the first torsional mode, which had natural periods of 2.222 and 1.724 s, respectively (Ghisbain et al., 2021). Four damper units, having a total mass ratio of about 0.5%, were installed (Schulz, 2015). These were divided into two sets, each comprising two damper units stacked vertically. One set of dampers was installed on the right side of the roof, and the other set was installed on the left side of the roof. As the torsional mode would also be controlled by the dampers, a sufficient gap between the two sets of ATDs was maintained. These ATD units were manufactured in Virginia and then transported to the construction site in New York by trucks. Using the construction crane, each unit was lifted to the roof in a single operation. To ensure functionality even in cold weather, a mixture of water and glycol was used as the damper liquid. A coating was provided on the corrugated PVC pipes to protect them from the UV rays of sunlight. The four ATD units were adjusted to different frequencies in order to cater for two specific vibration modes, the first short-direction translational mode and the first torsional mode, which were aimed to be controlled. Each individual ATD could also be effectively deactivated by increasing the stiffness of the air spring to a very high value, thereby preventing any significant interaction between the structural modes and the damper.

Since the commissioning of the damper system in February 2017, the ATDs are inspected and tested twice a year. The roof acceleration data recorded after the ATDs were fine-tuned indicate that the damper system has almost halved the roof acceleration of the building under wind excitation (https://hummingbirdkinetics.com/experience/).

8.11 AUSTRALIA 108

8.11.1 Overview

The Australia 108, topped out in 2021, is the tallest building of the southern hemisphere by roof height (Gold Coast's Q1 building is taller overall due to its spire). The supertall tower is located in the Southbank of Melbourne (see Figure 8.16). The initial proposal for the tower envisioned it as a 108 storey mixed-use building

FIGURE 8.16   The Australia 108 (the tallest one) among other buildings in the Southbank skyline of Melbourne (https://en.wikipedia.org/wiki/Australia_108#/media/File:Southbank_skyline,_November_2019.png).

with a total height of 388 m. However, it was subsequently modified to a residential tower with 100 stories, due to air-safety concerns related to the nearby Essendon Airport. The building comprises 1105 luxury apartments with associated amenities (https://fkaustralia.com/project/australia-108/). The sculptural glass tower includes two distinct golden levels that extend 8 m outward from the primary structure, forming a star-like shape. This architectural feature, known as the "Starburst," draws its inspiration from the Commonwealth Star depicted on the Australian flag. Positioned at approximately two-thirds of the building's height (at 70th and 71st floor levels), the "Starburst" section serves as a communal amenity space where swimming pools are suspended in the sky (Fender, 2015). Based on the site soil conditions, only a single basement storey was constructed in the building, leading to the development of an open-deck aboveground car parking.

The primary structural system comprises a concrete core outrigged by precast mega columns and mega frames. The concrete core is up to 1.95 m thick (https://www.worldconstructionnetwork.com/projects/australia-108-residential-tower/). The outriggers are 6 m deep, 0.9 m thick and are located at the 42–43 and the 68–69 storey levels (Muldowney, 2021). The outriggers contain 75 mm diameter ultra-high-strength reinforcement (MACALLOY and SAS bars) embedded within columns and the core. The floors of the building consist of post-tensioned two-way flat slabs (https://www.robertbird.com/rbg-projects/australia-108/). The building is founded on bored concrete piles, with a diameter of up to 2.1 m and a depth of up to 47 m, with 10–15 m of those piles rock socketed into the bedrock (Muldowney, 2021). Important information about the building is provided in Table 8.12.

The Australia 108 is located about 3 km from the shoreline of Port Phillip Bay. Thus, the governing dynamic loading on the building is wind, which may reach a velocity of 162 km/h. The form of the tower itself is a direct response to the formidable wind loads impacting the structure. Instead of sharp corners, a significant portion of the tower's perimeter features large curves that effectively disrupt the strong winds blowing off the bay. Further, the building structure is engineered to withstand this load safely.

TABLE 8.12

Salient Points on the Australia 108

Type of Building	:	Residential
Height /Number of storeys	:	318.7 m/100 + 1 basement
Total floor area	:	2,828 m²
Location	:	70 Southbank Boulevard, Melbourne, Australia
Material used in main structural system	:	Concrete
Developer	:	World Class Global
Architect	:	Fender Katsalidis Architects
Structural designer	:	Robert Bird Group
Project manager	:	Sinclair Brook
Main contractors	:	Multiplex
Year of completion	:	2021

However, under the design wind load, the lateral sway of the upper storeys of the building may reach up to 600 mm which would cause occupant discomfort (Muldowney, 2021). Thus, it was decided to employ a supplemental damping system to reduce the vibration of the structure and thereby preclude any possibility of occupant discomfort.

8.11.2 Description of Damper and Performance

A conventional TLCD, having 300,000 litre of water, is used to alleviate the wind-induced vibration of the building along the narrower direction of the building. However, the building does not require any supplemental damping along the wider direction. The damper is placed about 300 m from the grade level, at the 98th storey (Muldowney, 2021). This is for the first time a TLCD is installed in a super-tall building, which has a height of 300 m or more. The TLCD is made by assembling precast concrete walls and slabs (Oliver, 2020). Through an analytical study, it had been established that the TLCD would be able to reduce the lateral oscillation of the building within the limit acceptable from the occupant comfort point of view (Muldowney, 2021). However, data related to the full-scale performance assessment of the damper are not available to date.

REFERENCES

AIJ- GBV, 2004. *Guidelines for the Evaluation of Habitability to Building Vibration.* Tokyo, Architectural Institute of Japan.

Brazil, A., Joseph, L.M., Poon, D., Scarangello, T., 2006. Designing high rises for wind performance, in: Cross, B., Finke, J. (Eds.), *Structures Congress 2006: Structural Engineering and Public Safety.* ASCE, St. Louis, Missouri, USA. https://doi.org/10.1061/40889(201)184

Brenzel, K., 2015. *From NASA to Brooklyn: Here's Why Skyscraper Living Doesn't Make You Queasy.* https://therealdeal.com/new-york/2015/10/23/from-nasa-to-brooklyn-heres-why-sky-scraper-living-doesnt-make-you-queasy/ (accessed 18 October 23).

Cammelli, S., Li, Y.F., Mijorski, S., 2016. Mitigation of wind-induced accelerations using tuned liquid column dampers : experimental and numerical studies. *Jnl. Wind Eng. Ind. Aerodyn.* 155, 174–181. https://doi.org/10.1016/j.jweia.2016.06.002

Cho, B.-H., Jo, J.-S., Joo, S.-J., Kim, H., 2011. Dynamic parameter identification of secondary mass dampers based on full-scale tests. *Comput. Civ. Infrastruct. Eng.* 27, 155–230. https://doi.org/10.1111/j.1467-8667.2011.00740.x

Cho, B.H., Yu, E., Kim, H., 2016. Mitigation of wind-induced vibration of a tall residential building using liquid column vibration absorber. *J. Vibroeng.* 18, 1031–1040. https://doi.org/10.21595/jve.2015.16755

Das, A., Konar, T., Banerjee, A., 2023. Wind-induced vibration control of tall buildings by designing the overhead fire water tanks as compliant deep tank dampers-inerter. *Structures* 58, 105522. https://doi.org/10.1016/j.istruc.2023.105522

Farnsworth, D., 2014. Modular tall building design at Atlantic Yards B2, in: *CTBUH 2014 Shanghai Conference.* Shanghai, China, pp. 492–499.

Fender, K., 2015. Skyscraper citymaker, in: Wood, A., Sharry, A. (Eds.), *Asia & Australasia: A Selection of Written Works on the World's Tall Building Forefront.* Council on Tall Buildings and Urban Habitat, New York City, USA.

Fortner, B., 2001. Water tank damp motion in Vancouver high-rise. *Civ. Eng.* 71, 18.

Ganev, T., Yamazaki, F., Ishizaki, H., Kitazawa, M., 1998. Response analysis of the Higashi-Kobe bridge and surrounding soil in the 1995 Hyogoken-Nanbu earthquake. *Earthq. Eng. Struct. Dyn.* 27, 557–576. https://doi.org/10.1002/(SICI)1096-9845(19980 6)27:6<557::AID-EQE742>3.0.CO;2-Z

Ghisbain, P., Mendes, S., Pinto, M., Malsch, E., 2021. Innovative Liquid damper for wind-induced vibration of buildings: performance after 4 years of operation, and next iteration. *Int. J. High-Rise Build.* 10, 117–121. https://doi.org/10.21022/IJHRB.2021.10.2.117

Glanville, M.J., Kwok, K.C.S., 1995. Dynamic characteristics and wind induced response of a steel frame tower. *J. Wind Eng. Ind. Aerodyn.* 54–55, 133–149. https://doi.org/10.1016/0167-6105(94)00037-E

Glanville, M.J., Kwok, K.C.S., 1997. Wind-induced deflections of free-standing lattice towers. *Eng. Struct.* 19, 79–91. https://doi.org/10.1016/S0141-0296(96)00025-9

Glanville, M.J., Kwok, K.C.S., Denoon, R.O., 1996. Full-scale damping measurements of structures in Australia. *J. Wind Eng. Ind. Aerodyn.* 59, 349–364. https://doi.org/10.1016/0167-6105(96)00016-5

Hajjar, R., 2017. Sky Gate : an urban window of structural ingenuity from CCL. *Constr. Week.* https://www.constructionweekonline.com/projects-tenders/article-46469-sky-gate-an-urban-window-of-structural-ingenuity-from-ccl#:~:text=The%20Sky%20Gate%20project%20is,weight%20of%20construction%20above%20it

Henderson, J., 2010. Random House Tower Weitere Einzelheiten. https://de.wikipedia.org/wiki/Random_House_Tower#/media/Datei:Random_House_Tower_bluesky_jeh.JPG

Henry, C., 2011. Songdo International Business District /KPF. ArchDaily. https://www.archdaily.com/118790/songdo-international-business-district-kpf

Hesson, R., 2023. https://www.northernarchitecture.us/resisting-system/tuned-liquid-column-damper.html.

Hitchcock, P.A., Glanville, M.J., Kwok, K.C.S., Watkins, R.D., Samali, B., 1999. Damping properties and wind-induced response of a steel frame tower fitted with liquid column vibration absorbers. *J. Wind Eng. Ind. Aerodyn.* 83, 183–196. https://doi.org/10.1016/S0167-6105(99)00071-9

https://commons.wikimedia.org/wiki/File:Onewallcentre3.jpg?uselang=de#Lizenz, n.d.

https://commons.wikimedia.org/wiki/File:Skygate_-_Beirut_-_Nabil_Gholam_Architects_%28cropped%29.jpg, n.d. (accessed 16 June 24).

https://en.wikipedia.org/wiki/Australia_108#/media/File:Southbank_skyline,_November_2019.png, n.d. (accessed 19 October 23).

https://en.wikipedia.org/wiki/Comcast_Center#/media/File:Comcast_Philly.JPG, n.d. https://commons.wikimedia.org/wiki/File:Comcast_Philly.JPG?uselang=en#Licensing (accessed 10 October 23).

https://fkaustralia.com/project/australia-108/, n.d. (accessed 19 October 23).

https://new.manenterprise.com/our-work/skygate, n.d. (accessed 14 October 23).

https://upload.wikimedia.org/wikipedia/commons/b/b6/461Dean.jpg, n.d. (accessed 15 October 23).

https://www.cclint.com/case-studies/skygate-beirut-lebanon/, n.d. (accessed 14 October 23).

https://www.kunwon.com, n.d. https://www.kunwon.com/en/board/perform.read.php?BBS_ GUBUN=1&SC_LP=MASTER&SC_YEAR=&SC_SUPPLY=&SC_ LOCAL=2&SC_word=&page=1&BBS_IDX=281.

https://www.nabilgholam.com/project.8, n.d. (accessed 14 October 23).

https://www.ramsa.com/projects/project/comcast-center, 2023. https://www.ramsa.com/ projects/project/comcast-center

https://www.robertbird.com/rbg-projects/australia-108/, n.d. (accessed 21 October 23).

https://www.shoparc.com/projects/b2/, n.d. (accessed 23 October 23).

https://www.worldconstructionnetwork.com/projects/australia-108-residential-tower/, n.d. (accessed 21 October 23).

Hummingbird Kinetics, n.d. *Hummingbird Kinetics Damping System Installation*. https://humming birdkinetics.com/experience/ (accessed 18 October 23).

Irwin, P., Breukelman, B., 2001. Recent applications of damping systems for wind response, in: *6th World Congress of the Council on Tall Buildings and Urban Habitat. Council on Tall Buildings and Urban Habitat*, Melbourne, Australia.

Kaneko, S., Ishikawa, M., 1999. Modeling of tuned liquid dampers with submerged nets. *J. Press. Vessel Technol.* 121, 334–344. https://doi.org/10.1115/1.2883724

Kareem, A., Kijewski, T., Tamura, Y., 1999. Mitigation of motions of tall buildings with specific examples of recent applications. *Wind Struct.* 2, 201–251. https://doi.org/10.12989/ was.1999.2.3.201

Khalil, R., Ruta, M., 2023. Sky Gate, Nabil Gholam Architects, Beirut (in Italian). Arketipo Mag.

Kitazawa, M., Nishimori, K., Noguchi, J., Shimoda, I., 1992. Earthquake resistant design of a long-period cable-stayed bridge, in: *10th World Conferance on Earthquake Engineering*, Madrid, Spain. pp. 4797–4802.

Kitazawa, M., Noguchi, J., Yamagami, T., 1993. Design of the Higashi-Kobe Bridge, Japan. *Struct. Eng. Int.* 4, 226–228. https://doi.org/10.2749/101686693780607714

Konar, T., 2024a. Design of an overhead water tank as a passive tuned damper using an innovative support system adaptive to liquid depth fluctuation in the tank. *J. Earthq.* Tsunami in press. https://doi.org/10.1142/S1793431124500076

Konar, T., 2024b. Seismic vibration control of a building by overhead water tank designed as slender tuned sloshing damper. *Pract. Period. Struct. Des. Constr.* 29, 1–11. https://doi. org/10.1061/ppscfx.sceng-1393

Konar, T., Ghosh, A.D., 2023a. A review on various configurations of the passive tuned liquid damper. *J. Vib. Control* 29, 1945–1980. https://doi.org/10.1177/10775463221074077

Konar, T., Ghosh, A.D., 2023b. Adaptive design of an overhead water tank as dynamic vibration absorber for buildings by use of a stiffness-varying support arrangement. *J. Vib. Eng. Technol.* 11, 827–843. https://doi.org/10.1007/s42417-022-00611-y

Konar, T., Ghosh, A.D., Basu, B., 2024. Real-world installations of tuned liquid column dampers for wind- induced vibration control of buildings – some important case studies. *Struct. Infrastruct. Eng.* in press.

KPF, n.d. *Songdo First World Towers*. https://www.kpf.com/project/songdo-first-world-towers (accessed 23 October 23).

Kwok, K.C.S., 2013. Behaviour of tall buildings and structures in strong winds: dynamic properties, response characteristics and vibration mitigation. *Aust. J. Struct. Eng.* 14, 177–192. https://doi.org/10.7158/13287982.2013.11465131

Lago, A., Trabucco, D., Wood, A., 2019. Case studies of tall buildings with dynamic modification devices, in: Damping Technologies for Tall Buildings. Butterworth-Heinemann. Oxford, pp. 533–919. https://doi.org/10.1016/b978-0-12-815963-7.00001-4

Lord, J.F., Ventura, C.E., 2002. FE model calibration of a 48 storey building in vancouver using ambient vibration measurements, in: *4th Structural Specialty Conference Annual Conference of the Canadian Society for Civil Eng.* Montréal, Canada.

Mafi, N., 2016. *The World's Tallest Modular Skyscraper Welcomes Its First Residents.* https://www.architecturaldigest.com/story/worlds-tallest-modular-skyscraper-welcomes-first-residents

McKnight, J., 2016. *World's Tallest Modular High-Rise by SHoP Architects Opens in Brooklyn. Dezeen.* https://www.dezeen.com/2016/11/18/worlds-tallest-modular-prefabricated-apartment-tower-shop-architects-brooklyn-new-york/#:~:text=Rising%20to%20a%20height%20of,to%20the%20site%20by%20truck

Muldowney, S., 2021. *Engineering Australia 108, the Southern Hemisphere's Tallest Residential Building.* https://createdigital.org.au/engineering-australia-108-southern-hemispheres-tallest-residential-building/#:~:text=At%20close%20to%20319%20m%2C%20Australia%20108%20is%20the%20tallest,on%20levels%2070%20and%2071

Murayama, N., 2005. https://www.flickr.com/photos/12832970@N00/47617335.

Oder, N., 2015. *Forest City Touts New Technology to Keep B2 from Swaying, Previous Plan (Heavier, More Expensive) Unmentioned.* https://atlanticyardsreport.blogspot.com/2015/10/forest-city-touts-new-technology-to.html?m=1 (accessed 17 October 23).

Oliver, D., 2020. *Large Overturning Forces: How Australia 108 Stays Standing.* https://www.thisisconstruction.com.au/articles/large-overturning-forces-how-australia-108-stays-standing

One Wall Centre, n.d. https://glotmansimpson.com/project/one-wall-centre/ (accessed 2 October 23).

Peeters, B., De Roeck, G., 1999. Reference-based stochastic subspace identification for output-only modal analysis. *Mech. Syst. Signal Process.* 13, 855–878. https://doi.org/10.1006/mssp.1999.1249

Pettafor, E., 1999. Accor scores bargain in Tokyo hotel. *Aust. Financ. Rev.* https://www.afr.com/property/accor-scores-bargain-in-tokyo-hotel-19990323-k8mhe

RWDI, n.d. *A Unique Damping System for the Tallest Building in Philadelphia.* https://rwdi.com/en_ca/projects/comcast-center/ (accessed 2 May 24).

Sakai, F., Takaeda, S., Tamaki, T., 1991. Tuned liquid column dampers (TLCD) for cable-stayed bridges, in: *Proceddingas of Specialty Conference on Innovation in Cable-Stayed Bridges.* Fukuoka, Japan, pp. 197–205.

Saoka, Y., Sakai, F., Takaeda, S., Tamaki, T., 1988. On the suppression of vibrations by tuned liquid column dampers, in: *Annual Meeting of JSCE, JSCE.* Tokyo, Japan.

Schulz, D., 2015. *Atlantic Yards 'B2 Tower Employing Anti-Nausea Technology From NASA.* 6sqft. https://www.6sqft.com/atlantic-yards-b2-tower-employing-anti-nausea-technology-from-nasa/ (accessed 17 October 23).

Shimizu, K., Teramura, A., 1994. Development of vibration control system using U-shaped water tank, in: *1st International Workshop and Seminar on Behavior of Steel Structures in Seismic Areas.* Timisoara, Romania.

Simpson, R., 2001. Buildings: One Wall Centre, Vancouver. *Can. Consult. Eng.* 3–6. https://www.canadianconsultingengineer.com/features/buildings-one-wall-centre-vancouver/

Stephens, S., 2009. Robert A.M. Stern Architects raises the bar with Philadelphia's Comcast Center. *Archit. Rec.* 05.09 168–174.

Tamboli, A., Christoforou, C., Brazil, A., Joseph, L., Vadnere, U., Malmsten, B., 2005. Manhattan's mixed construction skyscrapers with tuned liquid and mass dampers, in: *CTBUH 2005–7th World Congress Renewing the Urban Landscape, Proceedings.* New York City, USA.

Tamboli, A., Joseph, L., Vadnere, U., Xu, X., 2008. Tall buildings : sustainable design opportunities, in: *CTBUH 2008 8th World Congress*. Dubai, UAE.

Teramura, A., Yoshida, O., 1996. Development of vibration control system using U-shaped water tank, in: *11th World Conferance on Earthquake Engineering*. Acapulco, Mexico, Paper No. 1343.

Thornton Tomasetti, n.d. *Comcast Center*. https://www.thorntontomasetti.com/project/comcast-center (accessed 11 October 23).

Ueda, T., Nakagaki, R., Koshida, K., 1992. Suppression of wind-induced vibration by dynamic dampers in tower-like structures. *J. Wind Eng. Ind. Aerodyn.* 43, 1907–1918. https://doi.org/10.1016/0167-6105(92)90611-D

Yamada, Y., Shiraishi, N., Toki, K., Matsumoto, M., Matsuhashi, K., Kitazawa, M., Ishizaki, H., 1991. Earthquake-resistant and wind-resistant design of the Higashi Kobe Brdge, in: *International Seminar on Cable Supported Bridges*, Yokohama, Japan. pp. 16–35.

Index

across-wind spectra 131–132
active dampers 10
active TLCD 200–211
    with a compliant mechanism 201–207
    with propellers 207–209
active tuned liquid column dampers *see* active
    TLCD
adaptive TLCD *see* Ad-TLCD
Ad-TLCD 105–106, 194–196
air tuned dampers 96–97
Airy's linear wave theory 182–184
along-wind spectra 132–133
aseismatic joint 5
ATLCD-C *see* active TLCD with a compliant
    mechanism
ATLCD-P *see* active TLCD with propellers
Australia 108, 258–260
    description of damper 260
    performance of damper 260

B2 tower *see* Dean Street, 461
base isolation 4–7
bi-directional CLCD 100
bi-directional LCVA 74
bi-directional TLCD 60–63, 238–240
bi-directional TLCD with period adjustment
    equipment 238–240
blocking ratio 48
Bretschneider spectrum 185
BTLCD *see* bi-directional TLCD
BTLCD-AR 62–63

central tuning ratio 50
circular TLCD 80–84
CLCD 97–100, 159–160, 162–164
closed circular TLCD 83–84
Clough and Penzien spectrum 157–158
coefficient of head-loss of orifice(s) 47–49, 144,
    170–171
Comcast Center 247–250
    description of damper 249–250
    performance of damper 250
compliant TLCD *see* CLCD
C-P model *see* Clough and Penzien spectrum

damped vibration absorber 15–16
Davenport spectrum 131
Dean Street, 461 255–258
    description of damper 257–258
    performance of damper 258
digital records of accelerograms 161

dynamic amplification factor 15
dynamic vibration absorber 11–13

earthquake-induced excitation 156–158, 160–161
edgewise vibration control of wind turbine
    blades 140
experimental studies 140–142, 165–169, 189
    for seismic vibration control 165–169
    for wave-induced vibration control 189
    for wind-induced vibration control 140–142

flow damping devices 20, 49
fluctuating wind speed 125, 129
Fraham's absorber *see* dynamic vibration
    absorber
Fraham's anti-rolling tank 22–24
frequency domain studies 44–46, 129–136,
    156–160, 184–185
    under earthquake loading 156–160
    under wind loading 129–136
frequency ratio 15
Froude-Krylov force 184

gas sealed TLCD with active pressure control
    209–211
gas-spring 93–94

H_∞ controller 207
Harris spectrum 131
hedge-algebra based controller 207
Higashi-Kobe Bridge 234–236
    description of damper 235
    performance of damper 235–236
Hotel Cosima 237–240
    description of damper 238–239
    performance of damper 240
Hotel Sofitel *see* Hotel Cosima
human perception levels to building acceleration
    127–128
hybrid dampers 10–11
hybrid dampers with semi-active TLCD
    226–227

inerter 103
inertia-based dampers 103

JONSWAP spectrum 185

Kaimal spectrum 132
Kanai-Tajimi spectrum 156–157
K-T model *see* Kanai-Tajimi spectrum

LCVA 71–74, 159, 162, 241–242, 251–253
length ratio 46–47
linear quadratic regulator controller 203–205
liquid column vibration absorbers *see* LCVA
LQR *see* linear quadratic regulator controller

magneto-rheological fluids 217
mass ratio 14, 41, 46, 143, 169–170
material-based dampers 7
mean wind speeds 125
Morison equation 184
MR fluids *see* magneto-rheological fluids
MTLCD 50
    under wind loading 135–137
    under earthquake loading 161–162
multi-layer TTLCD 68
multiple TLCD *see* MTLCD
multiresolution-based wavelet linear quadratic
        regulator controller 205–207

NatHaz online wind simulator 136–137
New Songdo City First World Towers 250–253
    description of damper 251–252
    performance of damper 252–253

offshore wind turbines 138–140, 185–190
Ohkuma and Kanaya spectrum 132
omnidirectional TLCDs 68–71, 189
One Pennsylvania Plaza *see* Comcast Center
One Wall Centre 242–244
    description of damper 243–244
    performance of damper 244
optimum damping ratio 16
optimum orifice head-loss coefficient 143–146,
        170–171
    under earthquake loading 170–171
    under wind loading 143–146
optimum TLCD parameters 143–146, 169–172
    under earthquake loading 169–172
    under wind loading 143–146
optimum tuning ratio 143–146, 170–172
    under earthquake loading 170–172
    under wind loading 143–146
orifice
    head loss coefficient *see* coefficient of head
        loss of orifice(s)
    opening ratio 47–49, 212
OTLCD *see* omnidirectional TLCDs

Park Imperial Apartments *see* Random
        House Tower
passive dampers 8–10
patents on TLCDs 233
pendulum TLCD 101–103
permissible lateral deflection 127
    of a civil engineering structure 127
    of building structures 127

Phillips constant 185
Pierson-Moskowitz spectrum 185
PLCD *see* pendulum TLCD
Polytropic index 93
power spectral density 45–46, 130–133, 156–158,
        222–223
    of the fluctuating wind speed 130–133
    of seismic excitation 156–158
pressurized tuned liquid column damper *see*
        sealed TLCD
prospect communication tower 240–242
    description of damper 241–242
    performance of damper 242

Random House Tower 244–247
    description of damper 247
    performance of damper 247
real-time hybrid testing 167–169
rolling-bearing device *see* aseismatic joint
RTHT *see* real-time hybrid testing

sealed TLCD 93–97
semi-active dampers 10
semi-active MR-TLCD 217–219
semi-active spring connected TLCD 219–225
semi-active TLCD 201, 211–225
Sétra guidelines 193
shake table test 141–142, 165–167, 255
shear wave velocity 175
Sheraton Vancouver Wall Centre Hotel *see* One
        Wall Centre
Simiu spectrum 132
Sky Gate 253–255
    description of damper 254–255
    performance of damper 255
slim tube 139
soil-structure interaction 172
    effect on TLCD performance 174–175
soil-structure-TLCD model 172–174
Songdo The Sharp First World *see* New Songdo
        City First World Towers
S-shaped TLCD 84–86
STLCD *see* semi-active spring connected TLCD
STLCD-APC *see* gas sealed TLCD with active
        pressure control
Stokes wave theories 183–184
storey drift ratio 155
structural vibration, causes and effects 2–3
structure-TLCD system 39–46
    equations of motion 39–43
    frequency domain representation of equations
        of motion 44–46
    transfer functions 45
supplemental damping devices 7

tank-pipe dampers 89–92
tank sloshing damper 18–22

time domain studies 136–140, 160–165
    under earthquake loading 160–165
    under wind loading 136–140
TLCBD 74–77, 190
TLCD
    mathematical modelling 39–43
    natural frequency 41–42
    performance under earthquake loading 158–169
    performance under pedestrian loading 192–194
    performance under train-induced vibration 194–196
    performance under wave loading 186–192
    performance under wind loading 133–142
    practical implementation 230–260
    with semi-active orifice control 211–215
    working mechanism 38
TLCDI 103–105
TLCSD 86–89
TLD 16–22
TLMCD 77–80
TMD 13–14
toroidal TLCD 63–68
torsional TLCD 81
torsional tuned liquid column gas damper 83
TPD *see* tank-pipe dampers
TTLCD *see* toroidal TLCD
tuned liquid column and sloshing dampers *see* TLCSD

tuned liquid column ball dampers *see* TLCBD
tuned liquid column ball gas damper 97
tuned liquid column damper inerters *see* TLCDI
tuned liquid column dampers *see* TLCD
tuned liquid column-gas damper *see* Sealed TLCD
tuned liquid column mass damper *see* CLCD
tuned liquid dampers *see* TLD
tuned liquid multi-column dampers *see* TLMCD
tuned mass dampers *see* TMD
tuning bandwidth 50
tuning ratio 15, 49–50

vertical sealed TLCD 106–107, 194
vertical TLCD 107
vertical vibration control 105–107, 194–196
vibration 1
Vickery and Clark Spectrum 132
von Kármán spectrum 131
VTLCD *see* vertical TLCD

walking force 192–193
wave frequency 182
wave-induced excitation 182–185
wave number 182
wind-induced excitation 125, 131–133, 136–137
wind thrust 125
wind tunnel tests 141–142

For Product Safety Concerns and Information please contact our EU
representative GPSR@taylorandfrancis.com
Taylor & Francis Verlag GmbH, Kaufingerstraße 24, 80331 München, Germany